COMPUTATIONAL OPTIMIZATION
A Tribute to Olvi Mangasarian
Volume I

edited by

Jong-Shi Pang
The Johns Hopkins University, U.S.A.

A Special Issue of
COMPUTATIONAL OPTIMIZATION AND APPLICATIONS
An International Journal
Volume 12, Nos. 1/2/3 (1999)

SPRINGER SCIENCE+BUSINESS MEDIA, LLC

COMPUTATIONAL OPTIMIZATION AND APPLICATIONS

An International Journal

Volume 12, Nos. 1/2/3, January 1999

Special Issue: COMPUTATIONAL OPTIMIZATION—A Tribute to Olvi Mangasarian, Part I
Guest Editor: Jong-Shi Pang

Computational optimization : a tribute to Olvi Mangasarian / edited by
Jong-Shi Pang.
p. cm.
"A special issue of Computational optimization and applications, an international journal, volume 12 nos. 1/2/3 (1999) and volume 13, nos. 1/2/3 (1999)."
Includes bibliographical references.
ISBN 978-1-4613-7367-4 ISBN 978-1-4615-5197-3 (eBook)
DOI 10.1007/978-1-4615-5197-3
1. Mathematical optimization--Data processing. 2. Nonlinear programming--Data processing. I. Mangasarian, Olvi L., 1934-
II. Pang, Jong-Shi.
QA402.5.C5587 1999
519.3'0285--dc21 99-13159
CIP

Printed on acid-free paper.

Computational Optimization and Applications 12, 5–12 (1999)

Guest Editorial

The two dozen papers in this collection were submitted by friends, former students, colleagues, and admirers of Professor Olvi L. Mangasarian in celebration of his sixty-fifth birthday on January 12, 1999. These papers were authored by forty-five international experts in the field of computational optimization who, directly and indirectly, have all been influenced and touched by Professor Mangasarian through his scholarly work, teaching and mentoring, and invaluable friendship. In appreciation of the fundamental and sustained contributions of this beloved and esteemed pioneer in our field, and the kindness and generosity of this wonderful friend and colleague, it is our great pleasure and honor to dedicate our works to you, Professor Mangasarian.

Four decades of contributions

The topics addressed by the papers in this collection cover many facets of Professor Mangasarian's forty years of research contributions. From the biography that follows this preface, we learn that Professor Mangasarian started his career as a doctoral student in Applied Mathematics at Harvard University. Several of his pioneering papers in 1960s paved the way for the fruitful development of the nonlinear programming field; others provided the foundation for his most recent entry into the fast growing areas of machine learning and data mining. Form his output in nonlinear programming, one must mention his joint work with Stan Fromovitz and the outstanding book titled *Nonlinear Programming*. The former introduced the famous Mangasarian-Fromovitz constraint qualification; the latter has served the field for thirty years and was recently recognized by the Society of Industrial and Applied Mathematics as a classic in applied mathematics. Another important work from this era is the paper on pseudo-convex functions which have played a fundamental role in the theory of generalized convexity.

In 1970s, Professor Mangasarian's research interests turned to algorithmic developments in nonlinear programming. Through his own work and that of his students, he was instrumental for the development of many well-known algorithms such as the tremendously successful family of sequential quadratic programming algorithms and the exact penalty approach to solving nonlinear programs. In the same period, he also began to publish his highly influential papers on the linear complementarity problem. In 1970, together with Ben Rosen and Klaus Ritter, Professor Mangasarian organized a major international conference on nonlinear programming that later turned out to be the first of four highly acclaimed conferences of this kind, with the latter three co-organized with Bob Meyer and Steve Robinson. As a young researcher, the Guest Editor was honored to be invited to speak at the fourth and last such conference held in 1980.

While maintaining his leading role in nonlinear programming and complementarity research, Professor Mangasarian broke ground, in the mid-1980s, on two new research frontiers: parallel optimization and error bounds for inequality systems. As with his previous works, he published some of the first papers on the latter two subjects. Today, the theory of error bounds has developed into a very fruitful topic within mathematical programming with vast connections to many interesting ideas and great potential for important application.

At the beginning this decade, Professor Mangasarian developed a renewed research interest that has occupied much of his time and energy till the present. This began with the subject of pattern classification and has led him to the novel areas of machine learning and data mining which, before his involvement, were virtually unknown to most researchers in the optimization community. Building on his early work in the mid-1960s, he has much expanded the domain of application and introduced new optimization methodology to deal with practical problems such as breast cancer diagnosis and prognosis. With the collaboration of his former and present students, he has been a strong advocate for applying optimization methods to important problems in data mining.

In addition to being a pioneer of our field, Professor Mangasarian has been a much loved teacher and mentor to several generations of young researchers. Many of his academic offspring are now established leaders in their respective areas of expertise. Today, he continues to teach and nurture the next generations of scholars and leaders. We all wish him good health and continued success in the next sixty-five years. Happy birthday, Olvi!

A personal tribute

As one rapidly gathers from the tributes in this collection of papers, many researchers have learned and benefited greatly from Professor Mangasarian. The Guest Editor is no exception. As a fresh Ph.D. graduate from Stanford University (supervised by Richard Cottle), I had the good fortune of first meeting Professor Mangasarian in Fall 1976 when I was an Assistant Research Scientist at the (then-called) Mathematics Research Center at the University of Wisconsin at Madison (I also thank Steve Robinson who was instrumental in bringing me to Madison). Since that time, I have always considered Professor Mangasarian as my second academic father. I have received invaluable support, encouragement, and inspiration from Professor Mangasarian throughout my career. When Bill Hager first suggested to me the idea of a special project in honor of Professor Mangasarian, I immediately embraced the suggestion and took on the task of Guest Editorship of these special issues of the journal COAP. My hope is to use this responsibility as an opportunity to express my deep appreciation to a great leader, a trusted mentor, an esteemed colleague, and a dear friend. Thank you, Olvi!

The papers in this festchrift

The collected papers cover a wide spectrum of computational optimization topics; these represent a good mix of familiar nonlinear programming topics and such novel paradigms as semidefinite programming and complementarity constrained nonlinear programs. Many new results are presented in these papers which are bound to inspire further research and generate new avenues for applications. An informal categorization of the papers is as follows.

Algorithmic advances for special classes of constrained optimization problems. De Leone, Meyer, and Zakarian present an ϵ relaxation algorithm for separable convex network flow problems; Di Pillo, Lucidi, and Palagi introduce a shifted-barrier primal-dual algorithm model for linearly constrained problems; Hager establishes new convergence results for stabilized sequential quadratic programming algorithms; Polak, Qi, and Sun describe first-order algorithms for semi-infinite min-max problems; Vanderbei and Shanno present extensive computational results on an interior pont algorithm for nonconvex programs; and Yamakawa and Fukushima report the results of testing parallel variable transformations.

Analysis of linear and nonlinear programs. Bertsekas presents an interesting approach to obtain error bounds for convex and nonconvex programs; Klatte and Kummer study generalized Kojima-functions and Lipschitz stability of critical points; Luo and Zhang obtain some extensions of the classical Frank-Wolfe Theorem to quadratically constrained quadratic programs; and Tseng analyzes the convergence and establishes new error bound results for linear programs under nonlinear perturbations.

Several papers deal with variational inequalities and the class of optimization problems with equilibrium constraints.

Algorithmic advances. Auslender, Teboulle, and Ben-Tiba describe a logarithmic-quadratic proximal method for variational inequalities and establish its global convergence under a very weak condition; Denault and Goffin report extensive computational results on a primal-dual analytic center cutting plane method for variational inequalities; Ferris and Munson describe the design, implementation, and usage of interfaces to the award-winning software PATH for solving mixed complementarity problems; Solodov presents some equivalent optimization reformations of the extended linear complementarity problem; Sun and Qi focus on various merit functions for the nonlinear complementarity problem; and Jiang and Ralph describe a MATLAB generator of test problems for mathematical programs with quadratic objectives and affine variational inequality constraints.

B-stationary points of mathematical programs with equilibrium constraints. Pang and Fukushima introduce some novel complementarity constraint qualifications under which verification for B-stationarity of this class of difficult mathematical programs can be simplified.

The remaining papers address various mathematical and applied topics related to computational optimization.

Applications of optimization. Alfakih, Khandani, and Wolkowicz apply the methodology of semidefinite programming to Euclidean distance matrix completion problems; Bennett and Bredensteiner explain a structured quadratic programming approach to multicategory classification by support vector machines; Powell discusses an interesting geometric problem and how it can be solved by a special optimization method.

Some mathematical topics. Cottle surveys some propositions arising from the literature of pure and applied mathematics whose validity is restricted by an associated natural number not exceeding four; Hiriart-Urruty and Lewis present computational formulas for the Clarke and Michel-Penot subdifferentials of the eigenvalues of a symmetric matrix.

Systems of nonlinear equations. Eaves and Rothblum discuss the continuation of regular points of formal parametric polynomial systems; and Gowda and Tawhid establish the existence and analyze the limiting behavior of trajectories associated with equations with P_0 defining functions.

Acknowledgments

The Editor-in-Chief of COAP, Bill Hager, suggested the idea of devoting a special issue of the journal dedicated to Olvi Mangasarian. We are thankful to all the authors for contributing their interesting papers to make this festchrift a success. Many of them have also acted as reviewers for others' papers. We have also relied on additional individuals to serve as referees for the papers published herein. Their efforts were particularly commendable; the editors and the authors are very grateful for their help.

Jong-Shi Pang
Baltimore, Maryland

Professor Olvi L. Mangasarian's Biography

Olvi Mangasarian was born in Baghdad, Iraq on January 12, 1934. His first two years of college, were at the American University of Beirut in Lebanon followed by three years at Princeton University where he obtained his BSE and MSE in 1954 and 1955. After Princeton, Olvi went to Harvard University where he completed his Ph.D. in Applied Mathematics in 1958 in the Division of Engineering and Applied Physics and stayed there for another year as a Research Fellow. In 1959 he joined the Applied Mathematics Department of Shell Development Company in Emeryville, California. In 1967 Olvi joined the Computer Sciences Department at the University of Wisconsin at Madison where he has been the John von Neumann Professor of Mathematics and Computer Science since 1982. During the period 1965–67 Olvi held an adjunct appointment in the Electrical and Industrial Engineering Departments at the University of California at Berkeley where he taught a course on nonlinear programming. The material taught in that course constituted the basis for his book, "Nonlinear Programming", published by McGaw-Hill in 1969 and re-published by SIAM in 1994.

In 1959 Olvi married Claire Garabedian of Worcester, Massachusetts, a graduate of Tufts University. They have three children. Leon, a journalist, and his wife Tatjana live in Bonn, Germany. Jeffrey, a computer scientist, his wife Cheryl and their son Tarrant live in El Granada, California. Aram, a biologist, lives currently in Geneva, Switzerland.

Olvi enjoys lap swimming on workdays. On weekends he likes to work in the garden when the temperature is above freezing. He also enjoys classical music, photography, historical biographies and discovering unusual World Wide Web sites.

Selected publications of Professor Olvi L. Mangasarian

1. "Linear and nonlinear separation of patterns by linear programming," Operations Research, vol. 13, pp. 444–452, 1965.
2. "Pseudo-convex functions," SIAM Journal on Control, vol. 3, pp. 281–290, 1965.
3. (with S. Fromovitz) "The Fritz John necessary optimality conditions in the presence of equality constraints," Journal of Mathematical Analysis and Applications, vol. 17, pp. 34–47, 1967.
4. "Multi-surface method of pattern separation," IEEE Transaction on Information Theory, vol. IT-14, pp. 801–807, 1968.
5. (with U.M. Garcia-Palomares) "Superlinearly convergent quasi-Newton algorithms for nonlinearly constrained optimization problems," Mathematical Programming, vol. 11, pp. 1–13, 1976.
6. "Equivalence of the complementarity problem to a system of nonlinear equations," SIAM Journal on Applied Mathematics, vol. 31, pp. 89–92, 1976.
7. "Solution of symmetric linear complementarity problems by iterative methods," Journal of Optimization Theory and Applications, vol. 22, pp. 465–485, 1977.
8. (with S.P. Han) "Exact penalty functions in nonlinear programming," Mathematical Programming, vol. 17, pp. 251–169, 1979.

9. "Locally unique solutions of quadratic programs, linear and nonlinear complementarity problems," Mathematical Programming, vol. 19, pp. 200–212, 1980.
10. "A condition number for linear inequalities and linear programs," in "Methods of Operations Research," Proceedings of 6th Symposium on Operations Research, Augsburg, September 1981, G. Bamberg and O. Opitz (Eds.), Verlagsgruppe Athenaum/Hain/Scriptor/Hanstein: Konigstein, 1981, pp. 3–15.
11. "A condition number for differentiable convex inequalities," Mathematics of Operations Research, vol. 10, pp. 175–179, 1985.
12. (with T.H. Shiau) "A variable-complexity norm maximization problem," SIAM Journal on Algebraic and Discrete Methods, vol. 7, pp. 455–461, 1986.
13. (with T.H. Shiau) "Error bounds for monotone linear complementarity problems," Mathematical Programming, vol. 36, pp. 81–89, 1986.
14. "A simple characterization of solution sets of convex programs," Operations Research Letters, vol. 7, pp. 21–26, 1988.
15. (with R. De Leone) "Serial and parallel solution of large scale linear programs by augmented Lagrangian successive overrelaxation," in Optimization, Parallel Processing and Applications, A. Kurzhanski, K. Neumann, and D. Pallaschke (Eds.), Lecture Notes in Economics and Mathematical Systems, vol. 304, Springer-Verlag, Berlin, 1988, pp. 103–124.
16. (with R. Setiono and W.H. Wolberg) "Pattern recognition via linear programming: Theory and application to medical diagnosis," in Large-Scale Numerical Optimization, T.F. Coleman and Y. Li (Eds.), SIAM Publications,: Philadelphia, 1990, pp. 22–30.
17. (with M.C. Ferris) "Minimum principle sufficiency," Mathematical Programming, vol. B 57, pp. 1–14, 1992.
18. (with K.P. Bennett) "Neural network training via linear programming," in Advances in Optimization and Parallel Computing, P.M. Pardalos (Ed.), North Holland: Amsterdam, 1992, pp. 56–67.
19. (with K.P. Bennett) "Robust linear programming discrimination of two linearly inseparable sets," Optimization Methods and Software, vol. 1, pp. 23–34, 1992.
20. (with M.V. Solodov) "Nonlinear complementarity as unconstrained and constrained minimization," Mathematical Programming, Series B, vol. 62, pp. 277–297, 1993.
21. (with M.V. Solodov) "Serial and parallel backpropagation convergence via nonmonotone perturbed minimization," Optimization Methods and Software, vol. 4, pp. 103–116, 1994.
22. (with M.C. Ferris) "Parallel variable distribution," SIAM Journal on Optimization, vol. 4, pp. 815–832, 1994.
23. "Misclassification minimization," Journal of Global Optimization, vol. 5, pp. 309–323, 1994.
24. (with J.S. Pang) "The extended linear complementarity problem," SIAM Journal on Matrix Analysis and Applications, vol. 16, pp. 359–368, 1995.
25. (with W.H. Wolberg and W.N. Street) "Breast cancer diagnosis and prognosis via linear programming," Operations Research, vol. 43, pp. 570–577, 1995.
26. (with C. Chen) "A class of smoothing functions for nonlinear and mixed complementarity problems," Computational Optimization and Applications, vol. 5, pp. 97–138, 1996.

27. (with P.S. Bradley and W.N. Street) "Feature selection via mathematical programming," INFORMS Journal on Computing, vol. 10, pp. 209–217, 1998.
28. "Mathematical programming in data mining," Journal of Data Mining and Knowledge Discovery, vol. 1, pp. 183–201, 1997.
29. (with P.S. Bradley and J.B. Rosen) "Parsimonious least norm approximation," Mathematical Programming Technical Report 97-03, March 1997, Computational Optimization and Applications, vol. 11, pp. 5–21, 1998.
30. (with P.S. Bradley) "Massive data discrimination via linear support vector machines," Mathematical Programming Technical Report 98-05, May 1998.

Computational Optimization and Applications 12, 13–30 (1999)

Solving Euclidean Distance Matrix Completion Problems Via Semidefinite Programming*

ABDO Y. ALFAKIH aalfakih@orion.math.uwaterloo.ca
University of Waterloo, Department of Combinatorics and Optimization, Waterloo, Ontario N2L 3G1, Canada

AMIR KHANDANI khandani@claude.uwaterloo.ca
University of Waterloo, Department of Electrical & Computer Engineering, Waterloo, Ontario N2L 3G1, Canada

HENRY WOLKOWICZ henry@orion.math.uwaterloo.ca
University of Waterloo, Department of Combinatorics and Optimization, Waterloo, Ontario N2L 3G1, Canada

Received May 5, 1998; Accepted July 24, 1998

Abstract. Given a partial symmetric matrix A with only certain elements specified, the Euclidean distance matrix completion problem (EDMCP) is to find the unspecified elements of A that make A a Euclidean distance matrix (EDM). In this paper, we follow the successful approach in [20] and solve the EDMCP by generalizing the completion problem to allow for approximate completions. In particular, we introduce a primal-dual interior-point algorithm that solves an equivalent (quadratic objective function) semidefinite programming problem (SDP). Numerical results are included which illustrate the efficiency and robustness of our approach. Our randomly generated problems consistently resulted in low dimensional solutions when no completion existed.

Keywords: Euclidean distance matrices, semidefinite programming, completion problems, primal-dual interior-point method

Dedication: (Henry) The first time that I came across Olvi's work was as a graduate student in the 70s when I studied from his book on Nonlinear Programming (now a SIAM classic) and also used the *Mangasarian-Fromovitz constraint qualification*. This is *the* constraint qualification (CQ) in nonlinear programming (NLP), since it guarantees the existence of Lagrange multipliers and is equivalent to stability of the NLP. This CQ has since been extended to various generalizations of NLP and plays a crucial role in perturbation theory.

In 1983 I was a visitor at The University of Maryland, College Park, and was teaching a course in the Business College. While walking through the halls one day I noticed the name Fromovitz on one of the doors. I could not pass this by and knocked and asked if this was *the* Fromovitz. The reply was "yes"; and, this is the story of the now famous CQ. Stan Fromovitz had just received his Ph.D. from Stanford and was working at Shell Development Company in the Applied Math Dept. Olvi needed a special Theorem of the Alternative for his work on a CQ. Stan went digging into the Math Library at Berkeley and came up with exactly what was needed: Motzkin's Theorem of the Alternative. The end result of this was the MF CQ.

I have followed Olvi's work very closely throughout my career. His work is marked by many beautiful and important results in various areas. Some of the ones I am aware of are: condition numbers for nonlinear programs; generalized convexity; complementarity problems; matrix splittings; solution of large scale linear programs.

It is a pleasure and an honour to be able to contribute to this special issue.

*Research supported by Natural Sciences Engineering, Research Council, Canada.

1. Introduction

Euclidean distance matrices have received a lot of attention in recent years both for their elegance and for their many important applications. Two other research areas of high interest currently, are semidefinite programming and interior-point methods. In this paper we solve the Euclidean distance matrix completion problem by generalizing the completion problem to allow for approximate completions, i.e., we find a weighted, closest Euclidean distance matrix. In particular, we introduce a primal-dual interior-point algorithm that solves an equivalent (quadratic objective) semidefinite programming problem.

An $n \times n$ symmetric matrix $D = (d_{ij})$ with nonnegative elements and zero diagonal is called a *pre-distance matrix* (or dissimilarity matrix). In addition, if there exist points $x^1, x^2, \ldots, x^n$ in $\Re^r$ such that

$$d_{ij} = \|x^i - x^j\|^2, \quad i, j = 1, 2, \ldots, n, \tag{1}$$

then D is called a *Euclidean distance matrix* (EDM). The smallest value of r is called *the embedding dimension* of D. Note that r is always $\leq n - 1$. Given a partial symmetric matrix A with certain elements specified, the Euclidean distance matrix completion problem (EDMCP) consists in finding the unspecified elements of A that make A a EDM. Alternatively, for the approximate EDMCP, let A be a pre-distance matrix, H be an $n \times n$ symmetric matrix with nonnegative elements, and let $\|A\|_F = \sqrt{\text{trace } A^t A}$ denote the *Frobenius norm* of A. Consider the objective function

$$f(D) := \|H \circ (A - D)\|_F^2,$$

where $\circ$ denotes *Hadamard product*, e.g., [19]. The weighted, *closest Euclidean distance matrix problem* is

$$(\text{CDM}_0) \quad \mu^* := \min \quad f(D) \quad \text{subject to} \quad D \in \mathcal{E},$$

where $\mathcal{E}$ denotes the cone of EDMs.

Applications of EDMs abound: e.g., molecular conformation problems in chemistry [9, 31]; multidimensional scaling and multivariate analysis problems in statistics [25, 26]; genetics, geography, and others [2]. Many of these applications require a low embedding dimension, e.g., $r = 3$.

Theoretical properties of EDMs can be found in, e.g., [7, 11, 15, 16, 21, 24, 34]. This includes characterizations as well as graph theoretic conditions for existence of completions. More information can be found in the recent survey article by Laurent [24]. (Generalizations of EDM arise in [36].)

An interesting discussion on algorithms for EDMCP appears in [41]. The point is made that there is no definitive general algorithm for EDMCP, i.e., one cannot provide an efficient decision rule for the question of whether a completion exists or not. However, there are many algorithms that find approximate completions. In [39–41], the author presents results on finding EDM completions based on spectral decompositions. In particular, the computationally hard problem of fixing the rank (the embedding dimension) is discussed. Some work on finding the closest EDM to a given symmetric matrix appears in [3, 12, 47]. Another approach based on global optimization and allowing for intervals for the distances is given in [28, 29] and also in [48]. We build a convex (tractable) model by relaxing the constraint on the embedding dimension (rank).

Semidefinite programming, SDP, is an extension of linear programming where nonnegativity constraints on vector variables is replaced by positive semidefiniteness constraints on matrix variables. This area has attracted a lot of interest recently because of the many applications as well as the elegant mathematics involved, see, e.g., the survey papers: [5, 13, 42, 43]. A lot of the interest in SDP is from the interior-point community who have completed so much successful work on linear programming. At the moment, interior-point methods are the most successful algorithms for general SDP problems, see, e.g., the above survey articles as well as the books [30, 46] and the recent theses [2, 4, 17, 22, 32, 33, 35]. The above references provide some evidence of the current high level of research activity in these areas.

The main contribution of this paper is a new approach to solving EDMCP. This approach is different from those in the literature in two ways. First we change the EDMCP into an approximation problem. This latter problem is a convex problem, i.e., our model is a tractable (polynomial time) model. Moreover, we further relax the original problem since we do not fix the embedding dimension. Thus, we do not solve problems that require a given dimension, though, we hope that this approach could be used as a first approximation to such problems. (A discussion on reducing the dimension is given in Section 5.) Our randomly generated tests consistently resulted in optimal solutions with low dimension, e.g., $r \leq 3$, when a completion did not exist. This is in contrast to the case where completions do exist, where the interior point method found the completion of maximum rank, since interior-point methods find optimal solutions in the relative interior of the optimal face. Second, we use a semidefinite programming formulation and a primal-dual interior-point algorithm to solve the approximation problem. And, we prove that the Slater constraint qualification holds for our model if and only if the graph of the matrix of weights is connected, see Corollary 2. Usually, the lack of the constraint qualification results in numerical difficulties due to unbounded optimal sets. However, in our case we can take advantage of a disconnected graph to replace the original problem by two smaller simpler problems.

As a side issue, we point out that our algorithm uses a new search direction for semidefinite programming introduced in [23]. This search direction is based on applying a Gauss-Newton approach to the optimality conditions. Our purpose is not to compare different search directions, and other public domain packages may be used for the problems that we have solved; though our tests show that our approach is comparable. We use this approach, rather than the more standard approaches already in the literature, since it is very well suited for our particular application. And, we think that having a program that is specific for this problem has many advantages for exploiting structure. Moreover, the details of the algorithm are self-contained in this paper; and, we provide a MATLAB program for those interested in our tests. Numerical results are included which illustrate the efficiency and robustness of the interior-point approach.

The paper is organized as follows. In Section 2 we introduce the basic results of EDMs. In Section 3 we present the optimality conditions for the problem. In Section 4 we derive the algorithm and the Slater constraint qualification result. We conclude with several remarks and numerical tests in Section 5. In addition, we include Section 5.1 with some technical details on the SDP algorithm.

2. Distance geometry

It is well known, e.g., [15, 16, 34, 38], that a pre-distance matrix D is a EDM if and only if D is negative semidefinite on

$$M := \{x \in \Re^n : x^t e = 0\},$$

the orthogonal complement of e, where e is the vector of all ones. Thus the set of all EDMs is a convex cone, which we denote by $\mathcal{E}$. We exploit this result to translate the cone $\mathcal{E}$ to the cone of semidefinite matrices in $\mathcal{S}_{n-1}$, the space of symmetric matrices of order $n-1$.

Define the $n \times n$ orthogonal matrix

$$Q := \left[\frac{1}{\sqrt{n}}e \mid V\right], \quad Q^t Q = I. \tag{2}$$

Thus $V^t e = 0$ and $V^t V = I$. Moreover, the subspace M can be represented as the range of the $n \times (n-1)$ matrix V and

$$J := VV^t = I - \frac{ee^t}{n} \tag{3}$$

is the orthogonal projection onto M.

Now define the *centered* and *hollow* subspaces of $\mathcal{S}_n$

$$\begin{aligned} \mathcal{S}_C &:= \{B \in \mathcal{S}_n : Be = 0\}, \\ \mathcal{S}_H &:= \{D \in \mathcal{S}_n : \text{diag}(D) = 0\}, \end{aligned} \tag{4}$$

where diag(D) denotes the column vector formed from the diagonal of D. Following [10], we define the two linear operators acting on $\mathcal{S}_n$

$$\mathcal{K}(B) := \text{diag}(B)\, e^t + e\, \text{diag}(B)^t - 2B, \tag{5}$$

and

$$\mathcal{T}(D) := -\frac{1}{2} JDJ. \tag{6}$$

The operator $-2\mathcal{T}$ is an orthogonal projection onto $\mathcal{S}_C$; thus it is a self-adjoint idempotent.

Theorem 1. *The linear operators satisfy*

$$\mathcal{K}(\mathcal{S}_C) = \mathcal{S}_H,$$
$$\mathcal{T}(\mathcal{S}_H) = \mathcal{S}_C,$$

and $\mathcal{K}_{|\mathcal{S}_C}$ *and* $\mathcal{T}_{|\mathcal{S}_H}$ *are inverses of each other.*

Proof: See [15, 21]. □

It can easily be verified that

$$\mathcal{K}^*(D) = 2(\text{Diag}(De) - D) \tag{7}$$

is the adjoint operator of $\mathcal{K}$, where $\text{Diag}(De)$ denotes the diagonal matrix formed from the vector De. In addition, a hollow matrix D is EDM if and only if $B = \mathcal{T}(D)$ is positive semidefinite, denoted by $B \succeq 0$. (We denote positive definiteness by $B \succ 0$.) Equivalently, D is EDM if and only if $D = \mathcal{K}(B)$, for some B with $Be = 0$ and $B \succeq 0$. In this case the embedding dimension r is given by the rank of B. Moreover, if $B = XX'$, then the coordinates of the points $x^1, x^2, \ldots, x^n$ that generate D are given by the rows of X and, since $Be = 0$, it follows that the origin coincides with the centroid of these points. For these and other basic results on EDM see [15, 16, 21, 24, 34].

We now introduce the composite operators

$$\mathcal{K}_V(X) := \mathcal{K}(VXV'), \tag{8}$$

and

$$\mathcal{T}_V(D) := V'\mathcal{T}(D)V = -\frac{1}{2}V'DV, \tag{9}$$

where V is defined in (2).

Lemma 1.

$$\mathcal{K}_V(\mathcal{S}_{n-1}) = \mathcal{S}_H,$$
$$\mathcal{T}_V(\mathcal{S}_H) = \mathcal{S}_{n-1},$$

and $\mathcal{K}_V$ *and* $\mathcal{T}_V$ *are inverses of each other on these two spaces.*

Proof: This immediately follows from Theorem 1 and the definition of V. □

From (7), we get that

$$\mathcal{K}_V^*(D) = V'\mathcal{K}^*(D)V \tag{10}$$

is the adjoint operator of $\mathcal{K}_V$. The following corollary summarizes the relationships between $\mathcal{E}$, the cone of Euclidean distance matrices of order n, and $\mathcal{P}$, the cone of positive semidefinite matrices of order $n-1$, which we will need for our model problem.

Corollary 1. *Suppose that V is defined as in* (2). *Then*:

$$\mathcal{K}_V(\mathcal{P}) = \mathcal{E},$$
$$\mathcal{T}_V(\mathcal{E}) = \mathcal{P}.$$

Proof: We saw earlier that D is EDM if and only if $D = \mathcal{K}(B)$ with $Be = 0$ and $B \succeq 0$. Let $X = V^t BV$, then since $Be = 0$ we have $B = VXV^t$. Therefore, $VXV^t \succeq 0$ if and only if $X \succeq 0$; and the result follows using (8) and Lemma 1. □

Note that the $n \times (n-1)$ matrix V as defined in (2) is not unique. In our code we use

$$V := \begin{bmatrix} y & y & \cdots & y \\ 1+x & x & \cdots & x \\ x & 1+x & \cdots & x \\ \cdots & \cdots & \ddots & \cdots \\ x & x & \cdots & 1+x \end{bmatrix}, \tag{11}$$

where $x = \frac{-1}{n+\sqrt{n}}$ and $y = \frac{-1}{\sqrt{n}}$. With this choice, it can be easily verified that $V^t e = 0$, $V^t V = I$, and $VV^t = J$ as required by (2).

2.1. *Program formulations*

Since $\operatorname{diag}(A) = \operatorname{diag}(D) = 0$, we can assume without loss of generality that $\operatorname{diag}(H) = 0$. Note that $H_{ij} = 0$ means that D_{ij} is free, while $H_{ij} > 0$ forces D_{ij} to be approximately equal to A_{ij}, i.e., A_{ij} is approximately fixed. If we want $D_{ij} = A_{ij}$ exactly, then we can add a linear constraint to the program (see below). Recall that the graph of H is *connected* if for all indices $i \neq j$ there is a *path* of indices $i_1, i_2, \ldots, i_k$ such that $H_{i,i_1} \neq 0$, $H_{i_1,i_2} \neq 0, \ldots, H_{i_{k-1},i_k} \neq 0$, $H_{i_k,j} \neq 0$, see [8]. Thus, we can assume that the graph of H is connected or the problem can be solved more simply as two smaller problems, see Lemma 3. In particular, we can assume that H does not have a row (hence a column) of all zeros; otherwise the corresponding row and column in A and D are free (independent) and the problem can be posed in a lower dimensional space.

By abuse of notation, let the function

$$f(X) := \|H \circ (A - \mathcal{K}_V(X))\|_F^2 = \|H \circ \mathcal{K}_V(B - X)\|_F^2,$$

where $B = \mathcal{T}_V(A)$. We now apply Corollary 1 and get the following problem, equivalent to (CDM_0).

$$\text{(CDM)} \quad \mu^* := \min \quad f(X) \quad \text{subject to} \quad \mathcal{A}X = b, \quad X \succeq 0.$$

We allow for an additional constraint using the linear operator $\mathcal{A}: \mathcal{S}_{n-1} \to \Re^m$, i.e., $b \in \Re^m$. The addition of this linear operator could represent some of the fixed elements in the given matrix A, e.g., adding the constraint $(K_V(X))_{ij} = A_{ij}$ fixes the ij element of D. Also, note that $X \in \mathcal{S}_{n-1}$. It is in this lower dimensional space that we solve the problem. We can recover the optimal distance matrix using the optimal X and the relation

$$D = \mathcal{K}_V(X).$$

Using finite precision, we can never solve the approximation problem exactly. In addition, we need to calculate the embedding dimension. The following lemma shows we lose little in the objective function if we choose a small embedding dimension using a numerical rank approach, i.e., if we only discard very small eigenvalues, then the objective function changes very little.

Lemma 2. *Suppose that X^* solves* (CDM). *Let $\bar{X}$ be the closest symmetric matrix to X^* with rank k, i.e., we set the smallest $n-k$ eigenvalues of X^* to 0, $\lambda_{k+1} = \ldots \lambda_n = 0$. Then,*

$$\sqrt{f(\bar{X})} \le \sqrt{f(X^*)} + 2\gamma(\sqrt{n}+1)\sqrt{\sum_{i=k+1}^{n} \lambda_i^2}. \tag{12}$$

where $\gamma := \max_{ij} H_{ij}$.

Proof: By the Cauchy-Schwartz inequality,

$$\sqrt{f(\bar{X})} \le \sqrt{f(X^*)} + \gamma \|\mathcal{K}(V(X^* - \bar{X})V')\|_F.$$

The result now follows from the definition of $\mathcal{K}$. More precisely, let $B = V(X^* - \bar{X})V'$. Then

$$\begin{aligned} \|\mathcal{K}(V(X^* - \bar{X})V')\|_F &= \|\mathcal{K}(B)\|_F \\ &\le 2\|B\|_F + 2\|\text{diag}(B)e'\|_F \\ &\le 2\|B\|_F + 2\sqrt{n}\|B\|_F. \end{aligned}$$

□

3. Duality and optimality

We now derive the optimality conditions and duality theory needed for a primal-dual interior-point approach.

For $\Lambda \in \mathcal{S}_{n-1}$ and $y \in R^m$, let

$$L(X, y, \Lambda) = f(X) + \langle y, b - \mathcal{A}(X)\rangle - \text{trace } \Lambda X \tag{13}$$

denote the *Lagrangian* of (CDM). It is easy to see that the primal program (CDM) is equivalent to

$$\mu^* = \min_{X} \max_{\substack{y \\ \Lambda \succeq 0}} L(X, y, \Lambda) = \min_{X \succeq 0} \max_{\substack{y \\ \Lambda \succeq 0}} L(X, y, \Lambda). \tag{14}$$

We assume that the generalized Slater's constraint qualification,

$$\exists X \succ 0 \text{ with } \mathcal{A}(X) = b,$$

holds for (CDM). Slater's condition implies that strong duality holds, i.e.,

$$\mu^* = \max_{\substack{y \\ \Lambda \succeq 0}} \min_{X} L(X, y, \Lambda) = \max_{\substack{y \\ \Lambda \succeq 0}} \min_{X \succeq 0} L(X, y, \Lambda), \tag{15}$$

and μ^* is attained for some y and $\Lambda \succeq 0$, see [44]. Since the semidefinite constraint on X can be treated as redundant, the inner minimization of the convex, in X, Lagrangian is unconstrained and we can differentiate to get the equivalent problem

$$\mu^* = \max_{\substack{\nabla f(X) - \mathcal{A}^* y = \Lambda \\ \Lambda \succeq 0}} f(X) + \langle y, b - \mathcal{A}(X)\rangle - \text{trace } \Lambda X. \tag{16}$$

We can now state the dual problem

$$\begin{array}{llll} \text{(DCDM)} & \mu^* := \max & f(X) + \langle y, b - \mathcal{A}(X)\rangle - \text{trace } \Lambda X & \\ & \text{subject to} & \nabla f(X) - \mathcal{A}^* y - \Lambda = 0 & (17) \\ & & \Lambda \succeq 0, (X \succeq 0). & \end{array}$$

We keep the semidefinite constraint on X in brackets to emphasize that it is a hidden constraint in the dual, though it is a given constraint in the primal. The above pair of primal and dual programs, (CDM) and (DCDM), provide an optimality criteria in terms of feasibility and complementary slackness. This provides the basis for many algorithms including primal-dual interior-point algorithms. In particular, we see that the duality gap, in the case of primal and dual feasibility, is given by the complementary slackness condition:

$$\text{trace } X\left(2\mathcal{K}_V^*\left(H^{(2)} \circ \mathcal{K}_V(\bar{X} - B)\right) - \mathcal{A}^*\bar{y}\right) = 0, \tag{18}$$

or equivalently

$$X\left(2\mathcal{K}_V^*\left(H^{(2)}\circ\mathcal{K}_V(\bar{X}-B)\right)-\mathcal{A}^*\bar{y}\right)=0,$$

where $H^{(2)} = H \circ H$.

Theorem 2. *Suppose that Slater's condition holds. Then $\bar{X} \succeq 0$, and $\bar{y}$, $\bar{\Lambda} \succeq 0$ solve (CDM) and $(DCDM)$, respectively, if and only if the following three equations hold.*

$$\begin{aligned} A(\bar{X}) &= b \quad \textit{primal feasibility} \\ 2\mathcal{K}_V^*\left(H^{(2)}\circ\mathcal{K}_V(\bar{X}-B)\right)-\mathcal{A}^*\bar{y}-\bar{\Lambda} &= 0 \quad \textit{dual feasibility} \\ \operatorname{trace}\bar{\Lambda}\bar{X} &= 0 \quad \textit{complementary slackness} \end{aligned}$$

In the case that there is no linear operator $\mathcal{A}$, we get a simplified variational principle. The statement that no feasible direction is a descent direction translates into the following characterization of an optimal solution $\bar{X}$ of (CDM):

$$\nabla f(\bar{X}) \in (\mathcal{P}-\bar{X})^+, \tag{19}$$

where $\mathcal{P}$ is the cone of positive semidefinite matrices and

$$S^+ = \{P : \operatorname{trace} QP \geq 0, \quad \forall Q \in S\}$$

is the *polar cone* of the set S. This yields the following characterization of optimality.

Theorem 3. *Suppose that (CDM) has no linear constraint $\mathcal{A}$. The matrix $\bar{X} \succeq 0$ solves (CDM) if and only if*

$$\operatorname{trace}\mathcal{K}_V^*\left(H^{(2)}\circ\mathcal{K}_V(\bar{X}-B)\right)(X-\bar{X}) \geq 0, \quad \forall X \succeq 0. \tag{20}$$

Proof: Note that the gradient acting on the symmetric matrix h, is

$$\begin{aligned} \langle\nabla f(X), h\rangle &= 2\operatorname{trace}\left(H^{(2)}\circ\mathcal{K}_V(\bar{X}-B)\right)\mathcal{K}_V(h) \\ &= 2\operatorname{trace}\mathcal{K}_V^*\left(H^{(2)}\circ\mathcal{K}_V(\bar{X}-B)\right)h. \end{aligned} \tag{21}$$

Therefore, the gradient of the objective function is

$$\nabla f(X) = 2\mathcal{K}_V^*\left(H^{(2)}\circ\mathcal{K}_V(\bar{X}-B)\right). \tag{22}$$

The result follows upon replacing h by the direction $X-\bar{X}$, for $X \succeq 0$, and applying the so-called Pshenichnyi condition (19). □

4. Primal-dual interior-point algorithm

We now derive a primal-dual interior point method using the log-barrier approach [18]. This is an alternative way of deriving the optimality conditions in Theorem 2. For simplicity, we treat the unconstrained problem, i.e., we consider (CDM) without the linear equality constraint $\mathcal{A}X = b$. In this case, the Slater constraint qualification holds for both primal and dual problems, as can be seen from the following lemma. (In the case that some of the elements of A are definitely fixed, then one needs to rederive the algorithm and include the constraint $\mathcal{A}X = b$.)

Lemma 3. *Let H be an $n \times n$ symmetric matrix with nonnegative elements and* 0 *diagonal such that the graph of H is connected. Then*

$$\mathcal{K}_V^*\left(H^{(2)} \circ \mathcal{K}_V(I)\right) \succ 0,$$

where $I \in \mathcal{S}_{n-1}$ is the identity matrix.

Proof: A simple calculation shows that $\mathcal{K}_V(I) = 2(ee^t - I)$. Thus $H^{(2)} \circ \mathcal{K}_V(I) = 2H^{(2)}$ and $\mathcal{K}_V^*(H^{(2)} \circ \mathcal{K}_V(I)) = 4V^t(\mathrm{Diag}(H^{(2)}e) - H^{(2)})V$. Note that e is an eigenvector of $\mathrm{Diag}(H^{(2)}e) - H^{(2)}$ with zero eigenvalue. We show that 0 is a simple eigenvalue. Assume to the contrary that there exists another nonzero eigenvector u, i.e., u is not a multiple of e. Then

$$u^t\left(\mathrm{Diag}\left(H^{(2)}e\right) - H^{(2)}\right)u = \sum_{i<j} H_{ij}^2(u_i - u_j)^2 = 0.$$

From this and since the graph of H is connected, it follows immediately that u_i = constant for all i, a contradiction. Since V is full column rank and $V^t e = 0$, we conclude that the columns of V form a basis for the range of $(\mathrm{Diag}(H^{(2)}e) - H^{(2)})$, i.e., $4V^t(\mathrm{Diag}(H^{(2)}e) - H^{(2)})V$ is full rank. □

Strict primal feasibility follows since every $X \succeq 0$ is feasible. Strict dual feasibility, i.e., $\Lambda \succ 0$, follows in the case of a connected graph, since $\alpha > 0$ can be chosen such that

$$X = B + \alpha I \succ 0; \tag{23}$$

we can then apply Lemma 3 to get

$$\mathcal{K}_V^*\left(H^{(2)} \circ \mathcal{K}_V(X - B)\right) \succ 0. \tag{24}$$

Corollary 2. *Slater's constraint qualification holds for the dual problem* (17) *if and only if the graph of H is connected.*

Proof: Sufficiency follows directly from Lemma 3.

To prove necessity, suppose that the graph is not connected. Then (CDM) can be solved as two smaller disconnected problems where the distances joining points for these two smaller problems are completely free, i.e., the set of optimal distances matrices for (CDM_0) is unbounded. Therefore, the set of optimal solutions X of (CDM) is unbounded. This implies that Slater's condition cannot hold in the dual, see, e.g., the proof of [14], Theorem 4 or [27]. In our case, it is rather simple to prove this result and we include it for completeness, i.e., we prove that Slater's condition for the dual implies that the optimal set for the primal is bounded.

Suppose that Slater's condition holds for the dual, i.e., there exists $\bar{\Lambda} \succ 0$ and $\bar{X} \succeq 0$ such that $\nabla f(\bar{X}) - \bar{\Lambda} = 0$. Equivalently, this means that $\nabla f(\bar{X}) \succ 0$. Let X^* be optimal for (CDM). Then convexity of f implies that

$$\langle \nabla f(\bar{X}), X^* - \bar{X} \rangle \leq 0.$$

Therefore, we get

$$\begin{aligned}\langle \nabla f(\bar{X}), \bar{X} \rangle &\geq \langle \nabla f(\bar{X}), X^* \rangle \\ &\geq \lambda_{\min}(\nabla f(\bar{X}))\operatorname{trace} X^* \\ &\geq \lambda_{\min}(\nabla f(\bar{X}))\lambda_{\max}(X^*),\end{aligned}$$

i.e., the norm of X^* is bounded by $\frac{n}{\lambda_{\min}(\nabla f(\bar{X}))}\langle \nabla f(\bar{X}), \bar{X} \rangle$. □

The log-barrier problem for (CDM) is

$$\min_{X \succ 0} B_\mu(X) := f(X) - \mu \log\det(X),$$

where $\mu \downarrow 0$. For each $\mu > 0$ we take one Newton step for solving the stationarity condition

$$\nabla B_\mu(X) = 2\mathcal{K}_V^*\big(H^{(2)} \circ \mathcal{K}_V(X - B)\big) - \mu X^{-1} = 0. \tag{25}$$

Let

$$C := 2\mathcal{K}_V^*\big(H^{(2)} \circ \mathcal{K}_V(B)\big) = 2\mathcal{K}_V^*\big(H^{(2)} \circ A\big). \tag{26}$$

Then the stationarity condition is equivalent to

$$\nabla B_\mu(X) = 2\mathcal{K}_V^*\left(H^{(2)} \circ \mathcal{K}_V(X)\right) - C - \mu X^{-1} = 0. \tag{27}$$

By equating $\Lambda = \mu X^{-1}$, and multiplying through by X, we get the optimality conditions, $F := \binom{F_d}{F_c} = 0$,

$$\begin{aligned}F_d &:= 2\mathcal{K}_V^*\big(H^{(2)} \circ \mathcal{K}_V(X)\big) - C - \Lambda = 0 \quad &&\text{dual feas.} \\ F_c &:= \Lambda X - \mu I = 0 &&\text{perturbed compl. slack.,}\end{aligned} \tag{28}$$

and the estimate of the barrier parameter

$$\mu = \frac{1}{n-1} \text{ trace } \Lambda X. \tag{29}$$

Following, we present the p-d i-p framework we used. This is a common framework for both linear and semidefinite programming, see [45]. We include a centering parameter σ_k (rather than the customary predictor-corrector approach) and let $\mathcal{F}^0$ denote the set of strictly feasible primal-dual points; F' denotes the derivative of the function of optimality conditions.

Algorithm 1 (p-d i-p framework).

Given $(X^0, \Lambda^0) \in \mathcal{F}^0$
for $k = 0, 1, 2, \dots$
solve *for the search direction* (*in a least squares sense*)

$$F'(X^k, \Lambda^k) \begin{pmatrix} \delta X^k \\ \delta \Lambda^k \end{pmatrix} = \begin{pmatrix} -F_d \\ \Lambda^k X^k + \sigma_k \mu_k I \end{pmatrix}$$

where σ_k *centering*, $\mu_k = \text{trace } X^k \Lambda^k / (n-1)$

$$(X^{k+1}, \Lambda^{k+1}) = (X^k, \Lambda^k) + \alpha_k(\delta X^k, \delta \Lambda^k), \quad \alpha_k > 0,$$

so that $(X^{k+1}, \Lambda^{k+1}) \succ 0$

end (for).

We use two approaches for solving the least squares problem. First we solve the large least squares problem; we call this the Gauss-Newton (GN) method. In the second approach, we restrict dual feasibility and substitute for $\delta \Lambda^k$ in the second equation; we call this the restricted Gauss-Newton (RGN) method. The numerical tests show that the second approach is significantly more efficient.

5. Conclusion and computational results

In this paper we have presented an interior-point algorithm for finding the weighted, closest Euclidean distance matrix; this provides a solution for the approximate Euclidean distance matrix completion problem. The algorithm has been extensively tested and has proven to be efficient and very robust, i.e., it has not failed on any of our test problems. In addition, an important observation is that the ranks of the optimal solutions X, for sparse problems where no completion existed, were typically between 1 and 3, i.e., a very small embedding dimension was obtained without any original rank restrictions.

Table 1. Data for closest distance matrix: dimension; tolerance for duality gap; density of nonzeros in H; rank of optimal X; number of iterations; cpu-time for one least squares solution of the GN and restricted GN directions.

					lss cpu-time	
dim	toler	H dens.	rank(X)	Iterations	GN	RGN
8	10^{-13}	0.8	2	25	0.16	0.1
9	10^{-13}	0.8	2	23	0.24	0.13
10	10^{-13}	0.8	3	25	0.34	0.18
12	10^{-9}	0.5	3	17	0.73	0.32
15	10^{-9}	0.5	2	20	2.13	0.79
18	10^{-9}	0.5	4	20	6.15	1.9
20	10^{-9}	0.3	2	20	11.35	3.3
24	10^{-9}	0.3	2	20	34.45	8.4
30	10^{-9}	0.3	4	20	138.0	31.5
35	10^{-9}	0.2	3	19	373.0	77.0
38	10^{-9}	0.2	3	19	634.0	127
40	10^{-8}	0.1	2	20	845.9	181.7
42	10^{-8}	0.1	4	18	1118.0	232.02

However, when a completion existed, then typically the embedding dimension was high. In this case, we can apply the technique presented in [1] to "purify", i.e., to iteratively move to an optimal solution of smaller rank on the optimal face (see also Pataki [32]).

We discuss more details of the Gauss-Newton approach in Section 5.1. The program was written in MATLAB. The tests were done on randomly generated problems; we used s SPARC 20 with 64 megs of RAM and SPECint 65.3, SPECfp 53.1. The documented MATLAB code, as well as ongoing test results, can be obtained with URL (or anonymous ftp)

ftp://orion.math.uwaterloo.ca/pub/henry/software/distance.d, or
http://orion.math.uwaterloo.ca/~hwolkowi/henry/software/distance.d.

In Table 1 we present a sample of test results. These results include problems with matrices up to dimension $n = 42$. We would like to emphasize that these are just preliminary test results. For example, we do not use a predictor-corrector approach, which has become the standard approach in interior-point methods, but rather use a centering parameter. Using the predictor-corrector approach should reduce the number of iterations from 20 to 30%. Tests with larger sized matrices are in progress.

We conclude with a specific example and its graph: the dimension is 11 generated with sparsity 0.5; the optimal value is 0.9256; the rank of X is 3; and the number of iterations is 25 to get 13 decimals accuracy. The matrices H and A and the optimal distance matrix D corresponding to the graph in figure 1 are, respectively:

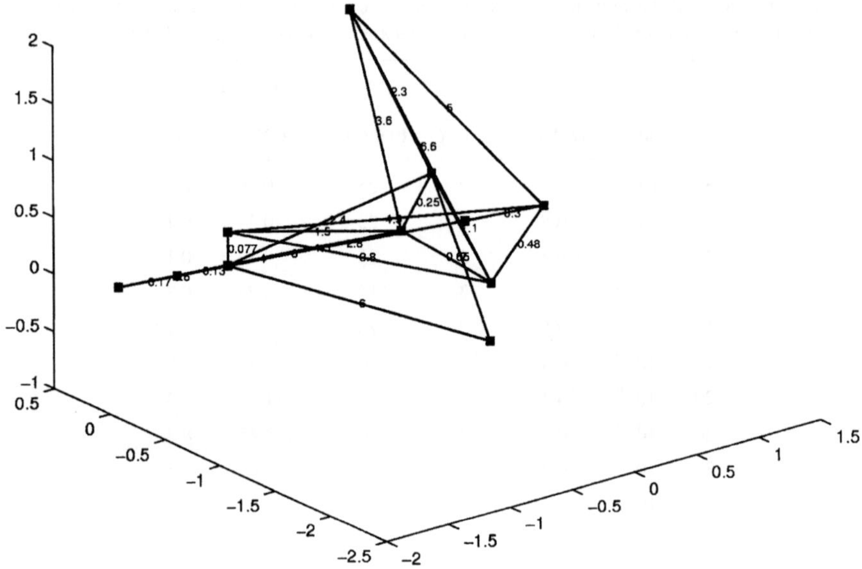

Figure 1. Approximate completion problem.

```
matrix H =

  0  0  0  0  0  5  2  5  0  7  0
  0  0  0  3  4  4  0  0  1  0  5
  0  0  0  0  0  0  1  0  0  0  2
  0  3  0  0  0  1  0  0  3  0  3
  0  4  0  0  0  2  7  0  3  0  0
  5  4  0  1  2  0  2  0  0  0  7
  2  0  1  0  7  2  0  6  0  1  2
  5  0  0  0  0  0  6  0  0  0  0
  0  1  0  3  3  0  0  0  0  3  0
  7  0  0  0  0  0  1  0  3  0  0
  0  5  2  3  0  7  2  0  0  0  0

matrix A =

  0  0  0  0  0  4  0  0  6  6  2
  0  0  0  8  4  0  2  0  0  0  0
  0  0  0  5  6  0  6  0  0  0  7
  0  8  5  0  0  4  1  4  5  4  0
  0  4  6  0  0  0  0  0  5  0  0
```

```
4  0  0  4  0  0  1  0  0  0  0
0  2  6  1  0  1  0  0  0  0  3
0  0  0  4  0  0  0  0  0  0  0
6  0  0  5  5  0  0  0  0  0  0
6  0  0  4  0  0  0  0  0  0  0
2  0  7  0  0  0  3  0  0  0  0
```

```
matrix D =
Columns 1 through 7
        0  6.9200  7.2655  6.9713  0.7190  3.9912  0.5989
   6.9200       0  8.1310  6.6123  3.8224  0.6523  3.5381
   7.2655  8.1310       0  9.3511  5.9827  6.4399  6.0000
   6.9713  6.6123  9.3511       0  3.7085  3.6117  4.7459
   0.7190  3.8224  5.9827  3.7085       0  1.4994  0.0775
   3.9912  0.6523  6.4399  3.6117  1.4994       0  1.4981
   0.5989  3.5381  6.0000  4.7459  0.0775  1.4981       0
   0.1684  4.9773  6.4452  5.6419  0.2296  2.5198  0.1321
   8.9782  0.4804  9.0262  5.0302  4.8968  0.9976  4.9394
   5.9737  0.3183  7.3517  4.0355  2.7767  0.1996  2.7897
   5.2815  1.0844  7.0000  2.3417  2.1414  0.2465  2.3892

Columns 8 through 11
   0.1684  8.9782  5.9737  5.2815
   4.9773  0.4804  0.3183  1.0844
   6.4452  9.0262  7.3517  7.0000
   5.6419  5.0302  4.0355  2.3417
   0.2296  4.8968  2.7767  2.1414
   2.5198  0.9976  0.1996  0.2465
   0.1321  4.9394  2.7897  2.3892
        0  6.6871  4.1359  3.5984
   6.6871       0  0.3050  0.7459
   4.1359  0.3050       0  0.2315
   3.5984  0.7459  0.2315       0
```

5.1. *Gauss-Newton direction*

The linear system for the search direction in Algorithm 1 is overdetermined. Therefore, it is not clear what is meant by solving this system. There are many different search directions that have been used for SDP, see [37]. Our problem is not a standard SDP since it has a quadratic objective and no constraints. Therefore, the standard public domain packages do not apply directly.[2] In our approach we use the Gauss-Newton direction introduced in [23], i.e., we linearize (28) and find the least squares solution of the resulting overdetermined linear system. This direction has many good properties, e.g.: it always exists; the linear systems for both GN and RGN are nonsingular at each iteration and in the limit; quadratic convergence steps are expected since the optimal value of the nonlinear least squares problems are 0. (For the details of these and other properties see [23].)

Notes

1. This report is available by anonymous ftp at orion.math.uwaterloo.ca in the directory pub/henry/reports or with URLs: ftp://orion.uwaterloo.ca/pub/henry/reports/distmat.ps.gz or http://orion.math.uwaterloo.ca:80/~hwolkowi/henry/reports/distmat.ps.gz.
2. It has been pointed out to us that the package SDPpack [6], will solve our problem if we replace our original objective function with the variable t and add a constraint $\|W\|_F \leq t$, where W represents the appropriate quantity in our original objective function. The numerical tests with this approach appear to be comparable to our approach. We thank Mike Overton for this observation.

References

1. A. Alfakih and H. Wolkowicz, "On the embeddability of weighted graphs in Euclidean spaces," Technical Report CORR Report 98-12, University of Waterloo, 1998. Submitted—URL: ftp://orion.uwaterloo.ca/pub/henry/reports/embedEDM.ps.gz.
2. S. Al-Homidan, "Hybrid methods for optimization problems with positive semidefinite matrix constraints," Ph.D. Thesis, University of Dundee, 1993.
3. S. Al-Homidan and R. Fletcher, "Hybrid methods for finding the nearest Euclidean distance matrix," in Recent Advances in Nonsmooth Optimization, World Sci. Publishing River Edge, NJ, 1995, pp. 1–17.
4. F. Alizadeh, "Combinatorial optimization with interior point methods and semidefinite matrices," Ph.D. Thesis, University of Minnesota, 1991.
5. F. Alizadeh, "Interior point methods in semidefinite programming with applications to combinatorial optimization," SIAM Journal on Optimization, vol. 5, pp. 13–51, 1995.
6. F. Alizadeh, J.-P. Haeberly, M.V. Nayakkankuppam, and M.L. Overton, "Sdppack user's guide—version 0.8 beta," Technical Report TR1997-734, Courant Institute of Mathematical Sciences, NYU, New York, March 1997.
7. M. Bakonyi and C.R. Johnson, "The Euclidean distance matrix completion problem," SIAM J. Matrix Anal. Appl., vol. 16, no. 2, pp. 646–654, 1995.
8. R.A. Brualdi and H.J. Ryser, Combinatorial Matrix Theory, Cambridge University Press: New York, 1991.
9. G.M. Crippen and T.F. Havel, Distance Geometry and Molecular Conformation, Wiley: New York, 1988.
10. F. Critchley, "On certain linear mappings between inner-product and squared distance matrices." Linear Algebra Appl., vol. 105, pp. 91–107, 1988.
11. R.W. Farebrother, "Three theorems with applications to Euclidean distance matrices," Linear Algebra Appl., vol. 95, pp. 11–16, 1987.
12. W. Glunt, T.L. Hayden, S. Hong, and J. Wells, "An alternating projection algorithm for computing the nearest Euclidean distance matrix." SIAM J. Matrix Anal. Appl., vol. 11, no. 4, pp. 589–600, 1990.

13. M.X. Goemans, "Semidefinite programming in combinatorial optimization," Mathematical Programmings, vol. 79, pp. 143–162, 1997.
14. E.G. Gol'stein, Theory of Convex Programming, American Mathematical Society: Providence, RI, 1972.
15. J.C. Gower, "Properties of Euclidean and non-Euclidean distance matrices," Linear Algebra Appl., vol. 67, pp. 81–97, 1985.
16. T.L. Hayden, J. Wells, W-M. Liu, and P. Tarazaga, "The cone of distance matrices," Linear Algebra Appl., vol. 144, pp. 153–169, 1991.
17. C. Helmberg, "An interior point method for semidefinite programming and max-cut bounds," Ph.D. Thesis, Graz University of Technology, Austria, 1994.
18. C. Helmberg, F. Rendl, R.J. Vanderbei, and H. Wolkowicz, "An interior point method for semidefinite programming," SIAM Journal on Optimization, pp. 342–361, 1996. URL: ftp://orion.uwaterloo.ca/pub/henry/reports/sdp.ps.gz.
19. R.A. Horn and C.R. Johnson, Matrix Analysis, Cambridge University Press: New York, 1985.
20. C. Johnson, B. Kroschel, and H. Wolkowicz, "An interior-point method for approximate positive semidefinite completions," Computational Optimization and Applications, vol. 9, no. 2, pp. 175–190, 1998.
21. C.R. Johnson and P. Tarazaga, "Connections between the real positive semidefinite and distance matrix completion problems," Linear Algebra Appl., vol. 223/224, pp. 375–391, 1995.
22. E. De Klerk, "Interior point methods for semidefinite programming," Ph.D. Thesis, Delft University, 1997.
23. S. Kruk, M. Muramatsu, F. Rendl, R.J. Vanderbei, and H. Wolkowicz, "The Gauss-Newton direction in linear and semidefinite programming," Technical Report CORR 98-16, University of Waterloo, Waterloo, Canada, 1998. Detailed Web Version at URL: ftp://orion.uwaterloo.ca/pub/henry/reports/gnsdplong.ps.gz.
24. M. Laurent, "A tour d'horizon on positive semidefinite and Euclidean distance matrix completion problems," in Topics in Semidefinite and Interior-Point Methods, The Fields Institute for Research in Mathematical Sciences, Communications Series, vol. 18, American Mathematical Society: Providence, RI, 1998.
25. J. De Leeuw and W. Heiser, "Theory of multidimensional scaling," in Handbook of Statistics, P.R. Krishnaiah and L.N. Kanal (Eds.), North-Holland, 1982, vol. 2, pp. 285–316.
26. S. Lele, "Euclidean distance matrix analysis (EDMA): Estimation of mean form and mean form difference," Math. Geol., vol. 25, no. 5, pp. 573–602, 1993.
27. F. Lempio and H. Maurer, "Differential stability in infinite-dimensional nonlinear programming," Appl. Math. Optim., vol. 6, pp. 139–152, 1980.
28. J.J. Moré and Z. Wu, "Global continuation for distance geometry problems," Technical Report MCS-P505-0395, Applied Mathematics Division, Argonne National Labs, Chicago, IL, 1996.
29. J.J. Moré and Z. Wu, "Distance geometry optimization for protein structures," Technical Report MCS-P628-1296, Applied Mathematics Division, Argonne National Labs, Chicago, IL, 1997.
30. Y.E. Nesterov and A.S. Nemirovski, Interior Point Polynomial Algorithms in Convex Programming, SIAM Publications, SIAM: Philadelphia, USA, 1994.
31. P.M. Pardalos, D. Shalloway, and G. Xue (Eds.), "Global minimization of nonconvex energy functions: Molecular conformation and protein folding," DIMACS Series in Discrete Mathematics and Theoretical Computer Science, vol. 23, American Mathematical Society: Providence, RI, 1996. Papers from the DIMACS Workshop held as part of the DIMACS Special Year on Mathematical Support for Molecular Biology at Rutgers University, New Brunswick, New Jersey, March 1995.
32. G. Pataki, "Cone programming and eigenvalue optimization: Geometry and algorithms," Ph.D. Thesis, Carnegie Mellon University, Pittsburgh, PA, 1996.
33. M.V. Ramana, "An algorithmic analysis of multiquadratic and semidefinite programming problems," Ph.D. Thesis, Johns Hopkins University, Baltimore, MD, 1993.
34. I.J. Schoenberg, "Remarks to Maurice Frechet's article: Sur la definition axiomatique d'une classe d'espaces vectoriels distancies applicables vectoriellement sur l'espace de Hilbert," Ann. Math., vol. 36, pp. 724–732, 1935.
35. J. Sturm, "Primal-dual interior point approach to semidefinite programming," Ph.D. Thesis, Erasmus University Rotterdam, 1997.
36. P. Tarazaga, T.L. Hayden, and J. Wells, "Circum-Euclidean distance matrices and faces," Linear Algebra Appl., vol. 232, pp. 77–96, 1996.

37. M. Todd, "On search directions in interior-point methods for semidefinite programming," Technical Report TR1205, School of OR and IE, Cornell University, Ithaca, NY, 1997.
38. W.S. Torgerson, "Multidimensional scaling. I. Theory and method," Psychometrika, vol. 17, pp. 401–419, 1952.
39. M.W. Trosset, "Applications of multidimensional scaling to molecular conformation," Technical Report, Rice University, Houston, Texas, 1997.
40. M.W. Trosset, "Computing distances between convex sets and subsets of the positive semidefinite matrices," Technical Report, Rice University, Houston, Texas, 1997.
41. M.W. Trosset, "Distance matrix completion by numerical optimization," Technical Report, Rice University, Houston, Texas, 1997.
42. L. Vandenberghe and S. Boyd, "Positive definite programming," in Mathematical Programming: State of the Art, 1994, The University of Michigan, 1994, pp. 276–308.
43. L. Vandenberghe and S. Boyd, "Semidefinite programming," SIAM Review, vol. 38, pp. 49–95, 1996.
44. H. Wolkowicz, "Some applications of optimization in matrix theory," Linear Algebra and its Applications, vol. 40, pp. 101–118, 1981.
45. S. Wright, Primal-Dual Interior-Point Methods, SIAM: Philadelphia, PA, 1996.
46. Y. Ye, Interior Point Algorithms: Theory and Analysis, Wiley-Interscience Series in Discrete Mathematics and Optimization, John Wiley & Sons: New York, 1997.
47. F.Z. Zhang, "On the best Euclidean fit to a distance matrix," Beijing Shifan Daxue Xuebao, vol. 4, pp. 21–24, 1987.
48. Z. Zou, R.H. Byrd, and R.B. Schnabel, "A stochastic/perturbation global optimization algorithm for distance geometry problems," Technical Report, Department of Computer Science, University of Colorado, Boulder, CO, 1996.

Computational Optimization and Applications 12, 31–40 (1999)

A Logarithmic-Quadratic Proximal Method for Variational Inequalities*

ALFRED AUSLENDER auslen@poly.polytechnique.fr
Laboratoire d' Econometrie de L'Ecole Polytechnique, 1 Rue Descartes, Paris 75005, France

MARC TEBOULLE teboulle@math.tau.ac.il
School of Mathematical Sciences, Tel-Aviv University, Ramat-Aviv 69978, Israel

SAMI BEN-TIBA bentiba@poly.polytechnique.fr
Laboratoire d' Econometrie de L'Ecole Polytechnique, 1 Rue Descartes, Paris 75005, France

Received December 22, 1997; Accepted May 2, 1998

Abstract. We present a new method for solving variational inequalities on polyhedra. The method is proximal based, but uses a very special logarithmic-quadratic proximal term which replaces the usual quadratic, and leads to an interior proximal type algorithm. We allow for computing the iterates approximately and prove that the resulting method is globally convergent under the sole assumption that the optimal set of the variational inequality is nonempty.

Keywords: variational inequalities, nonlinear complementarity, proximal-like methods, maximal monotone operators, global convergence, interior point methods, saddle point computation

Dedication: (Auslender) It is quite natural to acknowledge those who have been for you a source of inspiration. This was the case for me with Olvi Mangasarian, who inspired the origin of some of my works. Indeed, the article of Olvi on the condition number for differentiable functions, led J.P. Crouzeix and myself to study sharp error bounds and develop the notion of well behaved asymptotic functions. Likewise, the paper of Chen and Mangasarian on smoothing methods for complementarity, motivated a work by Cominetti, Haddou and myself which allowed us to deepen the role of recession functions in numerical methods for optimization problems. It has been always fruitful for me to read the works of Olvi, and I am delighted to contribute to this volume in honor of Olvi Mangasarian 65th birthday.

(Teboulle) It is always exciting in our profession to meet with people who are passionate by their work, are creative and who have original ideas. This was the case with Olvi, whom I met on several occasions in professional meetings and with whom a conversation always leaves enlightening traces. It is with great pleasure that I dedicate this paper to Olvi Mangasarian on the occasion of his 65th birthday in appreciation to his pioneering work in our field.

1. Introduction

Let C be a polyhedral set on $\mathbb{R}^m$ defined by

$$C := \{x \in \mathbb{R}^m : Ax \leq b\}$$

*The research of the second author was partially supported by The Israeli Ministry of Science under Grant No. 9636-1-96.

where A is an (p, m) matrix, $b \in \mathbb{R}^p$, $p \geq m$. We suppose that the matrix A is of maximal rank, i.e., rank $A = m$ and that int $C = \{x : Ax < b\}$ is nonempty. Let T be a maximal monotone set valued map such that dom $T \cap \text{int}\, C \neq \emptyset$(dom $T := \{x : T(x) \neq \emptyset\}$). Recall that a set valued map T is said to be a monotone operator if

$$\langle y' - y, x' - x \rangle \geq 0, \quad \forall y' \in T(x'), \forall y \in T(x),$$

$\forall x, x' \in \text{dom}\, T$, and a monotone operator is said to be maximal if its graph

$$G(T) := \{(x, y) \in \mathbb{R}^m \times \mathbb{R}^m : y \in T(x)\}$$

is not properly contained in the graph of any other monotone operator.

We consider the variational inequality

(VI) "Find a point $x^* \in C$ and $g^* \in T(x^*)$ satisfying $\langle g^*, x - x^* \rangle \geq 0 \quad \forall x \in C$,"

which includes as a special case the nonlinear complementarity problem by choosing $C = \mathbb{R}^m_+$.

A well known method for solving (VI) is the proximal algorithm, see e.g., [8, 14]. Given $\lambda_k \geq \lambda > 0$, the inexact version of the proximal point algorithm [14] generates iteratively sequences $\{x^k\} \subset C$, $\{e^k\} \in \mathbb{R}^m$ satisfying:

$$g^k + \lambda_k^{-1}(x^k - x^{k-1}) = e^k, \quad g^k \in \hat{T}(x^k), \tag{1.1}$$

where $\hat{T}(\cdot) := T(\cdot) + N_C(\cdot)$, N_C denotes the normal cone, [13] and e^k is the error sequence. The exact form of the method corresponds to $e^k = 0$, $\forall k$.

A number of recent papers have concentrated on generalization of the proximal point algorithm replacing the linear term $x^k - x^{k-1}$, by some nonlinear functionals $r(x^k, x^{k-1})$ based on entropic proximal terms arising from appropriately-formulated Bregman functions [1, 4–6], or entropic φ-divergence [16] and leading to "interior point" proximal methods for variational inequality problems.

In all the above papers, convergence was proved under restrictive assumptions on the problem data. In fact the challenge remains to generate an *interior proximal method* which is globally convergent to a solution of (VI), under the only assumption that the set of solutions of (VI) is nonempty. Indeed, until now this objective was not attained. In [16] only ergodic convergence is proved and in [1, 4–6], it is assumed that the map T is paramonotone that is: (x, y), $(x', y') \in G(T)$ and $(x - x', y - y') = 0$ collectively imply that $(x, y') \in G(T)$. This notion, introduced by Bruck [3] to prove convergence of an iterative method, is very close to strict monotonicity and quite restrictive. For example, it is not satisfied by the maximal monotone operator associated with the saddle function of a convex program.

For solving our problem (VI), we propose in the same spirit a very new proximal based method which not only overcomes this difficulty but also seems very promising from a numerical point of view. Indeed, the "proximal logarithmic-quadratic term" we introduce in Section 2, and which is neither Bregman-based nor an entropic φ-divergence, satisfies also self-concordance properties, a notion introduced by Nesterov and Nemirovski [10] and which appears to be fundamental if we want to solve the intermediate problem by an efficient Newton method.

The structure of the paper is simple. In Section 2, we present the method and prove its convergence. In Section 3, we give an existence theorem for the intermediate subproblem and in Section 4 we briefly discuss some extensions of the method and numerical considerations.

2. The logarithmic-quadratic proximal method

Let $\nu > \mu > 0$ by any given fixed parameters. For $v \in \mathbb{R}^p_{++}$ define:

$$d(u, v) = \begin{cases} \displaystyle\sum_{i=1}^{p} \frac{\nu}{2}(u_i - v_i)^2 + \mu\left(v_i^2 \log \frac{v_i}{u_i} + u_i v_i - v_i^2\right) & \text{if } u \in \mathbb{R}^p_{++} \\ +\infty & \text{otherwise} \end{cases}$$

It is easy to verify that $d(\cdot, v)$ is a closed proper convex function, nonnegative and $d(u, v) = 0$ if and only if $u = v$.

One of the motivation behind the specific form of the functional $d(\cdot, \cdot)$ is as follows: the first quadratic term is a usual *regularization* term used in a proximal method, while the second expression is added to enforce the method to become an interior one, i.e., to generate iterates staying in the positive orthant (see also Section 4 for other motivation).

From now on, for simplicity of exposition we will use $\nu = 2$, $\mu = 1$. Then, simple algebra shows that d given above can be conveniently written as:

$$d(u, v) = \begin{cases} \displaystyle\sum_{i=1}^{p} u_i^2 - u_i v_i - v_i^2 \log \frac{u_i}{v_i} & \text{if } u \in \mathbb{R}^p_{++} \\ +\infty & \text{otherwise} \end{cases}$$

Let a_i denotes the rows of the matrix A, and define the following quantities:

$$\begin{aligned} l_i(x) &= b_i - \langle a_i, x \rangle, \\ l(x) &= (l_1(x), l_2(x), \ldots, l_p(x))^T, \\ D(x, y) &= d(l(x), l(y)). \end{aligned}$$

For each $x \in \operatorname{int} C$, $y \in \operatorname{int} C$, we have

$$\nabla_x D(x, y) = -\sum_{i=1}^{p} a_i \left(2l_i(x) - l_i(y) - \frac{l_i(y)^2}{l_i(x)}\right), \tag{2.1}$$

Throughout the paper we assume that $\operatorname{dom} T \cap \operatorname{int} C \neq \emptyset$.

The LQP (Logarithmic-Quadratic Proximal) method to solve (VI) is defined as follows:

Start with $x^0 \in \operatorname{int} C$ and generate $\{x^k\} \subset \operatorname{int} C$, and $\{e^k\} \subset \mathbb{R}^m$ satisfying

$$g^k + \lambda_k^{-1} \nabla_x D(x^k, x^{k-1}) = e^k \quad \text{with } g^k \in T(x^k), \tag{2.2}$$

where $\lambda_k \geq \lambda > 0$.

To analyze the LQP method, it is important to know under which conditions there exist sequences $\{x^k, e^k\}$ satisfying the iteration (2.2). This will be obtained as a special case of a more general existence result, which is of interest in its on sake, and proven in Section 3.

Proposition 1. *For each $\lambda_k > 0$, $e^k \in \mathbb{R}^m$, $x^{k-1} \in \operatorname{int} C$, there exists a unique $x^k \in \operatorname{int} C$ satisfying* (2.2).

Proof: See Section 3. □

In order to ensure convergence we have to assume additional conditions on the sequence of errors $\{e^k\}$. We suppose that

$$\sum_{k=0}^{\infty} \lambda_k \|e^k\| < +\infty, \tag{2.3}$$

$$\sum_{k=0}^{\infty} \lambda_k \langle e^k, x^k \rangle < +\infty. \tag{2.4}$$

This kind of conditions for the error sequence was given in [6] and similar forms can be found for example in [7]. Condition (2.4) might appear somewhat unnatural since it involves the iterates x^k which are apriori unknown. However, as it was noticed in [6, p. 10], (2.4) is easily enforcable in practice, and also implied by the more easily verified condition

$$\sum_{k=0}^{\infty} \lambda_k \|e^k\| \|x^k\| < +\infty \tag{2.5}$$

which is satisfied in particular when $e^k = 0$ for each k or when C or $\operatorname{dom} T$ is bounded (in addition, with (2.3)). Before proving our main convergence result, we will also need the following technical result.

Lemma 1. *Let $\{v_k\}$ and $\{\beta_k\}$ be nonnegative sequences of real numbers satisfying*
(i) $v_{k+1} \leq v_k + \beta_k$,
(ii) $\sum_{k=1}^{\infty} \beta_k < \infty$.
Then the sequence $\{v_k\}$ converges.

Proof: Elementary. See also [11, Chapter 2]. □

The next Lemma will play an important role in the analysis of the proposed algorithm.

Lemma 2. *For any $s > 0$, $t > 0$ and $u \geq 0$ we have*

$$(t-u)\left(2t - s - \frac{s^2}{t}\right) \geq \frac{3}{2}((u-t)^2 - (u-s)^2) + \frac{1}{2}(t-s)^2. \tag{2.6}$$

Proof: Let δ be the left-hand side of (2.6), then developing and regrouping terms we obtain:

$$\begin{aligned}\delta &= 2t^2 - st - s^2 - u(2t-s) + u\frac{s^2}{t}\\ &\geq 2t^2 - st - s^2 - u(2t-s) + u(2s-t),\end{aligned}$$

where the inequality above follows from the property $\frac{s^2}{t} \geq 2s - t$. By simple calculations we then obtain

$$\begin{aligned}\delta &\geq s(t-s) + 2t(t-s) - 3u(t-s)\\ &= (s-t)(3u - 2t - s). \qquad (2.7)\\ &= (t-s)(s-u) + 2(t-s)(t-u) \qquad (2.8)\end{aligned}$$

Using in (2.8) the identities

$$\begin{aligned}2(t-s)(s-u) &= ((u-t)^2 - (u-s)^2 - (t-s)^2),\\ 2(t-s)(t-u) &= ((u-t)^2 - (u-s)^2 + (t-s)^2),\end{aligned}$$

we obtain (2.6). □

Theorem 1. *Let $\{x^k\}$ be the sequence generated by LQP and suppose that* (2.3) *and* (2.4) *hold. Suppose in addition that the set of solutions of* (VI) *denoted by S, is nonempty. Then the sequence $\{x^k\}$ converges to a solution $x^* \in S$.*

Proof: Define $\tilde{T} = T + N_C$ where N_C is the normal cone operator, i.e.,

$$N_C(x) = \begin{cases}\{y : \langle y, u\rangle \leq \langle y, x\rangle,\ \forall u \in C\} & \text{if } x \in C\\ \emptyset & \text{otherwise.}\end{cases}$$

Since $\operatorname{dom} T \cap \operatorname{int} C \neq \emptyset$, it is well known [12] that $\tilde{T}$ is a maximal monotone operator and that

$$x^* \in S \quad \text{iff } 0 \in \tilde{T}(x^*).$$

Furthermore, since for $x^k \in \operatorname{int} C$, it holds that $N_C(x^k) = \{0\}$ and we have

$$g^k \in \tilde{T}(x^k). \tag{2.9}$$

From (2.1) and (2.2) we have

$$g^k - e^k = \lambda_k^{-1}\left(\sum_{i=1}^{p} a_i(2l_i(x^k) - l_i(x^{k-1}) - \frac{l_i(x^{k-1})^2}{l_i(x^k)}\right) \tag{2.10}$$

Using the definition of l_i and the monotonicity of $\tilde{T}$, it follows from (2.10) that $\forall(x, g) \in G(\tilde{T})$,

$$\lambda_k\langle x - x^k, g - e^k\rangle \geq \sum_{i=1}^{p}(l_i(x^k) - l_i(x))\left(2l_i(x^k) - l_i(x^{k-1}) - \frac{l_i(x^{k-1})^2}{l_i(x^k)}\right). \tag{2.11}$$

Take now in (2.11) $(x, g) = (x^*, 0)$ with $x^* \in S$. Applying Lemma 2 componentwise with $s = l_i(x^{k-1})$, $t = l_i(x^k)$ and $u = l_i(x^*)$, and summing over $i = 1, \ldots, p$, we then obtain

$$\lambda_k\langle x^* - x^k, -e^k\rangle \geq \frac{3}{2}(\|A(x^* - x^k)\|^2 - \|A(x^* - x^{k-1})\|^2) + \frac{1}{2}\|A(x^k - x^{k-1})\|^2. \tag{2.12}$$

As a consequence, we get

$$\lambda_k\langle x^* - x^k, -e^k\rangle \geq \frac{3}{2}(\|A(x^* - x^k)\|^2 - \|A(x^* - x^{k-1})\|^2).$$

Then from (2.3) and (2.4), (recall that $\lambda_k \geq \lambda > 0$) it follows from Lemma 1 that the sequence $\{\|A(x^* - x^k)\|\}$ converges. Since A is of maximal rank the function $u \to \|u\|_A := \|Au\|$ is a norm on $\mathbb{R}^m$ and it follows that the sequence $\{\|x^* - x^k\|\}$ converges. Using again (2.12), it follows that $\{\|A((x^k - x^{k-1})\|\}$ converges to zero and as a result we have

$$\lim_{k\to\infty} \|x^k - x^{k-1}\| = 0. \tag{2.13}$$

Now, let $c_{i,k}(x) := (l_i(x^k) - l_i(x))(2l_i(x^k) - l_i(x^{k-1}) - \frac{l_i(x^{k-1})^2}{l_i(x^k)})$. From (2.11), we have

$$\langle x - x^k, g - e^k\rangle \geq \lambda_k^{-1}\sum_{i=1}^{p} c_{i,k}(x). \tag{2.14}$$

Using (2.7) with $s = l_i(x^{k-1})$, $t = l_i(x^k)$, $u = l_i(x)$, $x \in C$ we obtain that

$$c_{i,k}(x) \geq (l_i(x^{k-1}) - l_i(x^k))(3l_i(x) - (l_i(x^{k-1}) + 2l_i(x^k))).$$

Since $\{x^k\}$ is bounded, $\{(l_i(x^{k-1}) - l_i(x^k))\}$ converges to zero and $\{\lambda_k\}$ is bounded, we obtain for each x

$$\lim_{k\to\infty}\inf \lambda_k^{-1}\sum_{i=1}^{p} c_{i,k}(x) \geq 0. \tag{2.15}$$

Let x^∞ be a limit point of the sequence $\{x^k\}$. Using (2.15), since $\{e^k\}$ converges to zero, passing to the limit in (2.14), it follows that

$$\langle x - x^\infty, g\rangle \geq 0 \quad \forall (x, g) \in G(\tilde{T}).$$

Since $\tilde{T}$ is maximal monotone operator, the above inequality implies that $0 \in \tilde{T}(x^\infty)$, i.e., $x^\infty \in S$. Then since the sequence $\{\|x^* - x^k\|\}$ converges for each $x^* \in S$, it follows that the whole sequence $\{x^k\}$ converges to x^∞. □

3. An existence result

Proposition 1 can be obtained as an immediate consequence of a more general result of independent interest. For this we consider a class of functions $f : \mathbb{R}^p \to \mathbb{R} \cup \{+\infty\}$ satisfying the following properties

(i) f is a closed proper convex function with dom f open,
(ii) f is differentiable on dom f,
(iii) $f_\infty(d) = +\infty, \quad \forall d \neq 0.$

Here dom $f = \{x : f(x) < +\infty\}$ and f_∞ is the recession function of f, see [13] for definition. We denote by $\Im$ the class of functions satisfying (i), (ii) and (iii).

For fixed $x^{k-1} \in C$, the function (see (2.2))

$$u \to \lambda_k^{-1} \sum_{i=1}^{p} u_i^2 - u_i l_i(x^{k-1}) - l_i(x^{k-1})^2 \log \frac{u_i}{l_i(x^{k-1})}$$

clearly satisfies properties (i)–(iii), and Proposition 1 will be a consequence of the following general result.

Proposition 2. *Let $f \in \Im$. Then,*

(1) *the gradient mapping ∇f is onto.*
(2) *Let A be a (p, m) matrix with $p \geq m$ of rank m, $b \in \mathbb{R}^p$ with $(b - A(\mathbb{R}^p)) \cap \operatorname{dom} f \neq \emptyset$, and set $h(x) := f(b - Ax)$. Then,*
   (i) $h \in \Im$.
   (ii) *Let T be a maximal monotone map such that* dom $T \cap$ dom $h \neq \emptyset$ *and set*

$$U(x) = \begin{cases} T(x) + \nabla h(x), & \forall x \in \operatorname{dom} T \cap \operatorname{dom} \nabla h, \\ \emptyset & \textit{otherwise}. \end{cases} \tag{3.1}$$

Then there exists at least a solution x of the generalized equation

$$0 \in U(x), \tag{3.2}$$

which is unique if in addition f is supposed to be strictly convex on its domain.

Proof:

1. Let $y \in \mathbb{R}^p$ and set $v(x) = f(x) - \langle y, x \rangle$. Since $v_\infty(d) = f_\infty(d) - \langle y, d \rangle$, we have

$$v_\infty(d) = +\infty \quad \forall d \neq 0. \tag{3.3}$$

   As a consequence of (3.3), if we minimize v on $\mathbb{R}^p$, the optimal set is nonempty and since $\operatorname{dom} v$ is open, each optimal solution x satisfies $\nabla f(x) = y$, so that ∇f is onto.
2. (i) h is obviously a closed proper convex function. Since $\operatorname{dom} h = A^{-1}(b - \operatorname{dom} f)$ (where A^{-1} denotes the inverse mapping), and since $\operatorname{dom} f$ is open then $\operatorname{dom} h$ is open and h is differentiable on $\operatorname{dom} h$ by the chain-rule differentiation theorem. Finally, since $h_\infty(d) = f_\infty(-Ad)$ and since A is of maximal rank, we have

$$h_\infty(d) = +\infty \quad \forall d \neq 0,$$

   and therefore $h \in \Im$. (ii) Let ∂g be the sub-differential of a closed proper convex function defined on $\mathbb{R}^p$ such that $\operatorname{dom} T \cap \operatorname{int} \operatorname{dom} g \neq \emptyset$. It has been proven in [4] that if ∂g is onto then $T + \partial g$ is onto. As a consequence of part 1 and 2 (i), we have that ∇h is onto. Then using this result it follows that the generalized Eq. (3.2) admits at least a solution.

If in addition f is strictly convex on its effective domain, since A is of maximal rank, this implies that h is strictly convex on its domain and then $T + \nabla h$ is strictly monotone which implies obviously uniqueness. □

4. Extensions and computational remarks

This section briefly discusses some extensions of the proposed method and some remarks on its numerical aspects.

First, it should be remarked after simple algebra, that the functional d (see Section 2) can be rewritten as

$$d(u, v) = \sum_{i=1}^{p} v_i^2 \varphi\left(u_i v_i^{-1}\right), \quad \forall u, v > 0, \tag{4.1}$$

with $\varphi(t) = t^2 - t - \log t$. An extension of our analysis with functionals of the form (4.1) and with other choices of the kernel φ to optimization problems, will be analyzed elsewhere in a companion paper [2].

Next, we show how to extend the LQP method to the polyhedral sets where there are unrestricted variables $z \in \mathbb{R}^r$. In this case, the feasible set K is given by

$$K = \mathbb{R}^r \times C$$

where C is defined in Section 1 with a matrix A of maximal rank. Now, T is a maximal monotone map defined on $\mathbb{R}^r \times \mathbb{R}^m$ such that $\operatorname{dom} T \cap \operatorname{int} K \neq \emptyset$ and we consider the variational inequality

(VIG) "find a point $(z^*, u^*) \in K$ and $g \in T(z^*, u^*)$ such that
$\langle g^*, (z, u) - (z^*, u^*)\rangle \geq 0 \quad \forall (z, u) \in K$".

Set

$$D((z, u), (y, w)) := \frac{1}{2}\|z - y\|^2 + d(l(u), l(w)).$$

Then the extension of LQP is as follows:
Start with $(z^0, u^0) \in \mathbb{R}^r \times \operatorname{int} C$ and generate $\{(z^k, u^k)\} \subset \mathbb{R}^r \times \operatorname{int} C$ satisfying

$$g^k + \lambda_k^{-1}\nabla_{(z,u)} D((z^k, u^k), (z^{k-1}, u^{k-1})) = e^k \quad \text{with } g^k \in T(z^k, u^k).$$

Existence theorems and convergence results can be obtained exactly as done for the model analyzed in Sections 2 and 3. The proof is left to the reader.

As we can remark, the unrestricted variable z is treated as in the usual quadratic proximal method introduced by Martinet [8], in contrast with the constrained variable u.

The use of such a model is of particular interest if we consider a usual convex program (P) with inequalities:

$(P)\quad \inf\{f(z) : g_i(z) \leq 0, i = 1, \ldots, m, z \in \mathbb{R}^r\}$,

where $f, g_1, \ldots, g_p$ are proper closed convex functions on $\mathbb{R}^r$. Let L be the associated Lagrangian for (P):

$$L(z, u) = \begin{cases} f(z) + \sum_{i=1}^{m} u_i g_i(z) & \text{if } u \geq 0 \\ -\infty & \text{if } u \not\geq 0 \end{cases}$$

Then $L(z, u)$ is a closed saddle function, convex in z and concave in u, and therefore the set-valued map Σ on $\mathbb{R}^{r+m}$ defined by:

$$\Sigma(z, u) := \partial_z L(z, u) \times \{-\hat{\partial}_u L(z, u)\},$$

is maximal monotone (see [13]), where $\hat{\partial}$ denotes the upper sub-differential. The pair $(z^*, u^*) \in \mathbb{R}^{r+m}$ is optimal if and only if

$$L(z^*, u) \leq L(z^*, u^*) \leq L(z, u^*), \quad \forall z \in \mathbb{R}^r, u \in \mathbb{R}^m,$$

that is if and only if $(0, 0) \in \Sigma(z^*, u^*)$. As shown in Rockafellar [15], the application of the proximal point algorithm to Σ leads to the so-called *proximal method of multipliers*. In the same spirit, we can apply the extension of the LQP just described above to produce an interior (for the dual variables u) proximal method of multipliers. Details of this approach and other methods for solving convex programs will be addressed in [2].

Finally, we would like to point several important remarks with respect to the numerical aspects of the proposed methods. The function $\lambda_k^{-1}D(\cdot, l(x^{k-1}))$ is self-concordant (see [10] for the theory of self-concordant functions). For optimization problems with differentiable data, when $T = \nabla f$, if f is a self-concordant function then the function $f + \lambda_k^{-1}D(\cdot, l(x^{k-1}))$ is also self-concordant and we can use for computing the point x^k satisfying (2.2) the damped Newton method which is commonly recognized as a standard approach particularly efficient in such cases.

Very recently, Nemirovski has developed in [9] a similar theory on self-concordant convex-concave functions. In [9], Nemirovski proposed a saddle Newton method, equipped with certain policy for decreasing the "homotopy parameter", which differs from the usual damped Newton routine. However, the result is very similar to the basic result on the Newton method for minimizing a convex function.

For the case of finding a saddle point of a quadratic convex-concave functions under linear constraints, the computation of the vector x^k satisfying (2.2) can be done by computing a saddle point of a self-concordant convex-concave function by the saddle Newton method proposed in [9]. Further research has to be done in this direction, but the above considerations lead us to believe that LQP is a promising method.

References

1. A. Auslender and M. Haddou, "An interior proximal method for convex linearly constrained problems and its extension to variational inequalities," Mathematical Programming, vol. 71, pp. 77–100, 1995.
2. A. Auslender, M. Teboulle, and S. Ben-Tiba, "Interior proximal and multiplier methods based on second order homogeneous functionals," Submitted 1998.
3. R.D. Bruck, "An iterative solution of a variational inequality for certain monotone operators in Hilbert space," Bulletin of the American Math. Soc., vol. 81, pp. 890–892, 1975 (With corrigendum, in vol. 82, p. 353, 1976).
4. R.S. Burachik and A.N. Iusem, "A generalized proximal point algorithm for the variational inequality problem in a Hilbert space," SIAM Journal on Optimization, vol. 8, pp. 197–216, 1998.
5. Y. Censor, A.N. Iusem, and S.A. Zenios, "An interior-point method with Bregman functions for the variational inequality problem with paramonotone operators," Working paper, University of Haifa, 1994.
6. J. Eckstein, "Approximate iterations in Bregman-function-based proximal algorithms," RRR, 12-96, January 97, Rutgers University.
7. Z.Q. Luo and P. Tseng, "Error bounds and convergence analysis of feasible descent methods: A general approach," Annals of Operations Research, vol. 46, pp. 157–178, 1993.
8. B. Martinet, "Regularisation d'inéquations variationnelles par approximations successives, Revue Francaise d'Automatique et Informatique recherche Opérationnelle," vol. 4, pp. 154–159, 1970.
9. A. Nemirovski, "On self-concordant convex-concave functions," Research Report 3/97, Optimization Laboratory Faculty of Industrial Engineering and Management, Technion Haifa.
10. Y. Nesterov and A. Nemirovski, Interior Point Polynomial Algorithms in Convex Programming, SIAM Publications: Philadelphia, PA, 1994.
11. B.T. Polyak, Introduction to Optimization, Optimization Software Inc.: New York, 1987.
12. R.T. Rockafellar, "On the maximality of sums of nonlinear monotone operators, Transactions of the American Mathematical Society," vol. 149, pp. 75–88, 1970.
13. R.T. Rockafellar, Convex Analysis, Princeton University Press: Princeton, NJ, 1970.
14. R.T. Rockafellar, "Monotone operators and the proximal point algorithm," SIAM J. of Control and Optimization, vol. 14, pp. 877–898, 1976.
15. R.T. Rockafellar, "Augmented Lagrangians and applications of the proximal point algorithm in convex programming," Mathematics of Operations Research, vol. 1, pp. 97–116, 1976.
16. M. Teboulle, "Convergence of proximal-like algorithms," SIAM J. of Optimization, vol. 7, pp. 1069–1083, 1997.

Computational Optimization and Applications 12, 41–51 (1999)

A Note on Error Bounds for Convex and Nonconvex Programs*

DIMITRI P. BERTSEKAS bertsekas@LIDS.MIT.EDU
Department of Electrical Engineering and Computer Science, M.I.T., Cambridge, MA, 02139

Received August 7, 1997; Accepted February 19, 1998

Abstract. Given a single feasible solution x_F and a single infeasible solution x_I of a mathematical program, we provide an upper bound to the optimal dual value. We assume that x_F satisfies a weakened form of the Slater condition. We apply the bound to convex programs and we discuss its relation to Hoffman-like bounds. As a special case, we recover a bound due to Mangasarian [11] on the distance of a point to a convex set specified by inequalities.

Keywords: convex programming, optimization, duality, error bounds

Dedication: It is very appropriate that this paper appears in a special issue honoring Olvi Mangasarian, because its idea was actually born while listening to one of Olvi's talks. This is only the latest in a string of several occasions where Olvi's research has substantially influenced mine. For many years, I have benefited a great deal from reading his works, and from interacting with him at meetings, where his presence made serious and idle talk both personally enjoyable and professionally rewarding.

1. Introduction

We consider the problem

$$\begin{array}{ll} \text{minimize} & f(x) \\ \text{subject to} & x \in X, \quad g_j(x) \le 0, \quad j = 1, \ldots, r, \end{array} \tag{1}$$

where X is a nonempty subset of $\Re^n$, and $f : \Re^n \mapsto \Re$, $g_j : \Re^n \mapsto \Re$ are given functions. We denote by $g(x)$ the vector of constraint functions

$$g(x) = (g_1(x), \ldots, g_r(x)),$$

and we write the constraints $g_j(x) \le 0$ compactly as $g(x) \le 0$. In our notation, all vectors are column vectors and a prime denotes transposition.

*Research supported by NSF under Grant 9300494-DMI.

Let f^* and q^* be the optimal primal and dual value, respectively:

$$f^* = \inf_{\substack{x \in X \\ g_j(x) \leq 0,\ j=1,\ldots,r}} f(x), \tag{2}$$

$$q^* = \sup_{\mu \geq 0} q(\mu), \tag{3}$$

where $q : \Re^r \mapsto [-\infty, +\infty)$ is the dual function given by

$$q(\mu) = \inf_{x \in X} \{ f(x) + \mu' g(x) \}. \tag{4}$$

Throughout the paper, we assume the following:

Assumption 1. *We have two vectors x_F and x_I from X such that:*
(a) *x_F is feasible*, i.e., $g(x_F) \leq 0$.
(b) *x_I is infeasible*, i.e., $g_j(x_I) > 0$ *for at least one j. Furthermore, its cost $f(x_I)$ is strictly smaller than the cost $f(x_F)$ of x_F.*

We note that by weak duality, we have $q^* \leq f(x_F)$. We will show that the value $f(x_I)$ can be used to improve this upper bound. In particular, we prove the following result in Section 3:

Proposition 1. *Under Assumption* 1, *there holds*

$$\frac{q^* - f(x_I)}{f(x_F) - f(x_I)} \leq \frac{\Gamma}{\Gamma + 1}, \tag{5}$$

where

$$\Gamma = \inf\{\gamma \geq 0 \mid g(x_I) \leq -\gamma g(x_F)\}. \tag{6}$$

If $\Gamma = \infty$ because there is no $\gamma \geq 0$ such that $g(x_I) \leq -\gamma g(x_F)$, the bound in Eq. (5) reduces to the trivial bound $q^* \leq f(x_F)$.

Note that we have $\Gamma < \infty$ if the Slater condition

$$g_j(x_F) < 0, \quad \forall j = 1, \ldots, r, \tag{7}$$

holds. More generally, we have $\Gamma < \infty$ if and only if the following weakened form of the Slater condition holds:

$$g_j(x_I) \leq 0 \quad \text{for all } j \text{ with } g_j(x_F) = 0. \tag{8}$$

If the above condition holds, we have

$$\Gamma = \max_{\{j \mid g_j(x_F) < 0\}} \frac{g_j(x_I)}{-g_j(x_F)}. \tag{9}$$

Figure 1 illustrates the idea underlying the bound (5), (6). In the case of a single constraint ($r = 1$) the bound reduces to

$$\frac{q^* - f(x_I)}{f(x_F) - f(x_I)} \leq \frac{\hat{f} - f(x_I)}{f(x_F) - f(x_I)} = \frac{g(x_I)}{g(x_I) - g(x_F)}, \tag{10}$$

where $\hat{f}$ is the point of intersection of the vertical axis of $\Re^2$ with the line segment connecting the vectors $(g(x_F), f(x_F))$ and $(g(x_I), f(x_I))$. When there are multiple constraints, this line segment can be projected on the two-dimensional subspace spanned by the vertical axis $(0, 1)$ of $\Re^{r+1}$ and the vector $(g(x_I), 0)$. The inequality (10) can then be applied on this subspace in a suitably modified form (see the proof in Section 3).

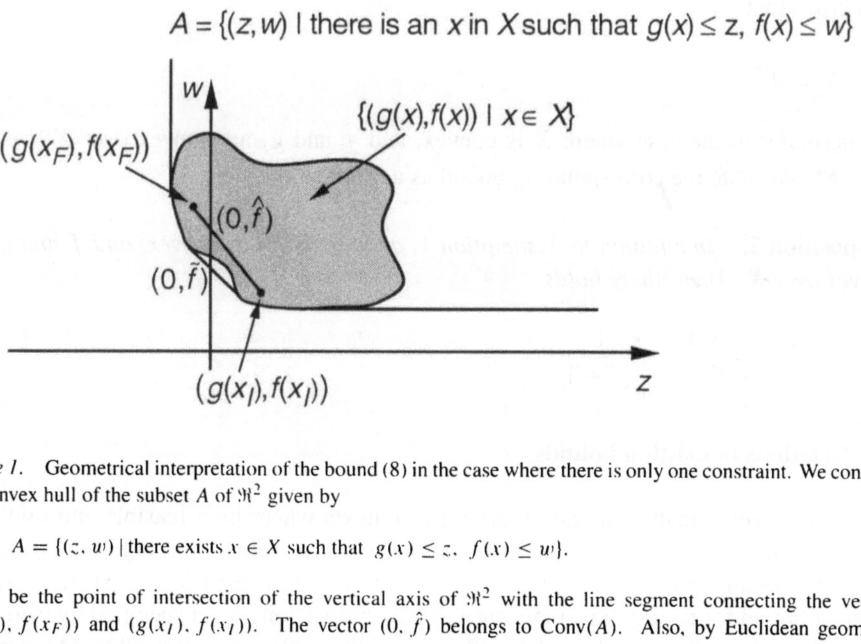

Figure 1. Geometrical interpretation of the bound (8) in the case where there is only one constraint. We consider the convex hull of the subset A of $\Re^2$ given by

$$A = \{(z, w) \mid \text{there exists } x \in X \text{ such that } g(x) \leq z,\ f(x) \leq w\}.$$

Let $\hat{f}$ be the point of intersection of the vertical axis of $\Re^2$ with the line segment connecting the vectors $(g(x_F), f(x_F))$ and $(g(x_I), f(x_I))$. The vector $(0, \hat{f})$ belongs to Conv(A). Also, by Euclidean geometry, we have

$$\frac{\hat{f} - f(x_I)}{f(x_F) - f(x_I)} = \frac{g(x_I)}{g(x_I) - g(x_F)},$$

and by the definition of q^* we have

$$q^* \leq \tilde{f} \leq \hat{f} \leq f^*,$$

where

$$\tilde{f} = \inf\{w \mid (z, w) \in \text{Conv}(A)\}.$$

Combining these two relations, the bound (5), (6) follows for the case of a single constraint.

Figure 1 also suggests the following slightly stronger version of our bound:

$$\frac{\tilde{f} - f(x_I)}{f(x_F) - f(x_I)} \leq \frac{\Gamma}{\Gamma + 1}, \tag{11}$$

where

$$\tilde{f} = \inf\{w \mid (z, w) \in \text{Conv}(A)\}, \tag{12}$$

the subset A of $\Re^2$ is given by

$$A = \{(z, w) \mid \text{there exists } x \in X \text{ such that } g(x) \leq z,\ f(x) \leq w\},$$

and $\text{Conv}(A)$ denotes its convex hull. Indeed, we prove this bound in Section 3, and we also prove that

$$\tilde{f} \leq f^*.$$

Furthermore, in the case where X is convex, and f and g_j are convex over X, we have $\tilde{f} = f^*$. We state the corresponding bound as a proposition:

Proposition 2. *In addition to Assumption* 1, *assume that X is convex, and f and g_j are convex over X. Then, there holds*

$$\frac{f^* - f(x_I)}{f(x_F) - f(x_I)} \leq \frac{\Gamma}{\Gamma + 1}. \tag{13}$$

2. Relations to existing bounds

There are several analytical and algorithmic contexts where both feasible and infeasible solutions are known in optimization problems (e.g., in primal-dual algorithms), and in which our bound may prove useful. As an illustration of one such context, let us derive an error bound for the Euclidean distance of a point to a convex set specified by inequality constraints. A similar error bound for this projection problem was derived by Mangasarian [11] using different methods, and was the inspiration for the present paper. In particular, let $y \in \Re^n$ be a given vector and consider the following projection problem

$$\begin{aligned} &\text{minimize} \quad \|y - x\| \\ &\text{subject to} \quad x \in X, \quad g_j(x) \leq 0, \quad j = 1, \ldots, r, \end{aligned}$$

where $\|\cdot\|$ denotes the standard Euclidean norm ($\|x\| = \sqrt{\sum_{i=1}^n x_i^2}$). Let us assume that X is a convex set and g_j are convex functions. Furthermore, let us assume the following two conditions:

(a) $y \in X$ and $g_j(y) > 0$ for at least one j.
(b) There exists a vector $x_F \in X$ such that

$$g_j(x_F) \leq 0, \quad \forall j = 1, \ldots, r,$$
$$g_j(y) \leq 0 \quad \text{for all } j \text{ with } g_j(x_F) = 0.$$

Then we can apply the error bound (13) with $f(x) = \|y - x\|$, f^* equal to the distance $d(y)$ of y from the convex set $X \cap \{x \mid g_j(x) \leq 0,\ j = 1, \ldots, r\}$, $x_I = y$, and $f(x_I) = 0$, $f(x_F) = \|y - x_F\|$. We have

$$d(y) \leq \frac{\Gamma}{\Gamma + 1} \|y - x_F\|, \tag{14}$$

where Γ is given by Eq. (9). It is easily seen that $\Gamma \leq \hat{\Gamma}$, where

$$\hat{\Gamma} = \frac{\max_{\{j \mid g_j(x_F) < 0\}} g_j(x_I)}{\min_{\{j \mid g_j(x_F) < 0\}} -g_j(x_F)}, \tag{15}$$

and the inequality (14) yields

$$d(y) \leq \frac{\hat{\Gamma}}{\hat{\Gamma} + 1} \|y - x_F\|, \tag{16}$$

or equivalently

$$d(y) \leq \frac{\max_{\{j \mid g_j(x_F) < 0\}} g_j(x_I)}{\max_{\{j \mid g_j(x_F) < 0\}} g_j(x_I) + \min_{\{j \mid g_j(x_F) < 0\}} -g_j(x_F)} \|y - x_F\|. \tag{17}$$

This bound coincides with the relative error bound derived by Mangasarian ([11], Theorem 2.2) under the assumption that $X = \Re^n$ and $g_j(x_F) < 0$ for all j. Note, however, that using Γ from Eq. (9) in place of $\hat{\Gamma}$ as in Eq. (16) yields a stronger bound.

For a generalization of the bound (14), let us consider replacing the distance $\|y - x\|$ with a more general metric. In particular, consider the problem

$$\begin{aligned} &\text{minimize} \quad f(x, y) \\ &\text{subject to} \quad x \in X, \quad g_j(x) \leq 0, \quad j = 1, \ldots, r, \end{aligned}$$

where X is a convex set, g_j is a convex function, and $f(\cdot, y)$ is a convex function satisfying

$$f(y, y) = 0, \quad f(x_F, y) > 0.$$

Then, if $f^*(y)$ is the optimal cost of this problem, the preceding analysis can be used to show that (cf. Eq. (14))

$$f^*(y) \leq \frac{\Gamma}{\Gamma + 1} f(x_F, y),$$

where Γ is given by

$$\Gamma = \max_{\{j \mid g_j(x_F)<0\}} \frac{g_j(y)}{-g_j(x_F)}.$$

In Mangasarian's paper [11], the bound (17) was used to derive an extension to Hoffman's error bound for the approximate solution of convex systems of inequalities. Hoffman's bound [4] and its extensions have been the subject of intensive investigation recently (see e.g., [2, 5–12], which give many additional references). It is thus interesting to inquire about the relation of our bound of Section 1 with Hoffman-like bounds. It turns out that the bound of Section 1 is fundamentally different. To see this, we note that Propositions 1 and 2 do not explicitly require the existence of a Lagrange multiplier for problem (1). By contrast, we will show shortly that Hoffman-like bounds essentially amount to assertions on the uniform boundedness of the Lagrange multipliers of some parametric convex program as the parameter vector ranges over some set. This connection, which is based on the theory of exact penalty functions, has apparently not been made explicit earlier and leads to slightly stronger versions of Hoffman's bound for convex systems of inequalities than some of those available in the literature (e.g., Mangasarian's Hoffman-like bound of [11]).

Indeed let Y be a subset of $\Re^n$, let y be a parameter vector taking values in Y, and consider the parametric program

$$\begin{aligned} &\text{minimize} \quad f(x, y) \\ &\text{subject to} \quad x \in X, \quad g_j(x, y) \le 0, \quad j = 1, \ldots, r, \end{aligned} \tag{18}$$

where X is a convex subset of $\Re^n$, and for each $y \in Y$, $f(\cdot, y)$ and $g_j(\cdot, y)$ are convex functions. We assume that for each $y \in Y$, this program has at least one feasible solution, and that its optimal value, denoted $f^*(y)$, is finite. Furthermore, we assume that when the constraints $g_j(x, y) \le 0$ are dualized, there is no duality gap; that is, the optimal value $q^*(y)$ of the dual problem

$$\begin{aligned} &\text{maximize} \quad q(\mu, y) \\ &\text{subject to} \quad \mu \ge 0 \end{aligned}$$

is equal to $f^*(y)$, where $q(\mu, y)$ is the dual function

$$q(\mu, y) = \inf_{x \in X} \{ f(x, y) + \mu' g(x, y) \}.$$

Consider a penalty function $P : \Re^r \mapsto \Re$ that is convex and satisfies

$$\begin{aligned} P(u) &= 0, \quad \forall u \le 0, \\ P(u) &> 0, \quad \text{if } u_j > 0 \text{ for some } j = 1, \ldots, r. \end{aligned}$$

Let $c > 0$ denote a penalty parameter. It is shown in [1] [Proposition 5.4.1(a)][1] that we have

$$f^*(y) = \inf_{x \in X} \{ f(x, y) + cP(g(x, y)) \}, \quad \forall y \in X, \tag{19}$$

if and only if

$$u'\mu^*(y) \le cP(u), \quad \forall u \in \Re^r, \ \forall y \in X,$$

for some dual optimal solution $\mu^*(y)$ (an optimal solution of the dual problem (14), which is also referred to as a Lagrange multiplier).

It is seen that Eq. (19) is equivalent to the bound

$$f^*(y) \le f(x, y) + \bar{c}P(g(x, y)), \quad \forall x \in X, \ y \in Y, \tag{20}$$

so this bound holds if and only if there exists a uniform bounding constant $c > 0$ such that

$$u'\mu^*(y) \le cP(u), \quad \forall u \in \Re^r, \ y \in Y. \tag{21}$$

For the above relation to hold, it is essential that the penalty function P be nondifferentiable, such as for example the Euclidean norm of the vector u^+ that has components $\max\{0, u_j\}$, $j = 1, \dots, r$,

$$P(u) = \|u^+\|,$$

or the corresponding l_1 or l_∞ norm

$$P(u) = \|u^+\|_1, \quad P(u) = \|u^+\|_\infty.$$

Given any of these choices, it is seen that Eq. (21), and the equivalent bound (20), hold if and only if for every $y \in Y$, it is possible to select a Lagrange multiplier $\mu^*(y)$ of the parametric problem (18) such that the set $\{\mu^*(y) \mid y \in Y\}$ is bounded.

Let us now specialize the preceding discussion to the parametric program

$$\begin{aligned} &\text{minimize} \quad f(x, y) = \|y - x\| \\ &\text{subject to} \quad x \in X, \quad g_j(x) \le 0, \quad j = 1, \dots, r, \end{aligned} \tag{22}$$

where $\|\cdot\|$ is the Euclidean norm, X is a convex subset of $\Re^n$, and g_j are convex functions. This is the projection problem considered in the beginning of this section. Let us take $Y = X$. If c satisfies Eq. (21), the bound (20) becomes

$$d(y) \le \|y - x\| + cP(g(x)), \quad \forall x \in X, \ y \in X,$$

and (by taking $x = y$) implies the bound

$$d(y) \le cP(g(y)), \quad \forall y \in X. \tag{23}$$

Thus, the Hoffman-like bound (23) holds if a Lagrange multiplier $\mu^*(y)$ of the projection problem (22) can be found such that Eq. (21) holds. In the case where P is given by

$$P(u) = \|u^+\|,$$

(or P is given by any one of a variety of other nondifferentiable functions of u) boundedness of the set $\{\mu^*(y) \mid y \in X\}$ is equivalent to the condition (21), and implies the Hoffman-like bound (23). It turns out that the reverse assertion also holds, as shown in the following proposition:

Proposition 3. *Let $d(y)$ be the optimal value of the projection problem*

$$\begin{aligned} &\textit{minimize} \quad \|y - x\| \\ &\textit{subject to} \quad x \in X, \quad g_j(x) \leq 0, \quad j = 1, \ldots, r, \end{aligned} \tag{24}$$

X is a convex subset of $\Re^n$, and the functions g_j are convex. The Hoffman-like bound

$$d(y) \leq c\|(g(y))^+\|, \quad \forall y \in X, \tag{25}$$

holds for some constant c if and only if the projection problem (24) has a Lagrange multiplier $\mu^*(y)$ such that the set $\{\mu^*(y) \mid y \in X\}$ is bounded.

Proof: The preceding discussion showed that the boundedness condition on the Lagrange multipliers implies the Hoffman-like bound (25), so there remains to prove the reverse assertion. Indeed, assume that Eq. (25) holds for some c, and to arrive at a contradiction, assume that there exist $x \in X$ and $y \in Y$ such that

$$d(y) > \|y - x\| + c\|(g(x))^+\|.$$

Then, using Eq. (25), we obtain

$$d(y) > \|y - x\| + d(x).$$

From this relation and the triangle inequality, it follows that

$$\begin{aligned} \inf_{z \in X,\, g(z) \leq 0} \|y - z\| &> \|y - x\| + \inf_{z \in X,\, g(z) \leq 0} \|x - z\| \\ &= \inf_{z \in X,\, g(z) \leq 0} \{\|y - x\| + \|x - z\|\} \\ &\geq \inf_{z \in X,\, g(z) \leq 0} \|y - z\|, \end{aligned}$$

which is a contradiction. Thus, Eq. (25) implies that we have

$$d(y) \leq \|y - x\| + c\|(g(x))^+\|, \quad x \in X,\ y \in X.$$

Using Proposition 5.4.1(a) of [1], this implies that there exists a Lagrange multiplier $\mu^*(y)$ such that

$$u'\mu^*(y) \leq c\|u^+\|, \quad \forall u \in \Re^r,\ y \in X.$$

This in turn implies the boundedness of the set $\{\mu^*(y) \mid y \in X\}$. □

Let us give two conditions under which the boundedness condition of Proposition 3 is satisfied and a Hoffman-like bound of the form (25) holds:

(a) $X = \Re^n$ *and* g_j *are linear* (this is the original Hoffman's bound [4]). For a simple way to prove this, let $g_j(x) = a_j'x - b_j$, where a_j is a vector in $\Re^n$ and b_j is a scalar. Then, because of the linearity of the constraints, the projection problem (24) has at least one Lagrange multiplier $\mu^*(y)$ (see e.g., [1], p. 437). This Lagrange multiplier satisfies

$$\frac{y - \hat{y}}{\|y - \hat{y}\|} = \sum_{j \in J(y)} \mu_j^*(y) a_j,$$

where $\hat{y}$ is the unique projection of y, and $J(y)$ is a subset of indices such that the set of vectors $\{a_j \mid j \in J(y)\}$ is linearly independent. (We assume here that $y \neq \hat{y}$, which is the case of interest.) Since the vector $\sum_{j \in J(y)} \mu_j^*(y) a_j$ has norm 1, it follows that $\mu^*(y)$ can be selected so that the set $\{\mu^*(y) \mid y \in \Re^n\}$ is bounded.

(b) *For each* $y \in X$, *a Slater condition holds*; *that is there exists a vector* $\bar{x}(y) \in X$ *such that* $g_j(\bar{x}(y)) < 0$ *for all* $j = 1, \dots, r$. *Furthermore, there is a constant* γ *such that*

$$\frac{\|y - \bar{x}(y)\| - d(y)}{\min_{j=1,\dots,r}\{-g_j(\bar{x}(y))\}} \leq \gamma, \quad \forall y \in X. \tag{26}$$

Mangasarian [11] used these two conditions (together with the additional condition $X = \Re^n$) to obtain a Hoffman-like bound of the form (25). For an alternative proof of this result, note that the Slater condition implies (see e.g., [1], p. 450 or [3], p. 313) that for each $y \in X$ there exists a Lagrange multiplier $\mu^*(y)$ with

$$\sum_{j=1}^{r} \mu_j^*(y) \leq \frac{\|y - \bar{x}(y)\| - d(y)}{\min_{j=1,\dots,r}\{-g_j(\bar{x}(y))\}}.$$

Thus, Eq. (26) implies the boundedness condition of Proposition 3.

3. Proof of Propositions 1 and 2

We consider the subset of $\Re^{r+1}$

$$A = \{(z, w) \mid \text{there exists } x \in X \text{ such that } g(x) \leq z,\ f(x) \leq w\},$$

and its convex hull $\mathrm{Conv}(A)$. The vectors $(g(x_F), f(x_F))$ and $(g(x_I), f(x_I))$ belong to A. In addition, the vector $(0, \tilde{f})$, where

$$\tilde{f} = \inf\{w \mid (z, w) \in \mathrm{Conv}(A)\},$$

is in the closure of Conv(A). Let us now show that $q^* \leq \tilde{f}$, as indicated by figure 1.

Indeed, for each $(z, w) \in \text{Conv}(A)$, there exist $\xi_1 \geq 0$ and $\xi_2 \geq 0$ with $\xi_1 + \xi_2 = 1$, and $x_1 \in X$, $x_2 \in X$ such that

$$\xi_1 g(x_1) + \xi_2 g(x_2) \leq z,$$
$$\xi_1 f(x_1) + \xi_2 f(x_2) \leq w.$$

Furthermore, by the definition of the dual function q, we have for all $\mu \in \Re^r$,

$$q(\mu) \leq f(x_1) + \mu' g(x_1),$$
$$q(\mu) \leq f(x_2) + \mu' g(x_2).$$

Combining the preceding four inequalities, we obtain

$$q(\mu) \leq w + \mu' z, \quad \forall\, (z, w) \in \text{Conv}(A),\ \mu \geq 0.$$

The above inequality holds also for all (z, w) that are in the closure of Conv(A), and in particular, for $(z, w) = (0, \tilde{f})$. It follows that

$$q(\mu) \leq \tilde{f}, \quad \forall \mu \geq 0,$$

from which, by taking the maximum over $\mu \geq 0$, we obtain $q^* \leq \tilde{f}$.

Let γ be any nonnegative scalar such that $g(x_I) \leq -\gamma g(x_F)$, and consider the vector

$$\Delta = -\gamma g(x_F) - g(x_I).$$

Since $\Delta \geq 0$, it follows that the vector

$$(-\gamma g(x_F), f(x_I)) = (g(x_I) + \Delta, f(x_I))$$

also belongs to the set A. Thus the three vectors

$$(g(x_F), f(x_F)), \quad (0, \tilde{f}), \quad (-\gamma g(x_F), f(x_I))$$

belong to the closure of Conv(A), and form a triangle in the plane spanned by the "vertical" vector $(0, 1)$ and the "horizontal" vector $(g(x_F), 0)$.

Let $(0, \hat{f})$ be the intersection of the vertical axis with the line segment connecting the vectors $(g(x_F), f(x_F))$ and $(-\gamma g(x_F), f(x_I))$ (there is a point of intersection because $\gamma \geq 0$). We have by Euclidean triangle geometry (cf. figure 1)

$$\frac{\hat{f} - f(x_I)}{f(x_F) - f(x_I)} = \frac{\gamma}{\gamma + 1}. \tag{27}$$

Since the vectors $(g(x_F), f(x_F))$ and $(-\gamma g(x_F), f(x_I))$ both belong to Conv(A), we also have $(0, \hat{f}) \in \text{Conv}(A)$. Therefore, there exist vectors $x_1, \ldots, x_m \in X$ and nonnegative scalars $\xi_1, \ldots, \xi_m$ with $\sum_{i=1}^m \xi_i = 1$, satisfying

$$\sum_{i=1}^m \xi_i g(x_i) \leq 0, \quad \sum_{i=1}^m \xi_i f(x_i) \leq \hat{f}.$$

Thus, if $\hat{f} < \tilde{f}$, we must have $\sum_{i=1}^{m} \xi_i f(x_i) < \tilde{f}$, contradicting the definition of $\tilde{f}$. It follows that $\tilde{f} \leq \hat{f}$ and since $q^* \leq \tilde{f}$, as shown earlier, from Eq. (27) we obtain

$$\frac{q^* - f(x_I)}{f(x_F) - f(x_I)} \leq \frac{\tilde{f} - f(x_I)}{f(x_F) - f(x_I)} \leq \frac{\gamma}{\gamma + 1}. \tag{28}$$

Taking the infimum over $\gamma \geq 0$, the error bound (5), (6) follows.

Assume now that X is convex, and f and g_j are convex over X. Then the set A is known to be convex under these assumptions (see e.g., [1], Proposition 5.3.1, p. 446), and we have $\tilde{f} = f^*$. Proposition 2 then follows from Eq. (28).

Note

1. Note that Proposition 5.4.1(a) of [1] includes a compactness assumption on the set X, but this assumption is unnecessary as long as finiteness of the optimal value $f^*(y)$ is assumed; see the discussion of p. 473 of [1].

References

1. D.P. Bertsekas, Nonlinear Programming, Athena Scientific, Belmont, MA, 1995.
2. J.V. Burke and P. Tseng, "A unified analysis of Hoffman's error bound via Fenchel duality," SIAM J. on Optimization, vol. 6, pp. 265–282, 1996.
3. J.-B. Hiriart-Urruty and C. Lemarechal, "Convex analysis and minimization algorithms," vols. I and II, Springer-Verlag: Berlin and NY, 1993.
4. A.J. Hoffman, "On approximate solutions of systems of linear inequalities," Natl. Bur. Standards, vol. 49, pp. 263–265, 1952.
5. D. Klatte and W. Li, "Asymptotic constraint qualifications and global error bounds for convex inequalities," Math. Programming, 1997, to appear.
6. A.S. Lewis and J.-S. Pang, "Error bounds for convex inequality systems," in Proceedings of the Fifth Symposium on Generalized Convexity, Luminy, J.P. Crouzeix, J.-E. Martinez-Legaz, and M. Volle (Eds.), June 1996, Kluwer Academic Publishers, 1998, pp. 75–110.
7. W. Li, "Abadie's constraint qualification, metric regularity, and error bounds for differentiable convex inequalities," SIAM J. on Optimization, vol. 7, pp. 966–978, 1997.
8. X.-D. Luo and Z.-Q. Luo, "Extension of Hoffman's error bound to polynomial systems," SIAM J. on Optimization, vol. 4, pp. 383–392, 1994.
9. Z.-Q. Luo and P. Tseng, "Error bound and convergence analysis of matrix splitting algorithms for the affine variational inequality problem," SIAM J. on Optimization, vol. 2, pp. 43–54, 1992.
10. Z.-Q. Luo and P. Tseng, 1992. "On the linear convergence of convex essentially smooth minimization," SIAM J. on Control and Optimization, vol. 30, pp. 408–425, 1992.
11. O.L. Mangasarian, "Error bounds for nondifferentiable convex inequalities under a strong Slater constraint qualification," Mathematical Programming, vol. 83, pp. 187–194, 1998.
12. J.-S. Pang, "Error bounds in mathematical programming," Math. Programming, vol. 79, pp. 299–332, 1997.

Computational Optimization and Applications, 12, 53–79 (1999)

Multicategory Classification by Support Vector Machines

ERIN J. BREDENSTEINER eb6@evansville.edu
Department of Mathematics, University of Evansville, Evansville, IN 47722

KRISTIN P. BENNETT bennek@rpi.edu
Department of Mathematical Sciences, Rensselaer Polytechnic Institute, Troy, NY 12180

Received December 8, 1997; Accepted July 10, 1998

Abstract. We examine the problem of how to discriminate between objects of three or more classes. Specifically, we investigate how two-class discrimination methods can be extended to the multiclass case. We show how the linear programming (LP) approaches based on the work of Mangasarian and quadratic programming (QP) approaches based on Vapnik's Support Vector Machine (SVM) can be combined to yield two new approaches to the multiclass problem. In LP multiclass discrimination, a single linear program is used to construct a piecewise-linear classification function. In our proposed multiclass SVM method, a single quadratic program is used to construct a piecewise-nonlinear classification function. Each piece of this function can take the form of a polynomial, a radial basis function, or even a neural network. For the $k > 2$-class problems, the SVM method as originally proposed required the construction of a two-class SVM to separate each class from the remaining classes. Similarily, k two-class linear programs can be used for the multiclass problem. We performed an empirical study of the original LP method, the proposed k LP method, the proposed single QP method and the original k QP methods. We discuss the advantages and disadvantages of each approach.

Keywords: support vector machines, linear programming, classification, data mining, machine learning.

Dedication: This work is a direct extension of my many collaborations with Olvi Mangasarian. It was my privilege to be his student in 1988 when the 1965 Multisurface Method of Pattern Recognition was resurrected as an approach to breast cancer diagnosis. This was the first of many cases where Olvi magically had already written a paper relevant to some new, interesting topic. Examining the parallels between Olvi's work and support vector methods, we can see the depth of insight found in his early work. Olvi's research in optimization approaches to learning exemplifies his willingness to tackle the latest problems, resulting in contributions in new areas. Optimization in learning is just one small part of his brilliant and prolific body of work. Probably there are other ideas that, like the Multisurface Method, will sit on the shelf for twenty years before we truly appreciate them. Only then will we fully understand the scope of his contributions. — Kristin Bennett

1. Introduction

We investigate the problem of discriminating large real-world datasets with more than two classes. Given examples of points known to come from $k > 2$ classes, we construct a function to discriminate between the classes. The goal is to select a function that will efficiently and correctly classify future points. This classification technique can be used for data mining or pattern recognition. For example, the United States Postal Service is interested in an efficient yet accurate method of classifying zip codes. Actual handwritten digits from zip codes collected by the United States Postal Service are used in our study. Each digit is represented by a 16 by 16 pixel grayscale map, resulting in 256 attributes for each sample number. Given the enormous quantities of mail the Postal Service sorts each day, the accuracy and efficiency in evaluation are extremely important.

In this paper, we combine two independent but related research directions developed for solving the two-class linear discrimination problem. The first is the linear programming (LP) methods stemming from the Multisurface Method of Mangasarian [16, 17]. This method and its later extension the Robust Linear Programming (RLP) approach [6] have been used in a highly successful breast cancer diagnosis system [31, 21]. The second direction is the quadratic programming (QP) methods based on Vapnik's Statistical Learning Theory [29, 30]. Statistical Learning Theory addresses mathematically the problem of how to best construct functions that generalize well on future points. The problem of constructing the best linear two-class discriminant can be posed as a convex quadratic program with linear constraints. The resulting linear discriminant is known as a Support Vector Machine (SVM) because it is a function of a subset of the training data known as *support vectors*. Specific implementations such as the Generalized Optimal Plane (GOP) method have proven to perform very well in practice [11]. Throughout this paper we will refer to the two different approaches as RLP and SVM.

The primary focus of this paper is how the two research directions have differed in their approach to solving problems with $k > 2$ classes. The original SVM method for multi-class problems was to find k separate two-class discriminants [29]. Each discriminant is constructed by separating a single class from all the others. This process requires the solution of k quadratic programs. When applying all k classifiers to the original multicategory dataset, multiply classified points or unclassified points may occur. This ambiguity has been avoided by choosing the class of a point corresponding to the classification function that is maximized at that point. The LP approach has been to directly construct k classification functions such that for each point the corresponding class function is maximized [5, 6]. The Multicategory Discrimination Method [5, 6] constructs a piecewise-linear discriminant for the k-class problem using a single linear program. We will call this method M-RLP since it is a direct extension of the RLP approach. We will show how these two different approaches can be combined to yield two new methods: k-RLP and M-SVM.

In Section 2, we will provide background on the existing RLP and SVM methods. While the k-class cases are quite different, the two-class linear discrimination methods for SVM and RLP are almost identical. The major difference is the regularization term used in the objective. According to the principle of structural risk minimization in Statistical Learning Theory, the regularization term in SVM acts as a form of capacity control that improves generalization both in theory and practice [29]. We use the regularized form of RLP proposed in [3] which is equivalent to SVM except that a different norm is used for the regularization term. For two-class linear discrimination, RLP generalizes equally well and is more computationally efficient than SVM. RLP exploits the fact that state-of-the-art commercial LP codes are far more efficient and reliable than QP codes.

The primary practical appeal of SVMs is that they can be simply and elegantly applied to nonlinear discrimination. With only minor changes, SVM methods can construct a wide class of two-class nonlinear discriminants by solving a single QP [29]. The basic idea is that the points are mapped nonlinearly to a higher dimensional space. Then the dual SVM problem is used to construct a linear discriminant in the higher dimensional space that is nonlinear in the original attribute space. By using kernel functions in the dual SVM problem, SVM can efficiently and effectively construct many types of nonlinear discriminant functions including polynomials, radial basis function machines, and neural networks. The successful polynomial-time nonlinear methods based on LP use multi-step approaches. The methods of Roy *et al.* [26, 25, 24] use clustering in conjunction with LP to generate neural networks in polynomial time. Another approach is to recursively construct piecewise-linear discriminants using a series of linear programs [17, 2, 19]. These approaches could also be used with SVM but we limit discussion to nonlinear discriminants constructed using the SVM kernel-type approaches.

After the introduction to the existing multiclass methods, M-RLP and k-SVM, we will show how the idea used in M-RLP can be adapted to construct multiclass SVMs using a single quadratic program. We adapt a problem formulation similar to the two-class case. In the two-class case, the initial problem is to construct a linear discriminant. The data points are then transformed to a higher dimensional feature space. A linear discriminant is constructed in the higher dimensional space. This results in a nonlinear classification function in the original feature space. The regularization term controls the capacity of the discriminant function preventing overfitting in the higher dimensional space. In Section 3, for the $k > 2$-class case, we begin by constructing a piecewise-linear discriminant function. A regularization term is added to avoid overfitting. This method is then extended to piecewise-nonlinear classification functions in Section 4. The variables are mapped to a higher dimensional space. Then a piecewise-linear discriminant function is constructed in the new space. This results in a piecewise-nonlinear discriminant in the original space. In Section 5, we extend the method to piecewise-inseparable datasets. We call the final approach the Multicategory Support Vector Machine (M-SVM). Depending on the choice of transformation, the pieces may be polynomials, radial basis functions, neural networks, etc. We concentrate our research on the polynomial classifier and leave the computational investigation of other classification functions as future work. Figure 1 shows a piecewise-second-degree polynomial separating three classes in two dimensions.

M-SVM requires the solution of a very large quadratic program. When transforming the data points into a higher dimensional feature space, the number of variables grows exponentially. For example, a second degree polynomial classifier in two dimensions requires the original variables x_1 and x_2 as well as the variables x_1^2, x_2^2, and x_1x_2. In the primal problem, the problem size will explode as the degree of the polynomial increases. The dual problem, however, remains tractable. The number of dual variables is $k-1$ times the number of points regardless of what transformation is selected. In the dual problem, the transformation appears as an inner product in the high dimensional space. Inexpensive techniques exist for computing these inner products. Each dual variable corresponds to a point in the original feature space. A point with a corresponding positive dual variable is referred to as a support vector. The goal is to maintain a high accuracy while using a small number of support vectors. Minimizing the number of support vectors is important for generalization and also for reducing the computational time required to evaluate new examples. Section 6 contains computational results comparing the two LP approaches k-RLP and M-RLP and the two QP approaches k-SVM and M-SVM. The methods were compared in terms of generalization (testing set accuracy), number of support vectors, and computational time.

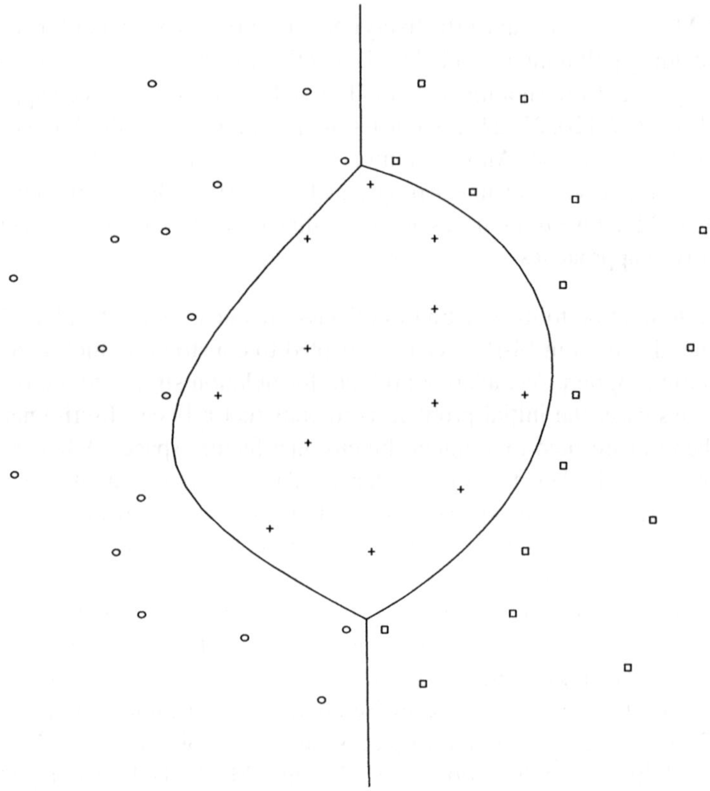

Figure 1. Piecewise-polynomial separation of three classes in two dimensions

The following notation will be used throughout this paper. Mathematically we can abstract the problem as follows: Given the elements of the sets, $\mathcal{A}^i, i = 1, \ldots, k$, in the n-dimensional real space R^n, construct a discriminant function that separates these points into distinct regions. Each region should contain points belonging to all or almost all of the same set. Let $\mathcal{A}^j$ be a set of points in the n-dimensional real space R^n with cardinality m_j. Let A^j be an $m_j \times n$ matrix whose rows are the points in $\mathcal{A}^j$. The ith point in $\mathcal{A}^j$ and the ith row of A^j are both denoted A^j_i. Let e denote a vector of ones of the appropriate dimension. The scalar 0 and a vector of zeros are both represented by 0. Thus, for $x \in R^n$, $x > 0$ implies that $x_i > 0$ for $i = 1, \ldots, n$. Similarly, $x \geq y$ implies that $x_i \geq y_i$ for $i = 1, \ldots, n$. For the vector x in R^n and the matrix A in $R^{n \times m}$, the transpose of x and A are denoted x^T and A^T respectively. The dot product of two vectors x and y is denoted $x^T y$ or $(x \cdot y)$. The 1-norm of x, $\sum_{i=1}^{n} |x_i|$, is denoted by $\|x\|_1$. The 2-norm or Euclidean norm of x, $\sqrt{\sum_{i=1}^{n} x_i^2} = \sqrt{x^T x}$, is denoted by $\|x\|$. The set of minimizers of $f(x)$ on the set S is

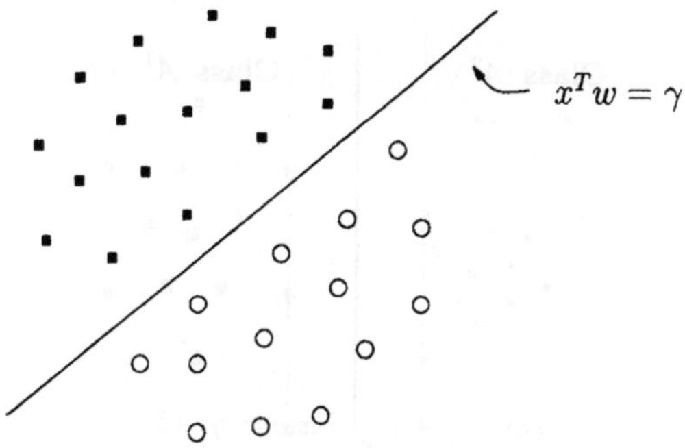

Figure 2. Two linearly separable sets and a separating plane

denoted by $arg \min_{x \in S} f(x)$. The sign of a scalar z, $sign(z) := 1$ if $z \geq 0$ and $sign(z) := -1$ otherwise.

2. Background

This section contains a brief overview of the existing RLP and SVM methods for classification. First we will discuss the two-class problem using a linear classifier. Then SVM for two classes will be defined. Then RLP will be reviewed. Finally, the piecewise-linear function used for multicategory classification in M-RLP will be reviewed.

2.1. *Two-Class Linear Discrimination*

Commonly, the method of discrimination for two classes of points involves determining a linear function that consists of a linear combination of the attributes of the given sets. In the simplest case, a linear function can be used to separate two sets as shown in Figure 2.

This function is the separating plane $x^T w = \gamma$ where w is the normal to the plane and γ is the distance from the origin. Let $\mathcal{A}^1$ and $\mathcal{A}^2$ be two sets of points in the n-dimensional real space R^n with cardinality m_1 and m_2 respectively. Let A^1 be an $m_1 \times n$ matrix whose rows are the points in $\mathcal{A}^1$. Let A^2 be an $m_2 \times n$ matrix whose rows are the points in $\mathcal{A}^2$. Let $x \in R^n$ be a point to be classified as follows:

$$\begin{array}{l} x^T w - \gamma > 0 \Rightarrow x \in \mathcal{A}^1 \\ x^T w - \gamma < 0 \Rightarrow x \in \mathcal{A}^2 \end{array} \tag{1}$$

The two sets of points, $\mathcal{A}^1$ and $\mathcal{A}^2$, are linearly separable if

$$A^1 w > \gamma e \quad \gamma e > A^2 w \tag{2}$$

where e is a vector of ones of the appropriate dimension. If the two classes are linearly separable, there are infinitely many planes that separate the two classes. The goal is to choose the plane that will generalize best on future points.

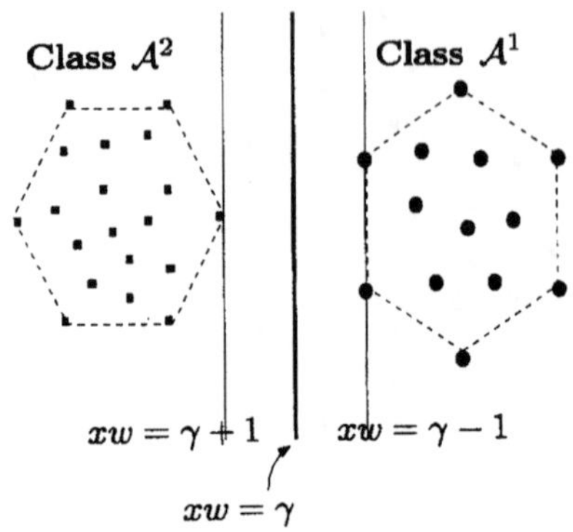

Figure 3. Two supporting planes and the resulting optimal separating plane

Both Mangasarian [16] and Vapnik and Chervonenkis [30] concluded that the best plane in the separable case is the one that minimizes the distance of the closest vector in each class to the separating plane. For the separable case the formulations of Mangasarian's Multi-surface Method of Pattern Recognition [17] and those of Vapnik's Optimal Hyperplane [29, 30] are very similar [3]. We will concentrate on the Optimal Hyperplane problem since it is the basis of SVM and it is validated theoretically by Statistical Learning Theory [29]. According to Statistical Learning Theory, the Optimal Hyperplane can construct linear discriminants in very high dimensional spaces without overfitting. The reader should consult [29] for full details of Statistical Learning Theory not covered in this paper.

The problem in the canonical form of Vapnik [29] becomes to determine two parallel planes $x^T w = \gamma + 1$ and $x^T w = \gamma - 1$ such that

$$A^1 w - \gamma e - e \geq 0 \qquad -A^2 w + \gamma e - e \geq 0. \tag{3}$$

and the margin or distance between the two planes is maximized. The margin of separation between the two supporting planes is $\frac{2}{\|w\|}$. An example of such a plane is shown in Figure 3.

The problem of finding the maximum margin becomes [29]:

$$\begin{aligned} \min_{w,\gamma} \quad & \tfrac{1}{2} w^T w \\ s.t. \quad & A^1 w - \gamma e - e \geq 0 \\ & -A^2 w + \gamma e - e \geq 0. \end{aligned} \tag{4}$$

In general it is not always possible for a single linear function to completely separate two given sets of points. Thus, it is important to find the linear function that discriminates best between the two sets according to some error minimization criterion. Bennett and Mangasarian [4] minimize the average magnitude of the misclassification errors in the construction of their following robust linear programming problem (RLP).

$$
\begin{array}{ll}
\min\limits_{w,\gamma,y,z} & \delta_1 e^T y + \delta_2 e^T z \\
s.t. & A^1 w - \gamma e + y - e \geq 0 \\
& -A^2 w + \gamma e + z - e \geq 0 \\
& y \geq 0 \quad z \geq 0
\end{array} \tag{5}
$$

where $\delta_1 > 0$ and $\delta_2 > 0$ are the misclassification costs. To avoid the null solution $w = 0$, use $\delta_1 = \frac{1}{m_1}$ and $\delta_2 = \frac{1}{m_2}$ where m_1 and m_2 are the cardinalities of $\mathcal{A}^1$ and $\mathcal{A}^2$ respectively. The RLP method is very effective in practice. The functions generated by RLP generalize well on many real-world problems. Additionally, the computational time is reasonably small because its solution involves only a single linear program. Note, however, that the RLP method no longer includes any notion of maximizing the margin. According to structural risk minimization in Statistical Learning Theory, maximizing the margin is essential for good generalization. For a fixed empirical misclassification rate, larger margins should lead to better generalization and prevent overfitting.

The SVM approach [11, 29] is a multiobjective quadratic program which minimizes the absolute misclassification errors and maximizes the separation margin by minimizing $\|w\|^2$.

$$
\begin{array}{ll}
\min\limits_{w,\gamma,y,z} & (1-\lambda)(e^T y + e^T z) + \frac{\lambda}{2} w^T w \\
s.t. & A^1 w - \gamma e + y - e \geq 0 \\
& -A^2 w + \gamma e + z - e \geq 0 \\
& y \geq 0 \quad z \geq 0
\end{array} \tag{6}
$$

where $0 < \lambda < 1$ is a fixed constant. Note that Problem (6) is equivalent to RLP with the addition of a regularization term $\frac{\lambda}{2} w^T w$ and with misclassification costs $\delta_1 = \delta_2 = 1$. Statistical Learning Theory shows that this regularization term is essential for good generalization.

A linear programming version of (6) can be constructed by replacing the norm used to minimize the weights w [3]. Recall that the SVM objective minimizes the square of the 2-norm of w, $\|w\|^2 = w^T w$. The 1-norm of w, $\|w\|_1 = e^T |w|$, can be used instead [3, 10, 9, 7]. Previous research found no empirical evidence that one norm was preferable in terms of generalization. The effects on Statistical Learning Theory caused by changing norms is an open question. The results in [20] can be used to show that minimizing $\|w\|_1$ corresponds to maximizing the seperation margin in terms of the infinity-norm. One benefit the 1-norm formulation has is that it is a sparse norm and automatically performs feature selection [9, 7]. Another benefit is that it can be solved using linear programming instead of quadratic programming. The linear programming problem can be constructed by removing the absolute value function through the introduction of the vector variable s and the constraints $-s \leq w \leq s$. The SVM objective is then modified by substituting $e^T s$ for $\frac{w^T w}{2}$. At optimality, $s_i = |w_i|$, $i = 1, \ldots, k$. The resulting LP is:

$$
\begin{array}{ll}
\min\limits_{w,\gamma,y,z,s} & (1-\lambda)(\delta_1 e^T y + \delta_2 e^T z) + \lambda e^T s \\
s.t. & A^1 w - \gamma e + y - e \geq 0 \\
& -A^2 w + \gamma e + z - e \geq 0 \\
& -s \leq w \leq s \\
& y \geq 0 \quad z \geq 0 \quad s \geq 0.
\end{array} \tag{7}
$$

We will refer to this problem as RLP since $\lambda = 0$ yields the original RLP method.

As in the SVM method, the RLP method minimizes both the average distance of the misclassified points from the relaxed supporting planes and a norm of the weight vectors. The main advantage of the RLP method over the SVM problem is that RLP is a linear program solvable using very robust algorithms such as the Simplex Method [14, 23]. SVM requires the solution of a quadratic program that is typically much more computationally costly for the same size problem. In [3], the RLP method was found to generalize as well as the linear SVM but with much less computational cost.

It is more efficient computationally to solve the dual RLP and SVM problems. The dual RLP problem is

$$\begin{array}{ll} \min\limits_{u,v} & e^T u + e^T v \\ s.t. & -\lambda e \le u^T A^1 - v^T A^2 \le \lambda e \\ & e^T u - e^T v = 0 \\ & 0 \le u \le (1-\lambda)\delta_1 \;\; 0 \le v \le (1-\lambda)\delta_2 \end{array} \tag{8}$$

In this paper we use $\delta_1 = \frac{1}{m_1}$ and $\delta_2 = \frac{1}{m_2}$ but δ_1 and δ_2 may be any positive weights for the misclassification costs. The dual SVM problem and its extension to nonlinear discriminants are given in the next section.

2.2. *Nonlinear Classifiers Using Support Vector Machines*

The primary advantage of the SVM (6) over RLP (7) is that efficient methods based on the dual of SVM (6) exist for constructing nonlinear discriminants [29, 11]. These methods with minor modifications can produce polynomial separators, radial basis functions, neural networks, etc. While such approaches are theoretically possible for RLP (7), their effective implementation is an open research question beyond the scope of this paper.

The basic idea is to map the original problem to a higher dimensional space and then to construct a linear discriminant in a higher dimensional space that corresponds to a nonlinear discriminant in the original space. So for example, to construct a quadratic discriminant for a two-dimensional problem, the input attributes $[x_1, x_2]$ are mapped into $[x_1^2, x_1^2, \sqrt{2}x_1x_2, x_1, x_2]$ and a linear discriminant function is constructed in the new five-dimensional space. Two examples of possible polynomial classifiers are given in Figure 4. The dual SVM is applied to the mapped points. The regularization term in the primal objective helps avoid overfitting the higher dimensional space. The dual SVM provides a practical computational approach through the use of generalized inner products or kernels.

The dual SVM is as follows:

$$\begin{array}{ll} \min\limits_{u,v} & \frac{1}{2}\left\| {A^1}^T u - {A^2}^T v \right\|^2 - e^T u - e^T v \\ s.t. & e^T u = e^T v \\ & \frac{(1-\lambda)}{\lambda} e \ge u \ge 0 \;\; \frac{(1-\lambda)}{\lambda} e \ge v \ge 0. \end{array} \tag{9}$$

To formulate the nonlinear case it is convenient to rewrite the problem in summation notation. Let $\mathcal{A}$ be the set of all points $\mathcal{A}^1$ and $\mathcal{A}^2$. Define $M = m_1 + m_2$ to be the total

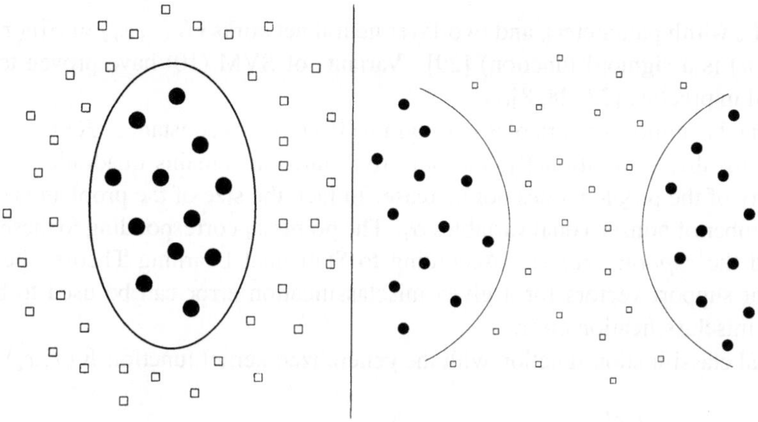

Figure 4. Two examples of second degree polynomial separations of two sets

number of points. Let $\alpha^T = [\alpha_1, \alpha_2, \ldots, \alpha_M] = [\frac{1}{\lambda}u^T \; \frac{1}{\lambda}v^T]$. Let $t \in R^M$ be such that for $x_i \in \mathcal{A}$ $t_i = \begin{cases} 1 & x_i \in A^1 \\ -1 & x_i \in A^2 \end{cases}$.

To construct the nonlinear classification function, the original data points x are transformed to the higher dimensional feature space by the function $\phi(x) : R^n \to R^{n'}$, $n' >> n$. The dot product of the original vectors $x_i^T x_j$ is replaced by the dot product of the transformed vectors $(\phi(x_i) \cdot \phi(x_j))$.

The first term of the objective function can then be written as the sum:

$$\frac{\lambda}{2}\sum_{i=1}^{M}\sum_{j=1}^{M} t_i t_j \alpha_i \alpha_j (\phi(x_i) \cdot \phi(x_j)).$$

Using this notation and simplifying the problem becomes:

$$\begin{aligned} \min_{\alpha} \quad & \tfrac{1}{2}\sum_{i=1}^{M}\sum_{j=1}^{M} t_i t_j \alpha_i \alpha_j (\phi(x_i) \cdot \phi(x_j)) - \sum_{i=1}^{M} \alpha_i \\ s.t. \quad & \sum_{i=1}^{M} \alpha_i t_i = 0 \\ & \frac{(1-\lambda)}{\lambda} e \geq \alpha \geq 0. \end{aligned} \tag{10}$$

In the support vector machine (SVM), Vapnik replaces the inner product $(\phi(x) \cdot \phi(x_i))$ with the inner product in the Hilbert space $K(x, x_i)$. This symmetric kernel function $K(x, x_i)$ must satisfy Theorem 5.3 in [29]. This theorem ensures $K(x, x_i)$ is an inner product in some feature space. The choice of $K(x, x_i)$ determines the type of classifier that is constructed. Possible choices include polynomial classifiers as in Figure 4 ($K(x, x_i) = (x^T x_i + 1)^d$, where d is the degree of the polynomial), radial basis function machines ($K_\gamma(|x - x_i|) = \exp\{-\gamma|x - x_i|^2\}$ where $|x - x_i|$ is the distance between two vectors and γ is the width parameter), and two-layer neural networks ($K(x, x_i) = S[v(x^T x_i) + c]$ where $S(u)$ is a sigmoid function) [29]. Variants of SVM (10) have proven to be quite successful in practice [27, 28, 8].

Note that the number of variables in Program (10) remains constant as $K(x, x_i)$ increases in dimensionality. Additionally, the objective function remains quadratic and thus the complexity of the problem does not increase. In fact, the size of the problem is dependent on the number of nonzero dual variables α_i. The points x_i corresponding to these variables are called the *support vectors.* According to Statistical Learning Theory, the expected number of support vectors for a given misclassification error can be used to bound the expected misclassification error.

The final classification function with the generalized kernel function $K(x, x_i)$ is:

$$f(x) = \text{sign}\left(\sum_{i=1}^{M} t_i \alpha_i K(x, x_i) - \gamma\right) \tag{11}$$

where $x \in \mathcal{A}^1$ if $f(x) = 1$, otherwise $x \in \mathcal{A}^2$.

2.3. *Multicategory Discrimination*

In multicategory classification a piecewise-linear separator is used to discriminate between $k > 2$ classes of m^i, $i = 1, \ldots, k$, points. We will examine two methods for accomplishing this. The first, used in SVM [29], is to construct a discriminant function to separate one class from the remaining $k - 1$ classes. This process is repeated k times. In the separable case, the linear discriminant for each class must satisfy the following set of inequalities. Find $(w^1, \gamma^1), \ldots, (w^k, \gamma^k)$, such that

$$A^i w^i - \gamma^i e > A^j w^i - \gamma^i e, \quad i, j = 1, \ldots, k, \quad i \neq j. \tag{12}$$

To classify a new point x, compute $f_i(x) = x^T w^i - \gamma^i$. The class of a point x is then determined from (w^i, γ^i), $i = 1, \ldots, k$ by finding i such that

$$f_i(x) = x^T w^i - \gamma^i \tag{13}$$

is maximized. Figure 5 shows a piecewise-linear function $f(x) = \max_{i=1,2,3} f_i(x)$ on R that separates three sets.

Note that either SVM (10) or RLP (8) can be used to construct the k two-class discriminants. For clarity, we will call this method used with SVM (10), k-SVM. We will denote this method used with RLP (8), k-RLP. The advantage of k-SVM is that it can be used for piecewise-nonlinear discriminants while k-RLP is currently limited to piecewise-linear discriminants. For both k-SVM and k-RLP to attain perfect training set accuracy, the following inequalities must be satisfied:

$$A^i w^i - \gamma^i e > A^i w^j - \gamma^j e, \quad i, j = 1, \ldots, k, \quad i \neq j.$$

This inequality can be used as a definition of piecewise-linear separability.

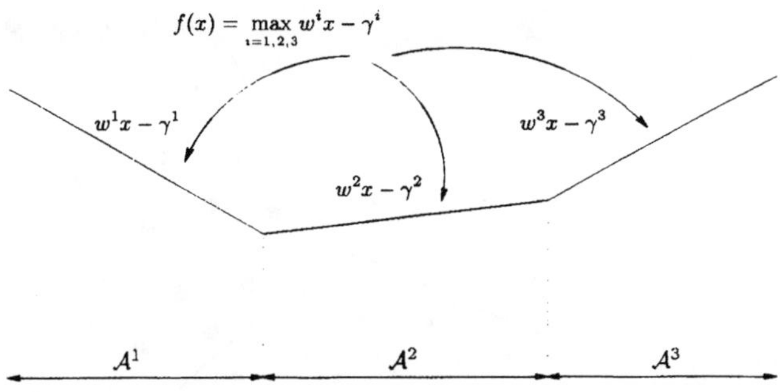

Figure 5. Piecewise-linear separation of sets $\mathcal{A}^1$, $\mathcal{A}^2$, and $\mathcal{A}^3$ by the convex piecewise-linear function $f(x)$.

DEFINITION 1 (PIECEWISE-LINEAR SEPARABILITY) *The sets of points $\mathcal{A}^i$, $i = 1, \ldots, k$, represented by the matrices $A^i \in R^{m_i \times n}$, $i = 1, \ldots, k$, are piecewise-linearly separable if there exist $w^i \in R^n$ and $\gamma^i \in R$, $i = 1, \ldots, k$, such that*

$$A^i w^i - \gamma^i e > A^i w^j - \gamma^j e, \; i, j = 1, \ldots, k, \; i \neq j. \tag{14}$$

Equivalent to Definition 1, finding the piecewise-linear separator involves solving the equation $A^i w^i - \gamma^i e \geq A^i w^j - \gamma^j e + e$, $i, j = 1, \ldots, k$, $i \neq j$. This can be rewritten as $0 \geq -A^i(w^i - w^j) + (\gamma^i - \gamma^j)e + e$, $i, j = 1, \ldots, k$, $i \neq j$. Figure 6 shows an example of a piecewise-linear separator for three classes in two dimensions.

The linear separating functions are represented by the quantities $(w^i - w^j, \gamma^i - \gamma^j)$, $i, j = 1, \ldots, k$, $j \neq i$, where $w^i \in R^{n \times 1}$ and $\gamma^i \in R^1$, $i = 1, \ldots, k$.

The M-RLP method[1] proposed and investigated in [5, 6] can be used to find (w_i, γ_i), $i = 1, \ldots, k$ satisfying Definition 1:

$$\min_{w^i, \gamma^i, y^{ij}} \left\{ \sum_{i=1}^{k} \sum_{\substack{j=1 \\ j \neq i}}^{k} \frac{e^T y^{ij}}{m^i} \;\middle|\; \begin{array}{c} y^{ij} \geq -A^i(w^i - w^j) + (\gamma^i - \gamma^j)e + e, \, y^{ij} \geq 0, \\ i \neq j, \; i, j = 1, \ldots, k \end{array} \right\} \tag{15}$$

where $y^{ij} \in R^{m_i \times 1}$. In M-RLP (15), if the optimal objective value is zero, then the dataset is piecewise-linearly separable. If the dataset is not piecewise-linearly separable, the positive values of the variables y^{ij} are proportional to the magnitudes of the misclassified points from the plane $x^T(w^i - w^j) = (\gamma^i - \gamma^j) + 1$. This program (15) is a generalization of the two-class RLP linear program (5) to the multicategory case. Like the original RLP (5), M-RLP does not include any terms for maximizing the margin and it does not directly permit the use of generalized inner products or kernels to allow extension to the nonlinear case. In the next section we will show how M-RLP and SVM can be combined by including margin maximization and generalized inner products in M-RLP.

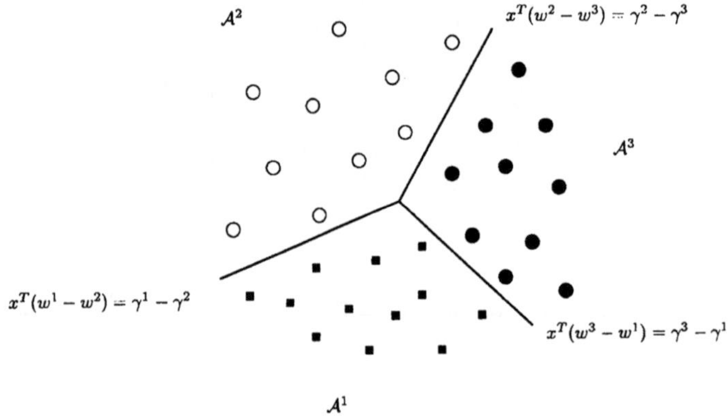

Figure 6. Three classes separated by a piecewise-linear function.

3. Formulation of M-SVM: Piecewise-linearly Separable Case

We now propose to construct piecewise-linear and piecewise-nonlinear SVMs using a single quadratic program. Analogous to the two-class case we start by formulating the "optimal" piecewise-linear separator for the separable case. Assume that the k sets of points are piecewise-linearly separable, i.e., there exist $w^i \in R^n$ and $\gamma^i \in R, i = 1, \ldots, k$, such that

$$A^i w^i - \gamma^i e > A^i w^j - \gamma^j e, \; i, \; j = 1, \ldots, k, \; i \neq j. \tag{16}$$

The class of a point x is determined from (w^i, γ^i), $i = 1, \ldots, k$ by finding i such that

$$f_i(x) = x^T w^i - \gamma^i \tag{17}$$

is maximized.

For this piecewise-linearly separable problem, infinitely many (w^i, γ^i) exist that satisfy (16). Intuitively, the "optimal" (w^i, γ^i) provides the largest margin of classification. So in an approach analogous to the two-class support vector machine (SVM) approach, we add regularization terms. The dashed lines in Figure 7 represent the margins for each piece $(w^i - w^j, \gamma^i - \gamma^j)$ of the piecewise-linear separating function. The margin of separation between the classes i and j, i.e., the distance between the supporting hyperplanes of the halfspaces

$$A^i(w^i - w^j) \geq (\gamma^i - \gamma^j)e + e$$
and
$$A^j(w^i - w^j) \leq (\gamma^i - \gamma^j)e - e,$$

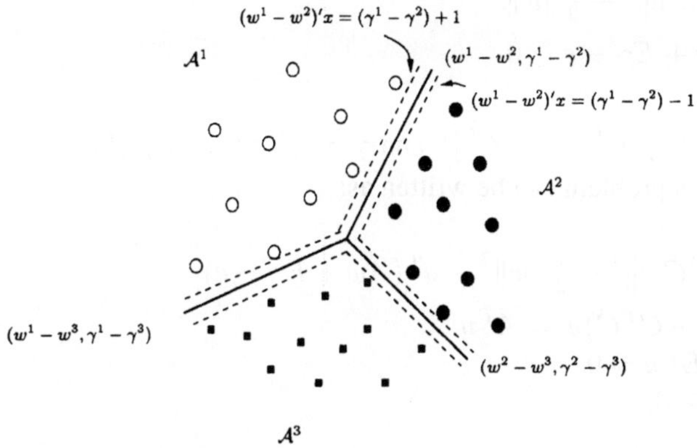

Figure 7. Piecewise-linear separator with margins for three classes.

is $\frac{2}{\|w^i - w^j\|}$. So we would like to minimize $\|w^i - w^j\|$ for all $i, j = 1, \ldots, k,\ i \neq j$. Also, we will add the regularization term $\frac{1}{2}\sum_{i=1}^{k} \|w^i\|^2$ to the objective.

For the piecewise-linearly separable problem we get the following:

$$\begin{array}{ll} \min\limits_{w^i, \gamma^i} & \frac{1}{2}\sum\limits_{i=1}^{k}\sum\limits_{j=1}^{i-1} \|w^i - w^j\|^2 + \frac{1}{2}\sum\limits_{i=1}^{k} \|w^i\|^2 \\ s.t. & A^i(w^i - w^j) - e(\gamma^i - \gamma^j) - e \geq 0 \\ & \qquad i, j = 1, \ldots, k \ \ i \neq j. \end{array} \tag{18}$$

To simplify the notation for formulation of the piecewise-linear SVM, we rewrite Problem (18) in matrix notation. See Appendix A.1 for complete matrix definitions for general k. For the three-class problem ($k = 3$) the following matrices are obtained:

Let

$$\bar{C} = \begin{bmatrix} I & -I & 0 \\ I & 0 & -I \\ 0 & I & -I \end{bmatrix}$$

where $I \in R^{n \times n}$ is the identity matrix.

Let

$$\bar{A} = \begin{bmatrix} A^1 & -A^1 & 0 \\ A^1 & 0 & -A^1 \\ -A^2 & A^2 & 0 \\ 0 & A^2 & -A^2 \\ -A^3 & 0 & A^3 \\ 0 & -A^3 & A^3 \end{bmatrix} \quad \bar{E} = \begin{bmatrix} -e^1 & e^1 & 0 \\ -e^1 & 0 & e^1 \\ e^2 & -e^2 & 0 \\ 0 & -e^2 & e^2 \\ e^3 & 0 & -e^3 \\ 0 & e^3 & -e^3 \end{bmatrix}$$

where $A^i \in R^{m_i \times n}, i = 1, \ldots, 3$, and $e^i \in R^{m_i \times 1}, i = 1, \ldots, 3$, is a vector of ones.

Using this notation for fixed $k > 2$ the program becomes:

$$\begin{array}{ll} \min\limits_{w,\gamma} & \frac{1}{2}\left\|\bar{C}w\right\|^2 + \frac{1}{2}\left\|w\right\|^2 \\ s.t. & \bar{A}w + \bar{E}\gamma - e \geq 0 \end{array} \tag{19}$$

where $w = [{w^1}^T, {w^2}^T, \ldots, {w^k}^T]^T$ and $\gamma = [\gamma^1, \gamma^2, \ldots, \gamma^k]^T$.
The dual of this problem can be written as:

$$\begin{array}{ll} \max\limits_{u,w,\gamma} & \frac{1}{2}\left\|\bar{C}w\right\|^2 + \frac{1}{2}\left\|w\right\|^2 - u^T(\bar{A}w + \bar{E}\gamma - e) \\ s.t. & (I + \bar{C}^T\bar{C})w = \bar{A}^T u \\ & -\bar{E}^T u = 0 \\ & u \geq 0. \end{array} \tag{20}$$

To eliminate the variables w and γ from this problem we will first show that the matrix $(I + \bar{C}^T\bar{C})$ is nonsingular.

PROPOSITION 1 (NONSINGULARITY OF $(I + \bar{C}^T\bar{C})$) *The inverse of matrix $(I+\bar{C}^T\bar{C})$ for $k > 2$ is*

$$(I_{kn} + \bar{C}^T\bar{C})^{-1} = \begin{bmatrix} \frac{2}{k+1}I_n & \frac{1}{k+1}I_n & \cdots & \frac{1}{k+1}I_n \\ \frac{1}{k+1}I_n & \ddots & \ddots & \vdots \\ \vdots & \ddots & \ddots & \frac{1}{k+1}I_n \\ \frac{1}{k+1}I_n & \cdots & \frac{1}{k+1}I_n & \frac{2}{k+1}I_n \end{bmatrix} \tag{21}$$

where I_n indicates the $n \times n$ identity matrix.

Proof. To show that $(I + \bar{C}^T\bar{C})$ is nonsingular for some $k > 2$, we will calculate its inverse. The matrix $\bar{C}$ as defined in Appendix A.1 has size $(n\sum_{i=2}^{k}(i-1) \times kn)$. Recall that n indicates the dimension of the feature space.

$$\bar{C}^T\bar{C} = \begin{bmatrix} (k-1)I_n & -I_n & \cdots & -I_n \\ -I_n & \ddots & \ddots & \vdots \\ \vdots & \ddots & \ddots & -I_n \\ -I_n & \cdots & -I_n & (k-1)I_n \end{bmatrix}$$

has size $kn \times kn$.

Therefore

$$I_{kn} + \bar{C}^T\bar{C} = \begin{bmatrix} kI_n & -I_n & \cdots & -I_n \\ -I_n & \ddots & \ddots & \vdots \\ \vdots & \ddots & \ddots & -I_n \\ -I_n & \cdots & -I_n & kI_n \end{bmatrix}.$$

Through simple calculations it can be shown that the inverse of this matrix is (21):

$$(I_{kn} + \bar{C}^T\bar{C})^{-1} = \begin{bmatrix} \frac{2}{k+1}I_n & \frac{1}{k+1}I_n & \cdots & \frac{1}{k+1}I_n \\ \frac{1}{k+1}I_n & \ddots & \ddots & \vdots \\ \vdots & \ddots & \ddots & \frac{1}{k+1}I_n \\ \frac{1}{k+1}I_n & \cdots & \frac{1}{k+1}I_n & \frac{2}{k+1}I_n \end{bmatrix}.$$

□

Using Proposition 1 the following relationship results:

$$(I + \bar{C}^T\bar{C})^{-1}\bar{A}^T = \frac{1}{k+1}\bar{A}^T. \tag{22}$$

It follows from Problem (20) and equation (22) that

$$w = (I + \bar{C}^T\bar{C})^{-1}\bar{A}^T u = \frac{1}{k+1}\bar{A}^T u. \tag{23}$$

Using this relationship, we eliminate w from the dual problem. Additionally, γ is removed because $-\bar{E}^T u = 0$.

After some simplification the new dual problem becomes:

$$\begin{array}{ll} \max\limits_{u} & e^T u - \frac{1}{2(k+1)} u^T \bar{A}\bar{A}^T u \\ s.t. & \bar{E}^T u = 0 \\ & u \geq 0. \end{array} \tag{24}$$

To construct the multicategory support vector machine, it is convenient to write this problem in summation notation. Let the dual vector

$$u^T = [u^{12^T}, u^{13^T}, \ldots, u^{1k^T}, u^{21^T}, u^{23^T}, \ldots, u^{k(k-1)^T}]$$

where $u^{ij} \in R^{m_i \times 1}$. The resulting dual problem for piecewise-linear datasets is:

$$\begin{array}{ll} \max\limits_{u} & \sum\limits_{i=1}^{k}\sum\limits_{\substack{j=1\\j\neq i}}^{k}\sum\limits_{l=1}^{m_i} u_l^{ij} - \frac{1}{2(k+1)}\sum\limits_{i=1}^{k}\sum\limits_{\substack{j=1\\j\neq i}}^{k}\sum\limits_{\substack{l=1\\l\neq i}}^{k}\left[\sum\limits_{p=1}^{m_i}\sum\limits_{q=1}^{m_i} u_p^{ij}u_q^{il}A_p^i A_q^{i^T}\right. \\ & \left. -2\sum\limits_{p=1}^{m_j}\sum\limits_{q=1}^{m_i} u_p^{ji}u_q^{il}A_p^j A_q^{i^T} + \sum\limits_{p=1}^{m_j}\sum\limits_{q=1}^{m_l} u_p^{ji}u_q^{li}A_p^j A_q^{l^T}\right] \\ s.t. & -\sum\limits_{\substack{j=1\\j\neq i}}^{k}\sum\limits_{l=1}^{m_i} u_l^{ij} + \sum\limits_{\substack{j=1\\j\neq i}}^{k}\sum\limits_{l=1}^{m_j} u_l^{ji} = 0 \text{ for } i = 1,\ldots,k \\ & u_l^{ij} \geq 0 \text{ for } i,j = 1,\ldots,k,\ i \neq j \text{ and } l = 1,\ldots,m_i \end{array} \tag{25}$$

where m_i is the number of points in class i.

Recall that for the piecewise-linear classification function, the class of a point x is determined by finding $i = 1,\ldots,k$, such that

$$f_i(x) = x^T w^i - \gamma^i \tag{26}$$

is maximized. From equation (23),

$$w = \begin{bmatrix} w^1 \\ w^2 \\ \vdots \\ w^k \end{bmatrix} = \frac{1}{k+1} \bar{A}^T u.$$

Solving for w^i in summation notation we get:

$$w^i = \frac{1}{k+1} \left[\sum_{\substack{j=1 \\ j \neq i}}^{k} \sum_{p=1}^{m_i} u_p^{ij} A_p^{i\,T} - \sum_{\substack{j=1 \\ j \neq i}}^{k} \sum_{p=1}^{m_j} u_p^{ji} A_p^{j\,T} \right].$$

Therefore,

$$f_i(x) = \frac{1}{k+1} \left[\sum_{\substack{j=1 \\ j \neq i}}^{k} \sum_{p=1}^{m_i} u_p^{ij} x^T A_p^{i\,T} - \sum_{\substack{j=1 \\ j \neq i}}^{k} \sum_{p=1}^{m_j} u_p^{ji} x^T A_p^{j\,T} \right] - \gamma^i.$$

4. Formulation of M-SVM: Piecewise-nonlinearly Separable Case

Just like in the two-class case, M-SVM can be generalized to piecewise-nonlinear functions. To construct the separating functions, $f_i(x)$, in a higher dimensional feature space, the original data points x are transformed by some function $\phi(x) : R^n \to R^{n'}$ [29, 11]. The function $f_i(x)$ is now related to the sum of dot products of vectors in this higher dimensional feature space:

$$f_i(x) = \frac{1}{k+1} \left[\sum_{\substack{j=1 \\ j \neq i}}^{k} \sum_{p=1}^{m_i} u_p^{ij} (\phi(x) \cdot \phi(A_p^{i\,T})) - \sum_{\substack{j=1 \\ j \neq i}}^{k} \sum_{p=1}^{m_j} u_p^{ji} (\phi(x) \cdot \phi(A_p^{j\,T})) \right] - \gamma^i.$$

According to [29], any symmetric function $K(x, x_i) \in L_2$ that satisfies Mercer's Theorem [12] can replace the dot product $(\phi(x) \cdot \phi(x_i))$. Mercer's Theorem guarantees that any eigenvalue λ_j in the expansion $K(x, x_i) = \sum_{j=1}^{\infty} \lambda_j (\phi_j(x) \cdot \phi_j(x_i))$ is positive. This is a sufficient condition for a function $K(x, x_i)$ to define a dot product in the higher dimensional feature space. Therefore we let $K(x, x_i) = (\phi(x) \cdot \phi(x_i))$.

Returning to dual Problem (25), the objective function contains the sum of dot products $A_p^j A_q^{i\,T}$ of two points in the original feature space. To transform the points A_p^j to a higher dimensional feature space we replace these dot products by $K(A_p^{j\,T}, A_q^{i\,T})$.

The resulting M-SVM for piecewise-linearly separable datasets is:

$$\begin{aligned}\max_{u}\ & \sum_{i=1}^{k}\sum_{\substack{j=1\\j\neq i}}^{k}\sum_{l=1}^{m_i} u_l^{ij} - \tfrac{1}{2(k+1)}\sum_{i=1}^{k}\sum_{\substack{j=1\\j\neq i}}^{k}\sum_{\substack{l=1\\l\neq i}}^{k}\Bigg[\sum_{p=1}^{m_i}\sum_{q=1}^{m_i} u_p^{ij}u_q^{il}K({A_p^i}^T,{A_q^i}^T) \\ & -2\sum_{p=1}^{m_j}\sum_{q=1}^{m_i} u_p^{ji}u_q^{il}K({A_p^j}^T,{A_q^i}^T) + \sum_{p=1}^{m_j}\sum_{q=1}^{m_l} u_p^{ji}u_q^{li}K({A_p^j}^T,{A_q^l}^T)\Bigg] \\ s.t.\ & -\sum_{\substack{j=1\\j\neq i}}^{k}\sum_{l=1}^{m_i} u_l^{ij} + \sum_{\substack{j=1\\j\neq i}}^{k}\sum_{l=1}^{m_j} u_l^{ji} = 0 \text{ for } i=1,\ldots,k \\ & u_l^{ij} \geq 0 \text{ for } i,j=1,\ldots,k,\ i\neq j \text{ and } l=1,\ldots,m_i.\end{aligned} \tag{27}$$

The points A_l^i corresponding to nonzero dual variables $u_l^{ij}, j = 1,\ldots,k, j \neq i$ are referred to as support vectors. It is possible for A_l^i to correspond to more than one nonzero variable $u_l^{ij}, j = 1,\ldots,k, j \neq i$. In Figure 8, support vectors are represented by a circle around the point. Some points have double circles which indicate that there are two dual variables $u_l^{ij} > 0, j = 1,\ldots,3, j \neq i$.

By the complementarity within the KKT conditions [18],

$$u_l^{ij} > 0 \Rightarrow A_l^i(w^i - w^j) = (\gamma^i - \gamma^j) + 1.$$

Consequently the support vectors are located "closest" to the separating function. In fact, the remainder of the points, those that are not support vectors, are not necessary in the construction of the separating function. The resulting nonlinear classification problem for a point x is to find $i = 1,\ldots,k$ such that the classification function

$$f_i(x) = \frac{1}{k+1}\left(\sum_{\substack{j=1\\j\neq i}}^{k}\left[\sum_{\substack{\text{support}\\\text{vectors}\in A^i}} u_p^{ij}K(x,{A_p^i}^T) - \sum_{\substack{\text{support}\\\text{vectors}\in A^j}} u_p^{ji}K(x,{A_p^j}^T)\right]\right) - \gamma^i \tag{28}$$

is maximized.

5. Formulation of M-SVM: Piecewise Inseparable Case

The preceding sections provided a formulation for the piecewise-linearly and piecewise-nonlinearly separable cases. To construct a classification function for a piecewise-linearly *inseparable* dataset, we must first choose an error minimization criterion. The technique, used in the preceding sections, of formulating the M-SVM for piecewise-linearly separable datasets can be combined with the 1-norm error criterion used in Problem (15) of Bennett and Mangasarian [6]. The result is the M-SVM for piecewise-linearly inseparable problems.

Using the same matrix notation as in Section 3, we add the terms $\frac{1}{2}\left\|\bar{C}w\right\|^2 + \frac{1}{2}\|w\|^2$ to the objective of Problem (15). The resulting primal problem is as follows:

$$\begin{aligned}\min_{w,\gamma,y}\ & \lambda(\tfrac{1}{2}\left\|\bar{C}w\right\|^2 + \tfrac{1}{2}\|w\|^2) + (1-\lambda)e^Ty \\ s.t.\ & \bar{A}w + \bar{E}\gamma - e + y \geq 0 \\ & y \geq 0\end{aligned} \tag{29}$$

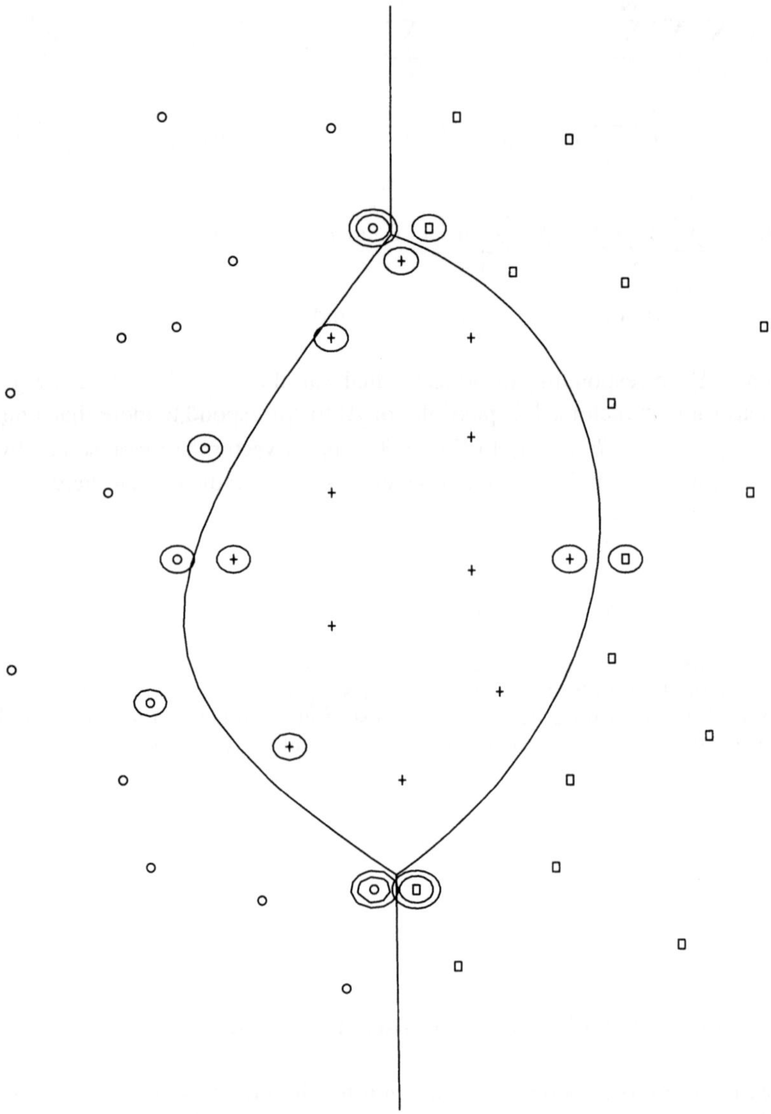

Figure 8. Piecewise-polynomial separation of three classes in two dimensions. Support vectors are indicated with circles.

where $y = [y_{12}^T, y_{13}^T, \ldots, y_{1k}^T, y_{21}^T, \ldots, y_{k(k-1)}^T]^T$ and $0 < \lambda < 1$.

Solving for the dual, substituting $w = \frac{1}{\lambda(k+1)}\bar{A}^T u$, and simplifying produces the following problem:

$$\begin{array}{ll} \max\limits_{u} & u^T e - \frac{1}{2(k+1)} u^T \bar{A}\bar{A}^T u \\ s.t. & \bar{E}^T u = 0 \\ & 0 \le u \le \frac{1-\lambda}{\lambda} e. \end{array} \tag{30}$$

As shown in Proposition 1, Problem (30) maximizes a concave quadratic objective over a bounded polyhedral set. Thus there exists a locally optimal solution that is globally optimal.

PROPOSITION 1 (CONCAVITY OF OBJECTIVE) *The function* $u^T e - \frac{1}{2(k+1)} u^T \bar{A}\bar{A}^T u$ *is concave.*

Proof. The matrix $\bar{A}\bar{A}^T$ is always positive semi-definite and symmetric. Thus the Hessian matrix $(-\frac{1}{(k+1)}\bar{A}\bar{A}^T)$ is negative semi-definite. Therefore, the objective is a concave function. □

Problem (30) is identical to Problem (24) in the piecewise-linearly separable case except the dual variables are now bounded by $\frac{1-\lambda}{\lambda}$. Therefore, transforming the data points A_l^i will proceed identically as in Section 4. Using the function $K(x, x_i)$ to denote the dot product in some feature space, the final M-SVM formulation is:

$$\begin{array}{ll} \max\limits_{u} & \sum\limits_{i=1}^{k}\sum\limits_{\substack{j=1\\ j\neq i}}^{k}\sum\limits_{l=1}^{m_i} u_l^{ij} - \frac{1}{2(k+1)} \sum\limits_{i=1}^{k}\sum\limits_{\substack{j=1\\ j\neq i}}^{k}\sum\limits_{\substack{l=1\\ l\neq i}}^{k} \left[\sum\limits_{p=1}^{m_i}\sum\limits_{q=1}^{m_i} u_p^{ij} u_q^{il} K(A_p^{i\,T}, A_q^{i\,T}) \right. \\ & \left. -2\sum\limits_{p=1}^{m_j}\sum\limits_{q=1}^{m_i} u_p^{ji} u_q^{il} K(A_p^{j\,T}. A_q^{i\,T}) + \sum\limits_{p=1}^{m_j}\sum\limits_{q=1}^{m_l} u_p^{ji} u_q^{li} K(A_p^{j\,T}, A_q^{l\,T}) \right] \\ s.t. & -\sum\limits_{\substack{j=1\\ j\neq i}}^{k}\sum\limits_{l=1}^{m_i} u_l^{ij} + \sum\limits_{\substack{j=1\\ j\neq i}}^{k}\sum\limits_{l=1}^{m_j} u_l^{ji} = 0 \text{ for } i = 1, \ldots, k \\ & 0 \le u_l^{ij} \le \frac{1-\lambda}{\lambda} \text{ for } i, j = 1, \ldots, k,\ i \neq j \text{ and } l = 1, \ldots, m_i. \end{array} \tag{31}$$

As in Sections 3 and 4, the class of a point x is determined by finding the maximum of the function

$$f_i(x) = \frac{1}{k+1} \left(\sum_{\substack{j=1\\ j\neq i}}^{k} \left[\sum_{\substack{\text{support}\\ \text{vectors} \in A^i}} u_p^{ij} K(x, A_p^{i\,T}) - \sum_{\substack{\text{support}\\ \text{vectors} \in A^j}} u_p^{ji} K(x, A_p^{j\,T}) \right] \right) - \gamma^i \tag{32}$$

for $i = 1, \ldots, k$.

To determine the threshold values $\gamma^i, i = 1, \ldots, k$, we solve the primal problem (29) with w fixed, where $\bar{A}w$ is transformed to the higher dimensional feature space. This problem is as follows:

$$
\begin{aligned}
\min_{\gamma,y} \quad & \sum_{i=1}^{k}\sum_{\substack{j=1\\ j\neq i}}^{k}\sum_{l=1}^{m_i} y_l^{ij} \\
s.t. \quad & -\gamma^i + \gamma^j + y_l^{ij} \geq \tfrac{1}{k+1}\sum_{\substack{l=1\\ l\neq i}}^{k}\sum_{q=1}^{m_i}\left[\sum_{r=1}^{m_i} K({A_q^i}^T, {A_r^i}^T)u_r^{il} - \sum_{r=1}^{m_l} K({A_q^i}^T, {A_r^l}^T)u_r^{li}\right] \\
& \quad - \tfrac{1}{k+1}\sum_{\substack{l=1\\ l\neq j}}^{k}\sum_{q=1}^{m_i}\left[\sum_{r=1}^{m_j} K({A_q^i}^T, {A_r^j}^T)u_r^{jl} - \sum_{r=1}^{m_l} K({A_q^i}^T, {A_r^l}^T)u_r^{lj}\right] + 1 \\
& y_l^{ij} \geq 0, \; i,j = 1,\ldots,k, \; i \neq j, \; l = 1,\ldots,m_i.
\end{aligned}
\tag{33}
$$

The right sides of the constraints are constant. Thus Problem (33) is a linear program and is easily solved.

6. Computational Experiments

In this section, we present computational results comparing M-SVM (32), M-RLP (15), k-SVM using SVM (10), and k-RLP using RLP (8). Several experiments on real-world datasets are reported. A description of each of the datasets follows this paragraph. Each of these methods was implemented using the MINOS 5.4 [23] solver. The quadratic programming problems for M-SVM and k-SVM were solved using the nonlinear solver implemented in MINOS 5.4. This solver uses a reduced-gradient algorithm in conjunction with a quasi-Newton method. In M-SVM, k-SVM and k-RLP, the selected values for λ are given. Better solutions may result with different choices of λ. Additionally, it is not necessary for the same value of λ to be used for both methods. The kernel function for the piecewise-nonlinear M-SVM and k-SVM methods is $K(x, x_i) = \left(\frac{x^T x_i}{n} + 1\right)^d$, where d is the degree of the desired polynomial.

Wine Recognition Data The Wine dataset [1] uses the chemical analysis of wine to determine the cultivar. There are 178 points with 13 features. This is a three-class dataset distributed as follows: 59 points in class 1, 71 points in class 2, and 48 points in class 3. This dataset is available via anonymous file transfer protocol (ftp) from the UCI Repository of Machine Learning Databases and Domain Theories [22] at ftp://ftp.ics.uci.edu/pub/machine-learning-databases.

Glass Identification Database The Glass dataset [15] is used to identify the origin of a sample of glass through chemical analysis. This dataset is comprised of six classes of 214 points with 9 features. The distribution of points by class is as follows: 70 float processed building windows, 17 float processed vehicle windows, 76 non-float processed building windows, 13 containers, 9 tableware, and 29 headlamps. This dataset is available via anonymous file transfer protocol (ftp) from the UCI Repository of Machine Learning Databases and Domain Theories [22] at ftp://ftp.ics.uci.edu/pub/machine-learning-databases.

US Postal Service Database The USPS Database [13] contains zip code samples from actual mail. This database is comprised of separate training and testing sets. There are

Table 1. Percent testing set accuracies and (total number of support vectors) for M-SVM and k-SVM. $\lambda = .05$ for k-RLP, M-SVM, and k-SVM.

Data	Method	Degree				
		1	2	3	4	5
Wine	M-RLP	91.01	-	-	-	-
	k-RLP	99.44	-	-	-	-
	M-SVM	97.19 (378)	97.19 (291)	97.75 (258)	96.63 (239)	96.63 (228)
	k-SVM	99.44 (537)	98.88 (424)	99.44 (405)	99.44 (394)	99.44 (413)
Glass	M-RLP	64.95	-	-	-	-
	k-RLP	54.67	-	-	-	-
	M-SVM	55.14 (1759)	62.15 (1660)	62.15 (1595)	66.82 (1533)	67.29 (1476)
	k-SVM	43.46 (1898)	55.61 (1854)	64.95 (1796)	70.56 (1769)	72.43 (1734)

7291 samples in the training set and 2007 samples in the testing set. Each sample belongs to one of ten classes: the integers 0 through 9. The samples are represented by 256 features.

Two experiments were performed. In the first, the datasets were normalized between -1 and 1. 10-fold cross validation was used to estimate generalization on future data. The second experiment was conducted on two subsets of the United States Postal Service (USPS) data. This data contains handwriting samples of the integers 0 through 9. The objective of this dataset is to quickly and effectively interpret zip codes. This data has separate training and testing sets, each of which consist of the 10 integer classes. We compiled two individual training subsets from the USPS training data. The first subset contains 1756 examples each belonging to the classes 3, 5, and 8. We call this set USPS-1 training data. The second subset contains 1961 examples each belonging to the classes 4, 6, and 7. We call this set USPS-2 training data. Similarly two subsets are created from the testing data. In all of these datasets, the data values are scaled by $\frac{1}{200}$. Testing set accuracies are reported for all four methods. The total numbers of unique support vectors in the resulting classification functions for the M-SVM and k-SVM methods are given.

Table 1 contains results for M-RLP, k-RLP, M-SVM, and k-SVM on the Wine and Glass datasets. As anticipated, adding the regularization term to the degree one problem in M-SVM produced better testing generalization than M-RLP on the Wine dataset. The Wine dataset is piecewise-linearly separable. Therefore, the M-RLP method has infinitely many optimal solutions. However, the testing accuracy for M-SVM with degree one on the Glass data was much lower than the M-RLP accuracy. This may indicate that the choice of λ is too large. As the degree increases, however, the accuracy of the M-SVM method improves and exceeds the M-RLP results. The k-SVM method generalized surprisingly well. The testing accuracies reported for k-SVM on the Wine dataset are higher than those of M-SVM. The linear k-RLP method performed just as well as the quadratic k-SVM program on the Wine dataset and better than the M-SVM and M-RLP methods. On the Glass data, as the degree increases, both methods, M-SVM and k-SVM, improve dramatically in testing accuracy. Using higher degree polynomials the M-SVM and k-SVM methods surpass

Table 2. Percent testing set accuracies and (total number of support vectors) for M-SVM and SVM. $\lambda = .05$ for k-SVM and $\lambda = .03$ for k-RLP and M-SVM.

Data	Method	Degree				
		1	2	3	4	5
USPS-1	M-RLP	80.69	-	-	-	-
	k-RLP	91.46	-	-	-	-
	M-SVM	91.26 (415)	91.87 (327)	92.28 (312)	92.07 (305)	92.28 (317)
	k-SVM	91.67 (666)	92.28 (557)	92.89 (514)	92.68 (519)	92.48 (516)
USPS-2	M-RLP	80.66	-	-	-	-
	k-RLP	96.13	-	-	-	-
	M-SVM	94.58 (228)	94.97 (185)	95.36 (167)	94.97 (166)	94.00 (180)
	k-SVM	96.13 (383)	96.52 (313)	96.13 (303)	95.16 (294)	94.58 (289)

Table 3. Total computational training time (in seconds) for M-RLP,k-RLP, M-SVM, and k-SVM on USPS-1.

	Degree				
	1	2	3	4	5
M-RLP	7922	-	-	-	-
k-RLP	341	-	-	-	-
M-SVM	20359	16978	11422	10402	12673
k-SVM	6883	5709	4819	4929	4932

the accuracies of M-RLP and k-RLP. This demonstrates the potential for polynomial and piecewise-polynomial classification functions over linear and piecewise-linear functions.

Table 2 contains results for the four methods on the USPS data subsets. Similar observations as above can be made. Both of these datasets are piecewise-linearly separable. The solution that M-RLP has found for each of these datasets has a significantly lower testing accuracy than the other methods. The k-SVM method generalizes slightly better than M-SVM. The k-RLP method reports accuracies similar to those of the k-SVM method. Additionally, it is solving linear programs rather than quadratic programs, so the computational training time is significantly smaller than that of the other methods. Changing the parameter λ may further improve generalization. The M-SVM method consistently finds classification functions using fewer support vectors than those of k-SVM. With fewer support vectors, a sample can be classified more quickly since only the dot-product of the sample with each support vector must be computed. Thus M-SVM would be a good method to choose when classification time is critical.

CPU times for training all four methods on the USPS-1 dataset are reported in Table 3. The times for all the datasets are not listed because the programs were run using a batch system on clusters of machines so the timing was not reliable. However, the trends were clear. The k-RLP method is significantly faster than the other methods. In the M-SVM and k-SVM methods, as the degree increased the computational time would decrease and then after a certain degree is reached it would increase. The degree of the polynomial for which it starts to increase varies by dataset. Surprisingly, for the USPS datasets the k-SVM method was faster than the M-RLP method. This was not the case for the Wine and Glass datasets. The M-RLP method had faster training times than k-SVM for these datasets. The times reported are for IBM RS6000 model 590 workstations with 128 MB RAM.

7. Conclusions

We have examined four methods for the solution of multicategory discrimination problems based on the LP methods of Mangasarian and the QP methods for SVM of Vapnik. The two-class methods, RLP and SVM, differ only in the norm of the regularization term. In the past two different approaches had been used for the $k > 2$-class case. The method we called k-SVM constructed k two-class discriminants using k quadratic programs. The resulting classifier was a piecewise-linear or piecewise-nonlinear discriminant function depending on what kernel function was used in the SVM. The original multicategory RLP for k classes constructed a piecewise-linear discriminant using a single linear program. We proposed two new hybrid approaches. Like the k-SVM method, k-RLP uses LPs to construct k two-class discriminants. We also formulated a new approach, M-SVM. We began the formulation by adding regularization terms to M-RLP. Then, like k-SVM with piecewise-nonlinear discriminants, the nonlinear pieces are found by mapping the original data points into a higher dimensional feature space. This transformation appeared in the dual problem as an inner product of two points in the higher dimensional space. A generalized inner product was used to make the problem tractable. The new M-SVM method requires the solution of a single quadratic program. We performed a computational study of the four methods on four datasets. In general we found that the k-SVM and k-RLP generalized better than M-RLP and M-SVM. M-SVM, however, used fewer support vectors – a counter-intuitive result since for the two-class case Statistical Learning Theory predicts that fewer support vectors produces a better bound on the expected generalization error. The theoretical justification of the better generalization of k-SVM and k-RLP and M-SVM and M-RLP is an open question. The k-RLP method provided accurate and efficient results on the piecewise-linearly separable datasets. The k-SVM method also tested surprisingly well but requires the solution of k quadratic programs. Overall we recommend k-RLP for faster accurate results when piecewise-linear discrimination is desired. For piecewise-nonlinear discrimination, k-SVM outperforms the other methods in every respect except the number of support vectors. M-SVM consistently finds solutions with the smallest number of support vectors.

Appendix

A.1. Matrix Representations for Multicategory Support Vector Machines

This appendix contains the definitions of the matrices used for the general k-class SVM formulation (18):

$$\begin{array}{ll} \min\limits_{w,\gamma} & \frac{1}{2}\left\|\bar{C}w\right\|^2 + \frac{1}{2}\left\|w\right\|^2 \\ s.t. & \bar{A}w + \bar{E}\gamma - e \geq 0 \end{array}$$

Let

$$\bar{C} = \begin{bmatrix} I & -I & 0 & 0 & \cdots & 0 \\ I & 0 & -I & 0 & \cdots & 0 \\ \vdots & \vdots & 0 & \ddots & 0 & \vdots \\ \vdots & \vdots & 0 & 0 & \ddots & 0 \\ I & 0 & \cdots & \cdots & 0 & -I \\ 0 & I & -I & 0 & \cdots & 0 \\ \vdots & \vdots & 0 & \ddots & 0 & \vdots \\ \vdots & \vdots & \vdots & 0 & \ddots & 0 \\ 0 & I & 0 & \cdots & 0 & -I \\ \vdots & \vdots & \vdots & \vdots & \vdots & \vdots \\ 0 & \cdots & 0 & I & -I & 0 \\ 0 & \cdots & 0 & I & 0 & -I \\ 0 & \cdots & \cdots & 0 & I & -I \end{bmatrix} \tag{A.1}$$

where $I \in R^{n \times n}$ is the identity matrix. The matrix $\bar{C}$ has $n \sum_{i=2}^{k} (i-1)$ rows and kn columns.

Let

$$\bar{A} = \begin{bmatrix} A^1 & -A^1 & 0 & 0 & \cdots & 0 \\ A^1 & 0 & -A^1 & 0 & \cdots & 0 \\ \vdots & \vdots & 0 & \ddots & 0 & \vdots \\ \vdots & \vdots & 0 & 0 & \ddots & \vdots \\ A^1 & 0 & \cdots & \cdots & 0 & -A^1 \\ -A^2 & A^2 & 0 & 0 & \cdots & 0 \\ 0 & A^2 & -A^2 & 0 & \cdots & 0 \\ \vdots & \vdots & 0 & \ddots & 0 & \vdots \\ \vdots & \vdots & \vdots & \cdots & \ddots & \vdots \\ 0 & A^2 & 0 & \cdots & \cdots & -A^2 \\ \vdots & \vdots & \vdots & \vdots & \vdots & \vdots \\ -A^k & 0 & \cdots & \cdots & 0 & A^k \\ 0 & -A^k & 0 & \cdots & 0 & A^k \\ \vdots & 0 & \ddots & 0 & \vdots & \vdots \\ \vdots & \vdots & 0 & \ddots & 0 & \vdots \\ 0 & \cdots & \cdots & 0 & -A^k & A^k \end{bmatrix} \tag{A.2}$$

where $A^i \in R^{m_i \times n}$. The matrix $\bar{A}$ has $(k-1)\sum_{i=1}^{k} m_i$ rows and kn columns.

Let

$$\bar{E} = \begin{bmatrix} -e^1 & e^1 & 0 & 0 & \cdots & 0 \\ -e^1 & 0 & e^1 & 0 & \cdots & 0 \\ \vdots & \vdots & 0 & \ddots & 0 & \vdots \\ \vdots & \vdots & 0 & 0 & \ddots & \vdots \\ -e^1 & 0 & \cdots & \cdots & 0 & e^1 \\ e^2 & -e^2 & 0 & 0 & \cdots & 0 \\ 0 & -e^2 & e^2 & 0 & \cdots & 0 \\ \vdots & \vdots & 0 & \ddots & 0 & \vdots \\ \vdots & \vdots & \vdots & \cdots & \ddots & \vdots \\ 0 & -e^2 & 0 & \cdots & \cdots & e^2 \\ & & & & & \\ \vdots & \vdots & \vdots & \vdots & \vdots & \vdots \\ & & & & & \\ e^k & 0 & \cdots & \cdots & 0 & -e^k \\ 0 & e^k & 0 & \cdots & 0 & -e^k \\ \vdots & 0 & \ddots & 0 & \vdots & \vdots \\ \vdots & \vdots & 0 & \ddots & 0 & \vdots \\ 0 & \cdots & \cdots & 0 & e^k & -e^k \end{bmatrix}$$

where $e^i \in R^{m_i \times 1}$ is a vector of ones. The matrix $\bar{E}$ has $(k-1)\sum_{i=1}^{k} m_i$ rows and kn columns.

Notes

1. The method was originally called Multicategory Discrimination

References

1. S. Aeberhard, D. Coomans, and O. de Vel, "Comparison of classifiers in high dimensional settings," Technical Report 92-02, Departments of Computer Science and of Mathematics and Statistics, James Cook University of North Queensland, 1992.
2. K. P. Bennett, "Decision tree construction via linear programming," in Proceedings of the 4th Midwest Artificial Intelligence and Cognitive Science Society Conference, Utica, Illinois, pages 97–101, 1992.
3. K. P. Bennett and E. J. Bredensteiner, "Geometry in learning," In C. Gorini, E. Hart, W. Meyer, and T. Phillips, editors, Geometry at Work, Washington, D.C., 1998. Mathematical Association of America, To appear.
4. K. P. Bennett and O. L. Mangasarian, "Neural network training via linear programming," in P. M. Pardalos, editor, Advances in Optimization and Parallel Computing, North Holland, pp. 56–67, 1992. .
5. K. P. Bennett and O. L. Mangasarian, "Multicategory discrimination via linear programming," Optimization Methods and Software, vol. 3, pp. 27–39, 1994.
6. K. P. Bennett and O. L. Mangasarian, "Serial and parallel multicategory discrimination," SIAM Journal on Optimization, Vol. 4, pp. 722–734, 1994.

7. K. P. Bennett, D. H. Wu, and L. Auslender, "On support vector decision trees for database marketing," R.P.I. Math Report No. 98-100, Rensselaer Polytechnic Institute, Troy, NY, 1998.
8. V. Blanz, B. Schölkopf, H. Bülthoff, C. Burges, V. Vapnik, and T. Vetter, "Comparison of view–based object recognition algorithms using realistic 3D models," in C. von der Malsburg, W. von Seelen, J. C. Vorbrüggen, and B. Sendhoff, editors, Artificial Neural Networks - ICANN'96, vol. 1112 of Lecture Notes in Computer Science, Springer: Berlin, 1996.
9. P. S. Bradley and O. L. Mangasarian, "Feature selection via concave minimization and support vector machines," Machine Learning Proceedings of the Fifteenth International Conference (ICML '98) (J. Shavlik, editor), Morgan Kaufmann: San Francisco, CA, pp. 82-90, 1998.
10. E. J. Bredensteiner and K. P. Bennett, "Feature minimization within decision trees," Computational Optimization and Applications, vol. 10, pp. 111–126, 1998.
11. C. Cortes and V. N. Vapnik, "Support vector networks," Machine Learning, vol. 20, pp. 273–297, 1995.
12. R. Courant and D. Hilbert, Methods of Mathematical Physics, J. Wiley: New York, 1953.
13. Y. Le Cun, B. Boser, J. S. Denker, D. Henderson, R. E. Howard, W. Hubbard, and L. J. Jackel, "Backpropagation applied to handwritten zip code recognition," Neural Computation, vol. 1, pp. 541–551, 1989.
14. G. B. Dantzig, Linear Programming and Extensions, Princeton University Press, Princeton, New Jersey, 1963.
15. I. W. Evett and E. J. Spiehler, "Rule induction in forensic science," Technical report, Central Research Establishment, Home Office Forensic Science Service, Aldermaston, Reading, Berkshire RG7 4PN, 1987.
16. O. L. Mangasarian, "Linear and nonlinear separation of patterns by linear programming," Operations Research, vol. 13, pp. 444–452, 1965.
17. O. L. Mangasarian, "Multi-surface method of pattern separation," IEEE Transactions on Information Theory, vol. IT-14, pp. 801–807, 1968.
18. O. L. Mangasarian, Nonlinear Programming, McGraw–Hill: New York, 1969.
19. O. L. Mangasarian, "Mathematical programming in machine learning," In G. DiPillo and F. Giannessi, editors, Proceedings of Nonlinear Optimization and Applications Workshop, pp. 283–295, Plenum Press: New York, 1996.
20. O. L. Mangasarian, "Arbitrary-norm separating plane," Mathematical Programming Technical Report 97-07, University of Wisconsin-Madison, 1997.
21. O. L. Mangasarian, W. N. Street, and W. H. Wolberg, "Breast cancer diagnosis and prognosis via linear programming," Operations Research, vol. 43, pp. 570–577, 1995.
22. P. M. Murphy and D. W. Aha, "UCI repository of machine learning databases," [http://www.ics.uci.edu/~mlearn/MLRepository.html], Department of Information and Computer Science, University of California, Irvine, California, 1994.
23. B. A. Murtagh and M. A. Saunders, "MINOS 5.4 user's guide," Technical Report SOL 83.20, Stanford University, 1993.
24. A. Roy, S. Govil, and R. Miranda, An algorithm to generate radial basis function (RBF)-like nets for classification problems, Neural Networks, vol. 8, pp. 179–202, 1995.
25. A. Roy, L. S. Kim, and S. Mukhopadhyay, "A polynomial time algorithm for the construction and training of a class of multilayer perceptrons," Neural Networks, vol. 6, pp. 535–545, 1993.
26. A. Roy and S. Mukhopadhyay, "Pattern classification using linear programming," ORSA Journal of Computing, vol. 3, pp. 66–80, 1990.
27. B. Schölkopf, C. Burges, and V. Vapnik, "Incorporating invariances in support vector machines," in C. von der Malsburg, W. von Seelen, J. C. Vorbrüggen, and B. Sendhoff, editors, Artificial Neural Networks - ICANN'96, vol. 1112 of Lecture Notes in Computer Science, pp. 47–52, Springer: Berlin 1996.
28. B. Schölkopf, K. Sung, C. Burges, F. Girosi, P. Niyogi, T. Poggio, and V. Vapnik, "Comparing support vector machines with gaussian kernels to radial basis function classifiers," AI Memo No. 1599; CBCL Paper No. 142, Massachusetts Institute of Technology, Cambridge, 1996.
29. V. Vapnik, The Nature of Statistical Learning Theory, Springer-Verlag, 1995.
30. V. N. Vapnik and A. Ja. Chervonenkis, Theory of Pattern Recognition, Nauka, Moscow, 1974. In Russian.
31. W. H. Wolberg and O. L. Mangasarian, "Multisurface method of pattern separation for medical diagnosis applied to breast cytology," Proceedings of the National Academy of Sciences U.S.A., vol. 87, pp. 9193–9196, 1990.

Computational Optimization and Applications 12, 81–105 (1999)

Quartic Barriers

RICHARD W. COTTLE cottle@soe.stanford.edu
Department of Engineering-Economic Systems and Operations Research, Stanford University, Stanford, CA 94305-4023

Received February 12, 1998; Accepted May 20, 1998

Abstract. Many mathematical propositions are expressed in terms of a natural number n for which they are either true or false. This expository paper surveys a collection of propositions $P(n)$ which are true if and only if $n \leq 4$.

Keywords: counterexamples, graphs, matrices, cones, volumes

Dedication: As I recall my first encounter with Olvi Mangasarian occurred in the early 1960s when he was a mathematician at the Shell Development Corportation in Emeryville, California and I was a doctoral student at Berkeley. Olvi's work on bimatrix games was one of interest to me at the time. Later his book *Nonlinear Programming* and his papers on duality and optimality conditions had a great impact on me. In the seventies, Olvi published several papers on solving certain kinds of linear complementarity problems by linear programming. I was fortunate enough to have an insight on the first of these papers and discussed it with a promising graduate student of mine named Jong-Shi Pang. This led us to a series of papers (essentially interleaved with Olvi's) on least-element solutions of LCPs. In later years, I was strongly influenced by Olvi's work on iterative and parallel computing methods for solving LCPs. I also fondly remember a stimulating but all-too-brief trimester that Olvi and his wife, Claire, spent at Stanford. Oh, how exciting it would be to repeat that experience!

Long ago a colleague of mine quipped that if you're an engineer and you get something wrong once, you're finished, whereas if you're a mathematician and get something right once, you're famous. Over the years, Olvi, the mathematician, has gotten many things right. Moreover, he has done so across an extremely impressive range of topics (see the opening of this issue). By virtue of his scientific accomplishments, kindness, and generosity, Olvi Mangasarian richly deserves this tribute from a group of his professional admirers. I am proud to participate in honoring a great researcher and a treasured friend.

Prologue

It is a well-known fact of remarkable importance in the history of mathematics that, in general, a real polynomial equation of degree n is solvable by radicals[1] if and only if $n \leq 4$. More formally, the proposition

> $P(n)$: *The general nth degree equation*
> $$x^n + a_1x^{n-1} + \cdots + a_{n-1}x + a_n = 0, \quad a_i \in \mathbb{R},\ n \in \mathbb{N}$$
> *is solvable by radicals*

is true for $n = 1, 2, 3, 4$ and false for $n \geq 5$.

The case $n = 1$ is trivial; the solution for $n = 2$ was known "algebraically" to the Babylonians about 2000 B.C. and "geometrically" to the ancient Greeks [15, pp. 36–37 and pp. 59–62]. Although the treatment of special cases was possible in classical antiquity, a considerable length of time elapsed before progress was made on the general third- and fourth-degree equations. As the historian D.E. Smith [44, p. 294] puts it, "the solution of the equations of third and fourth degrees was the chief mathematical achievement in Italian mathematics of the 16th Century." The formulas for the roots of the cubic and quartic equations are sometimes attributed to Cardano,[2] but this is an oversimplification. Their discovery was surrounded by controversy over priority, and several other names enter into the story.

The difficulty of going beyond the quartic equation was recognized by Lagrange as indicated in his work [29] of 1770. Lagrange's efforts influenced both Ruffini and Abel who, in the early 19th Century, proved that the general quintic equation cannot be solved by radicals, i.e., that $P(5)$ is false. For a fascinating account of this subject and this historical period, see [47].

These endeavors set the stage for the development of group theory and Galois theory where the insolvability of the quintic equation by radicals is demonstrated by the following line of reasoning: (1) a polynomial equation $p(x) = 0$ is solvable over the field K if and only if its Galois group over K is solvable; (2) there is a real quintic equation whose Galois group is the symmetric group on 5 letters; (3) the symmetric group on n letters is not solvable unless $n \leq 4$. This rests on the fact that the alternating group A_n is simple when $n \geq 5$. Of course, this chain of reasoning raises a question: exactly *why* is the alternating group A_n simple when $n \geq 5$? That is to say, what property of integers $n \geq 5$ implies this simplicity?

Purpose. It is the purpose of this survey to call attention to some *other* propositions, $P(n)$, indexed by the natural number n and which, like the one above, are true when $n \leq 4$ and false when $n \geq 5$. All the examples given here are drawn from the literature. In fact, most of them came to my attention while doing background reading for my teaching and my research in the field of mathematical programming. Brief comments on these connections are given along the way. In this expository paper, no attempt is made here to re-derive these results, but rather to assemble them and to pose (rather than answer) the question: *Are they instances of a general principle*?

It might be argued that many propositions "break down" at some positive integer,[3] hence new propositions can be "rigged" so that they, too, break down at 5. This is certainly true, but the examples offered here are *not* concocted in such a way. In each case, the dependence on n is completely natural, just as it is in the case of the degree of an equation or the order of a group.

A more sophisticated objection holds that what has been referred to as a "breakdown" in the truth of a proposition is biased in the wrong direction. To say that propositions break down is somehow to attribute merit to the cases where they are true and to treat these as if they were the *normal* state of affairs. Modern mathematics advances the opposite point of view: that it is normal for these propositions (as stated) to be false in general but true in a few special cases. Notwithstanding this viewpoint, the question raised above is not meaningless. It simply seeks to modify the proposition so as to reverse its truth value, i.e., to rephrase it to one that is true when $n \geq 5$ and false otherwise. The decision to state the propositions as originally indicated is deliberate. It reflects the naïve way we usually think about such matters. In addition to being a search for unity in an interesting and diverse collection of examples, this note implicitly renews the familiar warning about making unjustified assumptions, especially where higher dimensions are concerned.

Outline. Drawn from a fairly wide range of subject areas, the examples in this collection are organized under the following headings: (i) graphs, (ii) matrices and quadratic forms, (iii) cones, and (iv) volumes. Actually, these groupings are somewhat artificial since there are some connections between them.

1. Graphs

1.1. Planarity of complete graphs

Let $G=(V, E)$ denote a finite undirected graph with vertex set V and edge set E. The graph G is said to be *planar* if it can be represented on a plane so that the vertices are all distinct points, the edges are simple curves, and no two edges meet one another except at their endpoints.

Planar graphs have considerable relevance in the celebrated Four Color Problem [32, 33]. In operations research (and, more specifically, optimization theory) there are many applications graph-theoretic models involving planar graphs and their specialization to series-parallel and suspension graphs. For details on these models and the role planar graphs play in them, see for example [1, 14, 17, 48]. Taking account of these and many other applications, one can readily appreciate how important it is to be able to characterize planarity.

The graph G is *complete* if for each pair of distinct vertices i and $j \in V$ there exists an edge $e \in E$ having i and j as its endpoints. The complete graph having n vertices is denoted K_n. For the sake of completeness, we define K_1 as the graph having only one vertex and no edges. Figure 1 illustrates that K_n is planar for $n \leq 4$.

Note that to reveal the planarity of K_4 (which is sometimes called a Wheatstone Bridge) we have placed one of the edges in the part of the plane outside the convex hull of the vertices.

Success in exhibiting the planarity of K_n for $n \leq 4$ might lead one to hope for truth of the proposition

$P(n)$: *The graph* K_n *is planar.*

But this is known to be false. Indeed, a celebrated theorem[4] due to Kuratowski [28] states that *a graph is planar if and only if it contains no subgraph which is isomorphic* (*to within vertices of degree* 2) *to either of the graphs* K_5 *or* $K_{3,3}$ *shown in figure* 2.

We note that Kuratowski's theorem rules out the planarity of all the complete graphs K_n for $n \geq 5$.

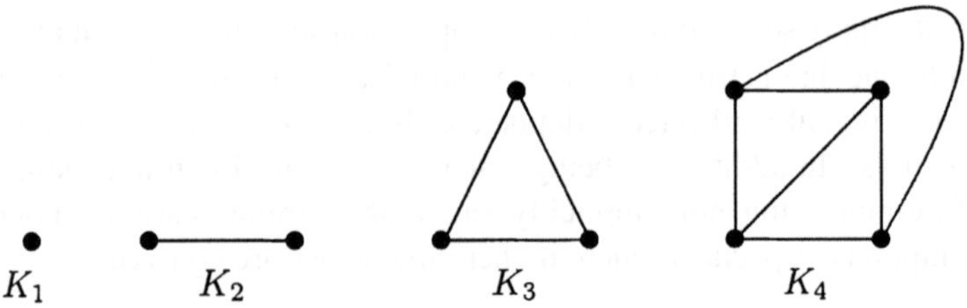

Figure 1. The planar graphs K_1, K_2, K_3 and K_4.

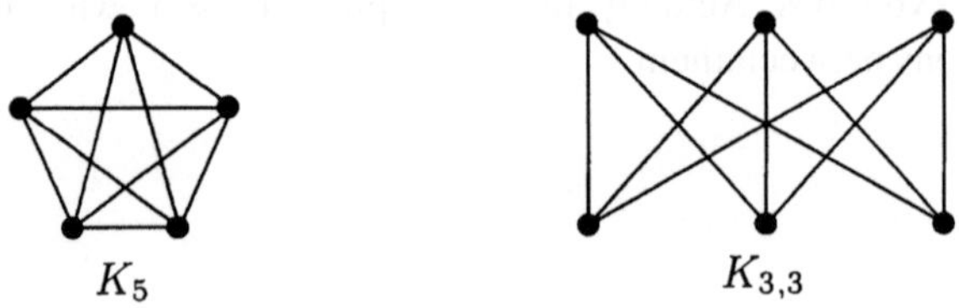

Figure 2. Two nonplanar graphs.

1.2. *n-Colorability: Unique and otherwise*

A *coloring* of a graph $G = (V, E)$ is an assignment[5] of colors to the vertices $v \in V$ such that adjacent vertices are colored differently. The *chromatic number* of G, denoted $\chi(G)$, is the smallest number n for which G can be n-colored. As a trivial example, it is clear that $\chi(K_n) = n$ for all $n \geq 1$. If $\chi(G) = n$, then G is called *n-chromatic.*

If γ is a $\chi(G)$-coloring of $G = (V, E)$, it induces a partition of V into *color classes.* Thus, two vertices belong to the same color class if they are colored the same. Suppose $\chi(G) = n$. If any two colorings of G induce the same partition of G, then G is said to be *uniquely n-colorable* or simply *uniquely colorable.* For $n \leq 4$, the graphs K_n are planar and uniquely n-colorable. In 1969, Chartrand and Geller [6] showed that there is no uniquely 5-colorable planar graph. It had long been known from a classical theorem of Heawood [24] that the chromatic number of a planar graph is at most 5. A powerful reason for the Chartrand-Geller result was given some five years later in the celebrated work of Appel and Haken [2] on the four-color conjecture. They showed that for $n \geq 5$ there simply are no n-chromatic planar graphs. To sum up, then, the proposition

$P(n)$: *There is a planar graph with chromatic number n.*

is true for all $n \leq 4$ and false for all $n \geq 5$.

1.3. *Checker jumping*

This example is based upon a chapter titled "A Problem in Checker-Jumping" in Ross Honsberger's book, *Mathematical Gems II* (see [25, Chapter 3]). Its underlying structure

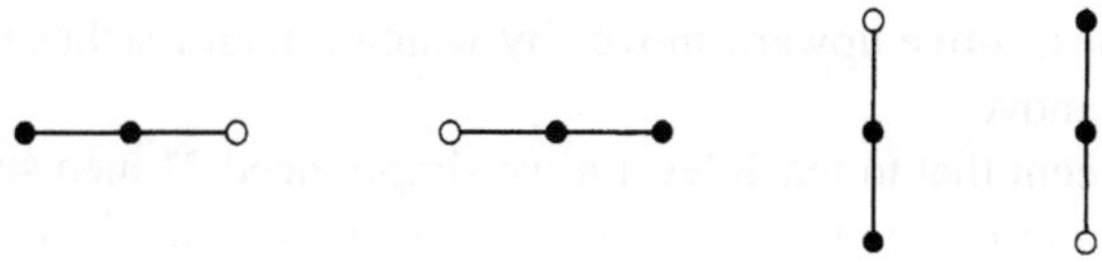

Figure 3. The segments that allow leap-frogging.

here is not graph-theoretic, but it is quite amusing and deserves to be included somewhere. This section is as good a place as any.

Suppose we think of the lattice of integer points in $\mathbb{R}^2$ as a very large checkerboard and say that a lattice point is either *occupied* or *unoccupied* by a "man".[6] Assume that the plane has finitely many men located at its lattice points and that they are all situated on or below the x-axis. Next consider a set of three consecutive lattice points on one of the lattice lines[7] having a man in the middle and another at one of the ends. The four possibilities are illustrated in figure 3.

From each of these configurations it is possible to make a *move* whereby the "occupancy status" of these three points is reversed. What happens in a move is that the man at an end of the segment jumps or "leapfrogs" the man in the middle. (In the game of Checkers, such jumping is done diagonally.) The man in the middle is then removed, and the man that did the jumping occupies the other end of the segment.

The problem here is to place the least number of men in the lower half-plane and reach a prescribed horizontal level in the upper half-plane by means of a sequence of such leapfrogging moves.

By *level n* we shall mean the set of lattice points (x, y) such that $y = n$, $n \in \mathbb{N}$. The proposition $P(n)$ is then

> $P(n)$: *Level n is reachable by a sequence of moves from a finite number of occupied points in the lower half plane.*

Reaching level 1 can be accomplished by placing men at two lattice points: $(x, 0)$ and $(x, -1)$ for some $x \in \mathbb{N}$. One upward move results in a man at $(x, 1)$.

With four men located at $(x-2, 0)$, $(x-1, 0)$, $(x, 0)$, and $(x, -1)$, it is possible to reach level 2 in three moves. The first move involves $(x, 0)$ and $(x, -1)$ just as in the case of $n = 1$. The second move goes horizontally, starting with men at $(x-2, 0)$ and $(x-1, 0)$ and ending with a man at $(x, 0)$. The third move is like the first, but from one level higher. It yields a man at $(x, 2)$.

When level 3 is the goal, eight men suitably placed in the lower half-plane will do the job. Indeed, the placement used for $n = 2$ is combined with the placement of four other men at

$$(x+1, 0), \quad (x+2, 0), \quad (x+1, -1), \quad \text{and} \quad (x+2, -1).$$

After the first four men are used to reach level 2, there remain two men on the x-axis and two men directly below them. The two possible leftward moves then yield a configuration that permits two consecutive upward moves by which a man reaches level 3. In all, this process takes seven moves.

So far, it would seem that to reach level n we simply need 2^n men suitably in the lower half-plane. But this conjecture fails when $n = 4$ where instead of 16 men, we need 20 of them positioned as follows:

$(x-2,0)$	$(x-1,0)$	$(x,0)$	$(x+1,0)$	$(x+2,0)$	$(x+3,0)$	$(x+4,0)$
$(x-2,-1)$	$(x-1,-1)$	$(x,-1)$	$(x+1,-1)$	$(x+2,-1)$		
$(x-2,-2)$	$(x-1,-2)$	$(x,-2)$	$(x+1,-2)$	$(x+2,-2)$	$(x+3,-2)$	
		$(x,-3)$	$(x+1,-3)$			

The idea behind the placement is to achieve a configuration like the one used for level 3, except that when it shows up, it is one level higher than the starting configuration for that case. Such a configuration can be generated in 12 moves. Once this occurs, seven moves like those of the case $n = 3$ finish the job.[8]

The preceding paragraphs should be enough to make the case that $P(n)$ is true for $n = 1, 2, 3, 4$. Remarkably, though, $P(5)$ is false: no matter how many men are deployed in the lower half-plane, there is no sequence of moves that will bring a man to level 5. Honsberger attributes this discovery to John Conway.[9] It clearly implies that $P(n)$ is false for all $n \geq 5$.

Let us follow the elementary, but ingenious proof (given in [25]) that $P(5)$ is false. We start from the observation that in analyzing the problem of reaching level 5, it is sufficient to distinguish an arbitrary yet definite lattice point of the form $P = (\bar{x}, 5)$. (The analogous statement would be true for reaching any level n above the x-axis.) Next, to each lattice point in the plane we associate a power of a particular real number, z. The power is determined by the (necessarily integral) ℓ_1-distance of the point (x, y) from the distinguished point $(\bar{x}, 5)$. Thus,

$$(x, y) \longleftrightarrow z^{|x-\bar{x}|+|y-5|}.$$

We note that

$$(\bar{x}, 5) \longleftrightarrow z^0 = 1.$$

Next we define the *value* of a set of men to be the sum of the values associated with their positions. After a move (as defined above) is made, only one of three lattice points making up the segment will be occupied. The ℓ_1-distance from this point to distinguished point P is either (i) less than, (ii) greater than, or (iii) equal to the ℓ_1-distance of the other endpoint of the segment. How do the moves of these three types affect the sets of men associated with the lattice points of the segment on which the move is made? In particular, what is the *change* in the value of the set of men on the segment (or any set of lattice points containing that segment)? The answer is given in Table 1.

Now choose z as a root of the equation $1 - z - z^2$, namely $(-1 + \sqrt{5})/2$. This positive number is less than 1. Using the fact that $z^2 = 1 - z$, it is easy to see that a move never

Table 1.

Type of move	Change in value
i (Closer)	$z^k - (z^{k+1} + z^{k+2}) = z^k(1 - z - z^2)$
ii (Farther)	$z^{k+2} - (z^{k+1} + z^k) = z^k(z^2 - z - 1)$
iii (Equidistant)	$z^{k+1} - (z^k + z^{k+1}) = -z^k$

increases the value of a set of men containing the move segment, and that there is no change in value when the move is of type i.

Since the point P is at a distance of 0 from itself, the value associated with this point is $z^0 = 1$. It follows that if some set S of men is capable of reaching P, its initial value must at least 1, for the value of the set cannot increase at any stage and yet must ultimately contain a man at the point P where the value must be 1. Now the set S, being finite, must have value less than that of the lower half-plane (with a man at every lattice point). To compute the latter value, first consider the set of lattice points $(\bar{x}, -k)$ for all integers $k \geq 0$. Because $P = (\bar{x}, 5)$, it follows that the value of this set (the "central column") is $z^5 + z^6 + z^7 + \cdots$. Since $0 < z < 1$, this geometric series sums to $z^5/(1 - z)$. Each pair of columns parallel to and at a distance d from the central column will contribute a value of $2z^{5+d}/(1 - z)$ to the value of the lower half-plane which then turns out to be

$$V = \frac{z^5}{1-z} + 2\left(\frac{z^6}{1-z} + \frac{z^7}{1-z} + \cdots\right) = \frac{z^5}{1-z} + \frac{2z^6}{1-z} \cdot \frac{1}{1-z}.$$

From the fact that $z^2 = 1 - z$, it is not difficult to deduce that $V = 1$. Hence no finite set of men in the lower half-plane can reach level 5 (or higher) by a sequence of these leapfrogging moves.

2. Matrices and quadratic forms

By a *real n-ary quadratic form* we mean a homogeneous second-degree polynomial in n variables, for example

$$\sum_{i=1}^{n}\sum_{j=1}^{n} a_{ij}x_ix_j.$$

In this instance, the coefficients a_{ij} and variables x_i, x_j are all real. It is customary to assemble the n^2 coefficients a_{ij} in an $n \times n$ array or *matrix*, say A. It is fairly customary[10] to assume that the matrix A is symmetric, i.e., that $a_{ij} = a_{ji}$ for all $i, j = 1, \ldots, n$. One can then write

$$\sum_{i=1}^{n}\sum_{j=1}^{n} a_{ij}x_ix_j = (x, Ax)$$

where $(\cdot, \cdot)$ represents the *inner product* in $\mathbb{R}^n$ for which the alternative notation $x^{\mathrm{T}}Ax$ is also used.

Caveat. In some cases, a class of square matrices is defined in terms of the function values associated with the corresponding quadratic form. Thus, for example, when $x^{\mathrm{T}}Ax \geq 0$ for all x, the matrix A and the quadratic form $x^{\mathrm{T}}Ax$ are called *positive semi-definite*. Other examples will be encounted below. In this section, we shall move freely between these matrix classes and the quadratic forms with which they are associated.

Quadratic forms (and quadratic functions) arise naturally in all sorts of (pure and applied) mathematical settings, especially in quadratic programming, the branch of optimization theory dealing with problems that can be posed as the minimization of a quadratic function subject to linear inequality constraints.

2.1. *Set intersection matrices*

In a short paper [19] published in 1941, Marshall Hall addressed what he called "a set partition problem." Since Hall's usage of the term "partition" seems at variance with its meaning today, we paraphrase his statement of the problem. In essence, we wish to find the minimum cardinality set X whose elements $x_1, \ldots, x_m$ (called *objects*) are to be associated with precisely n nonempty (but not necessarily distinct) subsets $S_1, \ldots, S_n$ of X so as to achieve a *prescribed pattern* of empty and nonempty pairwise intersections. In [19], the prescribed intersection pattern is expressed by a given symmetric $n \times n$ binary matrix $A = [a_{ij}]$, the entries of which have the interpretation

$$a_{ij} = \begin{cases} 1 & \text{if } S_i \cap S_j \neq \emptyset \\ 0 & \text{if } S_i \cap S_j = \emptyset \end{cases}$$

Reflecting the fact that the subsets $S_i \subseteq X$ are to be nonempty, the matrix A has the property that $a_{ii} = 1$ $(i = 1, \ldots, n)$. The sum of these diagonal elements is called the *trace* of A.

The case of $n = 1$ is completely uninteresting. For $n = 2$, there are just two matrices A of the type described above. They are

$$A_1 = \begin{bmatrix} 1 & 1 \\ 1 & 1 \end{bmatrix} \quad \text{and} \quad A_2 = \begin{bmatrix} 1 & 0 \\ 0 & 1 \end{bmatrix}.$$

For A_1 we require only $m = 1$ object which is associated with (belongs to) both sets S_1, S_2. Note that in this instance the sets are not distinct, but this is not ruled out. The other matrix A_2 summarizes the intersection pattern of two nonempty disjoint sets; for these a minimum of two objects are required. Notice that in either case, the set intersection pattern could have been realized by forming two nonempty subsets of a set X of cardinality no greater than 2.

For all $n \geq 2$, there are these two "extreme" cases: (i) where a single object belongs to each of the n subsets, corresponding to $a_{ij} = 1$ for all i, j, and (ii) where all n subsets are nonempty and disjoint corresponding to $a_{ii} = 1$ for all i and $a_{ij} = 0$ for all $i \neq j$. As we have seen, when $n = 2$, these are the only cases, but more possibilities arise when $n \geq 3$. For example, when $n = 3$, there are (up to principal rearrangement) four possible A matrices:

$$A_1 = \begin{bmatrix} 1 & 1 & 1 \\ 1 & 1 & 1 \\ 1 & 1 & 1 \end{bmatrix}, \quad A_2 = \begin{bmatrix} 1 & 1 & 0 \\ 1 & 1 & 1 \\ 0 & 1 & 1 \end{bmatrix}, \quad A_3 = \begin{bmatrix} 1 & 1 & 0 \\ 1 & 1 & 0 \\ 0 & 0 & 1 \end{bmatrix}, \quad A_4 = \begin{bmatrix} 1 & 0 & 0 \\ 0 & 1 & 0 \\ 0 & 0 & 1 \end{bmatrix}.$$

For the nonextreme cases (A_2 and A_3), the following subset-object incidence matrices indicate how the three subsets S_1, S_2, S_3 can be formed from just two distinct objects:

$$B_2 = \begin{bmatrix} 1 & 0 \\ 1 & 1 \\ 0 & 1 \end{bmatrix} \quad \text{and} \quad B_3 = \begin{bmatrix} 1 & 0 \\ 1 & 0 \\ 0 & 1 \end{bmatrix}.$$

We remark that, in general, for a subset-object incidence matrix B the product BB^{T} records the cardinalities of the pairwise intersections of the subsets. In a solution to the problem at hand, BB^{T} and A must have the same support[11] and B must have the smallest number of columns consistent with the given intersection pattern.

The case of $n = 4$ is more complicated than those for smaller n. As shown by Hall, there are 11 essentially distinct A matrices. Hall notes that only two of them require as many as four objects; these matrices are the 4×4 identity and

$$\begin{bmatrix} 1 & 0 & 1 & 1 \\ 0 & 1 & 1 & 1 \\ 1 & 1 & 1 & 0 \\ 1 & 1 & 0 & 1 \end{bmatrix}.$$

On the strength of these observations, one might be encouraged to formulate the proposition

> $P(n)$: *For every symmetric binary $n \times n$ matrix A with trace n there exists a binary $n \times m$ matrix B with $m \leq n$ such that* $\operatorname{supp} A = \operatorname{supp} BB^{\mathrm{T}}$.

But this proposition is false for all $n \geq 5$. Consider for instance the matrix

$$A = \begin{bmatrix} 1 & 0 & 0 & 1 & 1 \\ 0 & 1 & 0 & 1 & 1 \\ 0 & 0 & 1 & 1 & 1 \\ 1 & 1 & 1 & 1 & 0 \\ 1 & 1 & 1 & 0 & 1 \end{bmatrix}$$

and let

$$B = \begin{bmatrix} 1 & 1 & 0 & 0 & 0 & 0 \\ 0 & 0 & 1 & 1 & 0 & 0 \\ 0 & 0 & 0 & 0 & 1 & 1 \\ 1 & 0 & 1 & 0 & 1 & 0 \\ 0 & 1 & 0 & 1 & 0 & 1 \end{bmatrix}.$$

Then

$$BB^{\mathrm{T}} = \begin{bmatrix} 2 & 0 & 0 & 1 & 1 \\ 0 & 2 & 0 & 1 & 1 \\ 0 & 0 & 2 & 1 & 1 \\ 1 & 1 & 1 & 3 & 0 \\ 1 & 1 & 1 & 0 & 3 \end{bmatrix}$$

has the same support as A, but—as Hall shows—there is no subset-object incidence matrix B with *fewer* than six columns such that $\mathrm{supp}(BB)^{\mathrm{T}} = \mathrm{supp}(A)$.

In point of fact, Hall's paper gives two theorems. The first says, in effect, that every $n \times n$ binary matrix with trace n is the incidence matrix of a family of n subsets of a set having at most n objects when $n = 1, 2, 3, 4$ and of at most $n^2/4$ (n even and $n \geq 4$) or $(n^2 - 1)/4$ (n odd and $n \geq 5$). The second theorem shows that these bounds can be realized as follows. Let I_s denote the identity matrix of order s and let E_{st} be the $s \times t$ matrix of all ones. Now if $n = 2s + 1$ ($s \geq 2$), then

$$A = \begin{bmatrix} I_{s+1} & E_{s+1,s} \\ E_{s,s+1} & I_s \end{bmatrix}$$

will require that X contain $(n^2 - 1)/4$ objects, whereas when $n = 2s$ ($s \geq 3$),

$$A = \begin{bmatrix} I_s & E_{ss} \\ E_{ss} & I_s \end{bmatrix}$$

will require that X contain $n^2/4$ objects. When $n \geq 5$, these tight bounds are, respectively, larger than n.

It is interesting to note that these matrices are the node-node incidence matrices of the *complete bipartite graphs* $K_{s+1,s}$ and K_{ss}, respectively.

Some years later, Erdös et al. [13] considered a very similar problem couched in the language of graphs. They observe that a family of sets $S_1, S_2, \ldots$ gives rise to an undirected graph by regarding each set as a node and declaring that two nodes are adjacent if and only if the corresponding sets have a nonempty intersection. They then quote a result of E. Szpilrajn-Marczewski [46] which states that the converse is true, namely that for any graph

G there is a set X and a family of subsets $S_1, S_2, \ldots$ of X which can be put into one-to-one correspondence with the vertices of G in such a way that two vertices are adjacent if and only if the corresponding subsets of X have a nonempty intersection. Published in 1966, the paper by Erdös, Goodman and Pósa appears to have been written independently of Hall's paper [19], but it has the same objective: to determine the minimum number of elements in the set X. Erdös, Goodman and Pósa express the bound as follows.

Theorem 1. *If G is any graph with n vertices, then there is a set X and a family of n subsets of X such that two vertices are adjacent if and only if the corresponding subsets of X have a nonempty intersection. Further $[n^2/4]$ is the smallest such number (i.e., cardinality of X).*

In this statement, $[\cdot]$ denotes the bracket function. We note that

$$[n^2/4] \leq n \Leftrightarrow n \leq 4.$$

2.2. *Representation of copositive quadratic forms (matrices)*

A symmetric matrix $A \in \mathbb{R}^{n \times n}$ and its associated quadratic form $f(x) = x^{\mathrm{T}}Ax$ are said to be *copositive* if

$$x \geq 0 \Rightarrow x^{\mathrm{T}}Ax \geq 0.$$

Similarly, A and f are said to be *strictly copositive* if

$$0 \neq x \geq 0 \Rightarrow x^{\mathrm{T}}Ax > 0.$$

The concept of copositivity appears to have originated with T.S. Motzkin[12] [38] around 1952. Initial studies of copositivity centered on quadratic forms as applied to number theory. In 1965 the copositivity concept came up (minus the name) in the C.E. Lemke's seminal paper [31] on what was eventually to be called "the linear complementarity problem." Further discussion of this development is given in [8].

From the very definition, one is led to questions about how to test a matrix for (strict) copositivity; satisfactory answers were a while in coming. Every positive semi-definite matrix is copositive and so is every (elementwise) nonnegative matrix. In fact, so is the sum of any two such matrices. What about the converse statement? Obviously, a 1×1 copositive matrix is both positive semi-definite and nonnegative. It is not difficult to show that any copositive matrix of order 2 which is not nonnegative must be positive semi-definite. In other words, a copositive matrix is of one type or the other (and possibly both).

P.H. Diananda [10] provided proofs for the cases of 3×3 and 4×4 matrices and on the strength of this evidence conjectured that the converse statement is generally true, i.e., that for all n,

> $P(n)$: *Every copositive $n \times n$ matrix is the sum of a positive semi-definite matrix and a nonnegative matrix.*

Before long, Hall and Newman [22, p. 334] reported a counterexample (due to A. Horn) for $n = 5$. (A discussion of this example also appears in Hall's book [21, p. 273].) Since Horn's matrix

$$A = \begin{bmatrix} 1 & -1 & 1 & 1 & -1 \\ -1 & 1 & -1 & 1 & 1 \\ 1 & -1 & 1 & -1 & 1 \\ 1 & 1 & -1 & 1 & -1 \\ -1 & 1 & 1 & -1 & 1 \end{bmatrix}$$

can always be made into a principal submatrix of a copositive matrix of order $n > 5$ (by bordering it with nonnegative entries), it follows that Diananda's conjecture cannot hold in general for any $n \geq 5$.

2.3. *The matrix equation $X^{\mathrm{T}}X = A$*

It is quite evident that for any $m \times n$ matrix X, the product $A = X^{\mathrm{T}}X$ is symmetric and positive semi-definite. Moreover, if X happens to be (elementwise) nonnegative, then so is A. The converse of this assertion was investigated by Maxfield and Minc [34].

It is very well known that every symmetric positive semi-definite matrix A can be decomposed as $A = X^{\mathrm{T}}X$. What if $A \geq 0$, that is $a_{ij} \geq 0$ for all i, j? Does there exist a decomposition $A = X^{\mathrm{T}}X$ such that $X \geq 0$? As noted by Maxfield and Minc [34], an equivalent[13] statement of the problem is: Can a [symmetric] positive semi-definite matrix with nonnegative entries be expressed as the sum of rank 1 positive semi-definite matrices with nonnegative entries? Maxfield and Minc establish an affirmative answer for $n \leq 4$. So, for these values of n, we have the truth of the proposition

> $P(n)$: *If A is a symmetric nonnegative positive semi-definite matrix of order n, there exists a nonnegative matrix X such that $X^{\mathrm{T}}X = A$.*

For the case of $n \geq 5$, Maxfield and Minc exhibit a symmetric nonnegative positive semi-definite matrix due to M. Hall [20, p. 101], for which $P(n)$ is false,[14] namely

$$A = \begin{bmatrix} 1 & 0 & 0 & \frac{1}{2} & \frac{1}{2} \\ 0 & 1 & \frac{3}{4} & 0 & \frac{1}{2} \\ 0 & \frac{3}{4} & 1 & \frac{1}{2} & 0 \\ \frac{1}{2} & 0 & \frac{1}{2} & 1 & 0 \\ \frac{1}{2} & \frac{1}{2} & 0 & 0 & 1 \end{bmatrix}.$$

By taking the direct sum of this matrix with any other symmetric nonnegative positive semi-definite matrix, one sees that $P(n)$ is false for all $n \geq 5$.

Maxfield and Minc also make the following interesting remark.

There is a rather surprising geometrical consequence of this answer: In an n-dimensional space ($n \geq 5$) not every pencil of n vectors such that the angle between any two vectors does not exceed $\pi/2$ can be placed in the nonnegative orthant of coordinates by a rotation or by a rotation followed by a reflection.

Hall's argument that the 5×5 matrix A above is not expressible as $X^{\mathrm{T}}X$ for some matrix $X \geq 0$ is actually based on the equivalent formulation of the problem mentioned above. He shows that this particular quadratic form $x^{\mathrm{T}}Ax$ cannot be the sum of squares of linear forms each having nonnegative coefficients. We shall consider the aforementioned equivalence in the next section.

2.4. *Representation of zero by quadratic forms*

A classical problem in number theory is that of representation of numbers by quadratic forms. Typically, one has a symmetric $n \times n$ integer matrix $A = [a_{ij}]$ and asks what are the integers k for which there exists an integer n-vector x such that $x^{\mathrm{T}}Ax = k$? When the equation $x^{\mathrm{T}}Ax = k$ has an integer solution, the quadratic form $x^{\mathrm{T}}Ax$ is said to *represent* k. (See Davenport [9, Chapter 6] and O'Meara [41, Chapter 4] for discussions of this subject.)

Our attention will be confined to the representation of zero by quadratic forms,[15] specifically by *indefinite* quadratic forms.[16] Since there are no indefinite n-ary quadratic forms when $n = 1$, our proposition $P(n)$ will need to be expressed for $n \geq 2$.

It is clear when $n = 2$, some indefinite quadratic forms represent zero nontrivially. For instance the equation

$$x_1^2 - 4x_2^2 = 0$$

has the nontrivial solution $(x_1, x_2) = (2, 1)$. But there exist binary quadratic forms that represent zero only trivially as is the case for the quadratic form $x_1^2 - 3x_2^2$. Accordingly, we can assert *there exist indefinite quadratic forms in two variables that represent zero only trivially.*

The analogous statements for ternary and quaternary quadratic forms are also true. The criteria for the integer solvability of the equation

$$a_1x_1^2 + a_2x_2^2 + a_3x_3^2 = 0$$

were first given by A.-M. Legendre [30]. The case of $n = 4$ was analyzed by Meyer [35]. Discussions of these results can be found in Nagell [40, Section 61] and Mordell [36, Chapter 7].

These cases for two, three, and four variables might lead one to formulate the general proposition

> *$P(n)$: For all integers $n \geq 2$ there exist indefinite integral quadratic forms in n variables that represent zero only trivially.*

But this proposition also fits the pattern of this paper. Indeed, Meyer [35] established the remarkable fact that *any indefinite integral quadratic form in five or more variables represents zero nontrivially.*

2.5. *The field of values of a square complex matrix*

Let $\mathbb{C}$ denote the field of complex numbers. Then $\mathbb{C}^n$ denotes the vector space of complex n-vectors and $\mathbb{C}^{n\times n}$ denotes the vector space of complex matrices of order n. In this setting, the notation A^H denotes the *conjugate transpose* of $A \in \mathbb{C}^{n\times n}$.

For $u,\ v \in \mathbb{C}^n$ we write

$$(u, v) = \sum_{j=1}^{n} \bar{u}_j v_j \quad \text{and} \quad \|u\| = (u, u)^{1/2}.$$

Now for $A \in \mathbb{C}^{n\times n}$, define the function $f : \mathbb{C}^n \to \mathbb{C}$ by

$$f(z) = (z, Az).$$

We are interested in the restriction of f to the unit sphere.

Definition 1. The *field of values* of the matrix $A \in \mathbb{C}^{n\times n}$ is the set

$$F(A) = \{f(z) : \|z\| = 1\}.$$

The topic of the field of values is covered extensively in [27, Chapter 1] which can be consulted for proofs of (and insights on) the results mentioned in this section.

It is known (see [23, 27, p. 8]) that $F(A)$ is a convex set in the (complex) plane. The set $F(A)$ contains all the eigenvalues of A and (by convexity) their convex hull, which we denote Co(($\sigma(A)$)). That is:

$$\mathrm{Co}(\sigma(A)) \subseteq F(A) \quad \forall A \in \mathbb{C}^{n\times n}.$$

Next, we recall two more definitions from matrix theory.

The matrix $U \in \mathbb{C}^{n\times n}$ is said to be *unitary* if $U^H = U^{-1}$; the matrix $A \in \mathbb{C}^{n\times n}$ is said to be *normal* if $AA^H = A^H A$.

Since unitary transformations are length preserving, it follows that for any $A \in \mathbb{C}^{n\times n}$:

$$F(UAU^H) = F(A).$$

Moreover, when A is normal, it can be diagonalized by a unitary matrix [26, p. 101]:

$$UAU^H = \mathrm{Diag}(\lambda_1, \ldots, \lambda_n)$$

where $\lambda_1, \ldots, \lambda_n$ are the eigenvalues of A. For these reasons, we may assume

$$f(z) = \sum_{j=1}^{n} \lambda_j |z_j|^2 \quad (z \in S).$$

This means that $f(z) \in \mathrm{Co}(\sigma(A))$.

These arguments lead to the following result (see [39, 27, p. 11]):

Theorem 2. *If $A \in \mathbb{C}^{n\times n}$ is normal, then $F(A) = Co(\sigma(A))$.*

One can now pose the question: *Does the equality of $F(A)$ and $Co(\sigma(A))$ hold only when A is normal*? To facilitate the discussion, let

$$Q_n = \{A \in \mathbb{C}^{n\times n} : F(A) = \mathrm{Co}(\sigma(A))\},$$
$$N_n = \{A \in \mathbb{C}^{n\times n} : A \text{ is normal}\}.$$

In this notation, theorem above asserts that $N_n \subseteq Q_n$. The reverse inclusion would say that only normal matrices of order n have the property $F(A) = \mathrm{Co}(\sigma(A))$.

It is shown in [39, 27, p. 52] that the proposition

$$\boxed{P(n)\colon Q_n = N_n}$$

is true when $n \leq 4$ and false when $n \geq 5$.

This assertion can be deduced from the necessary and sufficient conditions for $A \in \mathbb{C}^{n\times n}$ to satisfy $F(A) = \mathrm{Co}(\sigma(A))$ given in [27, Theorem 1.6.8]. This criterion is the following

Theorem 3. *Let $A \in \mathbb{C}^{n\times n}$. Then $F(A) = Co(\sigma(A))$ if and only if either A is normal or A is unitarily similar to a matrix of the form*

$$A_1 \oplus A_2 = \begin{bmatrix} A_1 & 0 \\ 0 & A_2 \end{bmatrix}$$

where A_1 is normal and $F(A_2) \subset F(A_1)$.

Using this theorem, it is easy to illustrate the falsity of $P(n)$ for all $n \geq 5$. Consider, for instance, the 5×5 matrix $A_1 \oplus A_2$ with

$$A_1 = \begin{bmatrix} 0 & 1 & 0 \\ 0 & 0 & 1 \\ 1 & 0 & 0 \end{bmatrix} \quad \text{and} \quad A_2 \begin{bmatrix} 0 & 1 \\ 0 & 0 \end{bmatrix}.$$

A direct computation shows that A_1 is normal whereas A_2 is not, and for this reason, A is not normal. But $F(A_2) \subset F(A_1)$ since A_2 is a principal submatrix of A_1. Hence A satisfies the sufficient conditions for the equality of $F(A)$ and $\mathrm{Co}(\sigma(A))$. Thus N_5 is a proper subset of Q_5 and $P(5)$ is false. For all larger values of n, it suffices to border A_2 by an appropriate number of zeros as in

$$\tilde{A}_2 = \begin{bmatrix} A_2 & 0 \\ 0 & 0 \end{bmatrix}.$$

3. Cones

A nonempty subset of $\mathbb{R}^n$ (or any real vector space) is said to be *cone* if it contains all nonnegative scalar multiples of its elements. Symbolically, C is a cone if

$$C \neq \emptyset, \quad x \in C, \quad \text{and} \quad \lambda \geq 0 \Rightarrow \lambda x \in C.$$

This definition implies that the zero vector belongs to every cone.

Cones play an important role in many branches of mathematics. Their place in mathematical programming for instance was very well established while linear programming was still in its infancy. Cones are quite conspicuous in the study of optimality criteria for mathematical programs. The examples exhibited in this section stem from efforts to analyze the behavior of quadratic functions, primarily on polyhedral convex sets as they arise in quadratic programming or in linear complementarity theory [8]. Among the types of quadratic functions of special interest are those for which the quadratic part is a copositive quadratic form. Another related topic is the study of criteria for the quasi-convexity and pseudo-convexity of quadratic functions. In these cases, the quadratic forms involve cones in their definition and also form cones as a class.

3.1. The cone of completely positive quadratic forms

Although the examples exhibited in Sections 2.2 and 2.3 may look different, they are intimately linked through the theory of convex cones. This point is elaborated in the remainder of this subsection. It borrows heavily from the paper by Hall and Newman [22] cited in Section 2.2.

Recall that Section 2.2 dealt with three classes of symmetric matrices: copositive, positive semi-definite, and (elementwise) nonnegative. It is clear that for each $n \in \mathbb{N}$ the copositive matrices of order n form a convex cone. The analogous statements are also true for the positive semi-definite matrices and the nonnegative matrices. Moreover, each of these convex cones is *closed*, a fact of great importance for the application of the theory.

To the aforementioned collection of matrix classes, Hall and Newman add another, the class of completely positive matrices. They give the definition first in terms of quadratic forms. A symmetric quadratic form $x^{\mathrm{T}}Ax$ is called *completely positive* if it can be expressed as the sum of squares of linear forms each having nonnegative coefficients. An n-ary quadratic form of this kind can written as

$$x^{\mathrm{T}}Ax = \sum_{i=1}^{m} \left(\sum_{j=1}^{n} c_{ij} x_j \right)^2 .$$

It follows, then, that the coefficient matrix of a completely positive quadratic form is the sum of products $c^i \cdot (c^i)^{\mathrm{T}}$, $c^i \geq 0$, $i = 1, \ldots, m$. Notice that if $m = 1$ the matrix of coefficients of the quadratic form is the product of a nonnegative matrix and its transpose. Conversely, a matrix product $Y^{\mathrm{T}}Y$ in which $Y \geq 0$ (elementwise) gives rise to a completely positive quadratic form in which $m = n$ since

$$x^{\mathrm{T}}Y^{\mathrm{T}}Yx = \sum_{i=1}^{n}\left(\sum_{j=1}^{n} y_{ij}x_j\right)^2 .$$

It may be of interest to mention a convention used by Hall and Newman as well as by many other authors. Instead of regarding the real symmetric $n \times n$ matrix $A = [a_{ij}]$ as a point in n^2-space, they take advantage of its symmetry and associate it with the point $(a_{11}, \ldots, a_{nn}, \sqrt{2}a_{12}, \ldots \sqrt{2}a_{n-1,n})$ which has $m := n(n+1)/2$ coordinates. In so doing they achieve a one-to-one correspondence between points in n-ary quadratic forms and points in the Euclidean space $\mathbb{E}^m$.

Hall and Newman use the following notations[17] for these four classes of quadratic forms (or matrices):

B the class of completely positive quadratic forms
C the class of copositive quadratic forms
P the class of coefficientwise nonnegative quadratic forms
S the class of positive semi-definite quadratic forms

Now let $H \subset \mathbb{E}^m$ be given. The corresponding *dual set* is

$$H^* = \{x : (x, h) \geq 0 \text{ for all } h \in H\}.$$

It is easily seen that the process of taking the dual set is *inclusion reversing*:

$$H_1 \subseteq H_2 \Rightarrow H_1^* \supseteq H_2^*.$$

Hall and Newman cite the following well known properties:

(i) H^* is a closed convex cone.
(ii) $H^{**} \supseteq H$ and $H^{**} = H$ if H is a closed convex cone.
(iii) $(H + K)^* = H^* \cap K^*$.
(iv) $(H \cap K)^* = H^* + K^*$.

In (iii), $H + K = \{h + k : h \in H,\ k \in K\}$. When H itself is a closed convex cone, then (ii) says that H is the dual of H^*.

Hall and Newman then prove a key result.

Theorem 4. *The spaces of copositive quadratic forms and completely positive quadratic forms are both closed convex cones. Each is the dual of the other, that is*

$$C = B^* \quad \textit{and} \quad C^* = B.$$

Hall and Newman also recall (and prove) a well-known result to the effect that $P$and S are each self-dual closed convex cones:

$$P = P^* \quad \text{and} \quad S = S^*.$$

With these notations and results in place, we can express Diananda's conjecture as

$$C_n = P_n + S_n \quad \forall n \in \mathbb{N}$$

where the subscripts reflect the order n of the coefficient matrices involved. We know this equation holds only for $n \leq 4$. So in particular

$$C_n \supset P_n + S_n \quad \forall n \geq 5.$$

Since the aforementioned duality relationships hold for each n, we have

$$B_n = P_n \cap S_n \quad \forall n \leq 4$$

and

$$B_n \subset P_n \cap S_n \quad \forall n \geq 5.$$

In other words, the result of Maxfield and Minc described in Section 2.3 follows directly from this analysis.

3.2. *Homogeneous convex cones*

In [49] and [50], the Russian mathematician E.B. Vinberg discusses the classification of self-adjoint convex homogeneous cones. We begin by reviewing some of Vinberg's terminology.

A *cone* will be defined here to mean a region V of a finite-dimensional real vector space which is closed under multiplication by positive scalars. The group of all linear transformations leaving V invariant will be denoted $\mathcal{G}(V)$. If $\mathcal{G}(V)$ *acts transitively* on V, i.e., for each $v^1, v^2 \in V$ there exists a transformation $\sigma \in \mathcal{G}(V)$ such that $v^2 = \sigma(v^1)$, then the cone is said to be *homogeneous.* A cone V is called *convex* if it contains no line[18] and contains $V + V$.

Now given a convex cone $V \subset \mathbb{R}^n$, we introduce an arbitrary Euclidean metric in $\mathbb{R}^n$ and consider the set

$$V^* = \{a : \langle a, v\rangle > 0 \quad \forall v \in \bar{V}\backslash 0\}$$

where $\bar{V}$ is the closure of V. It is evident that V^* is a convex cone. It is called the *adjoint* of V. The convex cone V is called *self-adjoint* if for an appropriate choice of Euclidean metric, $V = V^*$. As an example, consider $V = \mathbb{R}^n_{++}$, the interior of the nonnegative orthant. Relative to the ordinary Euclidean metric on $\mathbb{R}^n$, we have $\bar{V}\backslash 0 = \{x \in \mathbb{R}^n : 0 \neq x \geq 0\}$ and $V^* = V$.

In terms of the theme of this paper, we have the proposition

An n-dimensional homogeneous convex cone is self-adjoint.

Vinberg [49, 50] states that this proposition is not true for all n and that the simplest counterexample occurs when $n = 5$. The word "simplest" means "lowest dimensional" [51]. Thus, according to Vinberg, $P(n)$ is another example of a proposition that is true when $n = 1, 2, 3, 4$ and false when $n \geq 5$.

It may be worth mentioning that the classification of convex homogeneous self-adjoint cones makes use of Jordan algebras which enter the picture in connections with $\mathcal{G}(V)$. For an additional reference on this subject see the monograph by Faraut and Korányi [16]. Unfortunately, these authors misquote Vinberg's result in their note [16, p. 23] on the issue discussed above.

4. Volumes

The study of volume dates from classical antiquity. Even its connection with extremal problems in the higher-dimensional geometry of convex bodies is over one hundred years old [4, p. 87], [54, p. 297]. The topics treated in this section were suggested by research on the problem of minimal triangulation of the unit cube in n-space. It was hoped that such triangulations would be beneficial in the development of efficient homotopy algorithms for computing fixed points.

4.1. *The volume of the unit ball in real n-space*

For each $n \in \mathbb{N}$, let

$$B^n = \{x \in \mathbb{R}^n : \|x\| \leq 1\}.$$

The set B^n is just the *closed unit ball* in R^n. It has a positive n-dimensional *volume*, V_n, the formula for which depends on whether n is even or odd. These formulas[19] can be found in [45, p. 136].

$$V_{2m} = \frac{\pi^m}{m!}$$

$$V_{2m+1} = \frac{\pi^m m! 2^{2m+1}}{(2m+1)!}$$

An examination of the first few values of V_n would suggest the validity of the proposition

$$\boxed{P(n) : V_n < V_{n+1}.}$$

It can be shown, however, that $P(n)$ is only true for $n \leq 4$. Indeed, the volume of B^n is maximum when $n = 5$. Thereafter, it decreases asymptotically to zero! (see figure 4.)

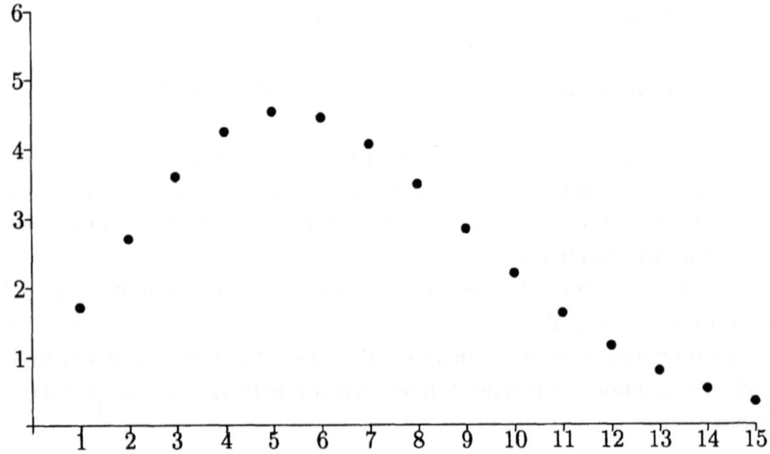

Figure 4. Volume of the unit ball B^n as a function of n.

4.2. *Volumes of certain simplices and determinants of binary matrices*

For each $n \in \mathbb{N}$, let $\mathbb{U}^n$ denote the *unit n-cube*, that is, n-fold cartesian product of the *unit interval*

$$\mathbb{U} = \{\theta : 0 \le \theta \le 1\}.$$

The 2^n vertices of the unit n-cube are binary vectors. A vertex of $\mathbb{U}^n$ is said to be *even* if its representation has an even number of ones. Otherwise it is said to be *odd*. This property of evenness or oddness is called the *parity* of the vertex. Exactly half the vertices of $\mathbb{U}^n$ are of even parity, and no two vertices of the same parity are *neighbors* (endpoints of an edge of $\mathbb{U}^n$).

For every vertex v of $\mathbb{U}^n$ there is a unique *antipodal point*

$$A(v) = e - v,$$

where $e = (1, \ldots, 1)$, and a set of neighbors, $N(v)$. Corresponding to v and $A(v)$ we distinguish two n-dimensional simplices. The first of these is, $\langle v, N(v)\rangle$, the convex hull of v and $N(v)$, is the *corner simplex* of $\mathbb{U}^n$ at v. The other, $\langle N(v), A(v)\rangle$, is called the *antipodal simplex* corresponding to v. Our interest now centers on the volume of the antipodal simplex. Before discussing this, we need a little more general apparatus.

In general, an n-simplex σ is the convex hull of $n+1$ points in general position. In our case, the points will be vertices of $\mathbb{U}^n$. Thus, we have

$$\sigma = \langle v^1, v^2, \ldots, v^{n+1}\rangle.$$

For convenience, we shall regard the coordinates of these points as row vectors. Then corresponding to σ is an $(n+1) \times (n+1)$ matrix S whose ith row is $S_{i\cdot} = (v^i, 1)$. When the v^i are vertices of the $\mathbb{U}^n$ (as in our case), the entries of the matrix S are all zeros and ones, so S is said to be a *binary matrix*. It is well known [45, p. 124] that the n-dimensional volume of σ is given by the formula

$$\mu_n(\sigma) = \frac{|\det S|}{n!}.$$

It can be shown that for every vertex $v \in \mathbb{U}^n$, the corresponding antipodal simplex has volume $(n-1)/n!$. The next proposition is the statement

> $P(n)$: *The antipodal simplices $\langle N(v), A(v)\rangle$ have the largest volume among all simplices whose vertices are also vertices of $\mathbb{U}^n$.*

The statement is more or less vacuous for $n = 1$, and it is not difficult to verify for $n = 2, 3$. The case of $n = 4$ is covered in Cottle [7]. As we shall see below, the statement $P(n)$ is false when $n \geq 5$.

The issue here is related to the question of how large the determinant of a binary matrix can be. Indeed, according to our conventions, the matrix corresponding to the simplex σ is

$$S = \begin{bmatrix} v_1^1 & v_2^1 & \cdots & v_n^1 & 1 \\ v_1^2 & v_2^2 & \cdots & v_n^2 & 1 \\ \vdots & \vdots & & \vdots & \vdots \\ v_1^n & v_2^n & \cdots & v_n^n & 1 \\ v_1^{n+1} & v_2^{n+1} & \cdots & v_n^{n+1} & 1 \end{bmatrix}.$$

By taking $(v_1^{n+1}, v_2^{n+1}, \ldots, v_n^{n+1}) = (0, 0, \ldots, 0)$ and then computing the determinant of S by cofactors of its last column, we see that to maximize the determinant of the $(n+1) \times (n+1)$ matrix S one wants to know $f(n)$, the maximum value of the determinant of an $n \times n$ binary matrix. Even though an all-purpose formula for $f(n)$ is not known, we have enough information to handle the question at hand. As asserted above, $f(4) = 3$. A result of K.W. Schmidt [43] states that

$$(n+1)f(n+1) \geq 2nf(n).$$

It follows that $f(5) \geq \frac{24}{5}$, and since $f(n)$ is always an integer, $f(5) \geq 5$. This value is attained by

$$\begin{bmatrix} 1 & 0 & 0 & 0 & 1 \\ 1 & 1 & 1 & 1 & 0 \\ 0 & 0 & 1 & 1 & 1 \\ 0 & 1 & 0 & 1 & 1 \\ 0 & 1 & 1 & 0 & 1 \end{bmatrix},$$

hence $P(5)$ is false. In fact, Schmidt's inequality can be used to show that $P(n)$ is false for all $n \geq 5$. Actually, for values of $n > 5$, the discrepancy between $n - 1$ and $f(n)$ grows dramatically.

4.3. Cross sections of simplices

It is intuitively obvious and not difficult to prove that the longest line segment within a triangular region must be one of the sides of the triangle. More formally,

$$\max_{M^1} \mu_1(M^1 \cap \sigma^2) = \max\{\mu_1(\sigma_1^1), \mu_1(\sigma_2^1), \mu_1(\sigma_3^1)\}.$$

where M^1 is a 1-flat (line), σ^2 is a 2-simplex (triangle), σ_i^1 $(i = 1, 2, 3)$ are the one-dimensional simplicial faces (sides) of σ^2, and μ_1 is 1-dimensional measure (length).

The corresponding three-dimensional case arose (twice) as an *American Mathematical Monthly* problem concerning the validity of the proposition

Every plane section of a tetrahedron is smaller (*in area*) *than the largest face.*

Paraphrased, the statement involves planes (2-flats), tetrahedra (3-simplices) and the areas of their intersections (cross sections). It says

$$\max_{M^2} \mu_2(M^2 \cap \sigma^3) = \max\{\mu_2(\sigma_1^2), \mu_2(\sigma_2^2), \mu_2(\sigma_3^2), \mu_2(\sigma_4^2)\}.$$

Proofs were provided by Brands and Laman [5] and by Eggleston [12]. We note in passing that in addition to the empty set, $M^2 \cap \sigma^3$ can be a point, a line segment, a triangle, or a quadrilateral (see figure 5).

On the strength of this evidence, one is tempted to generalize to the proposition

$P(n)$: *A hyperplane section of an n-simplex with maximum volume is a face of the simplex.*

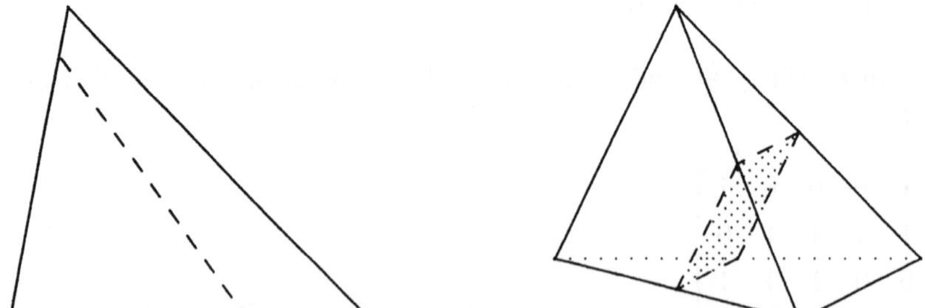

Figure 5. Cross sections of σ^2 and σ^3.

Let σ^n be an n-dimensional simplex with (($n-1$)-dimensional) faces σ_i^{n-1} ($i = 1, \ldots, n+1$). Let M^{n-1} be an ($n-1$)-dimensional hyperplane and let μ_{n-1} denote ($n-1$)-dimensional "volume". The proposition $P(n)$ above says that

$$\max_{M^{n-1}} \mu_{n-1}(M^{n-1} \cap \sigma^n) = \max_{1 \le i \le n+1} \mu_{n-1}(\sigma_i^{n-1}).$$

Historically, the *falsity* of $P(n)$ for *odd* integers $n \geq 5$ was established by Walkup [53]. The gaps (even integers greater than or equal to 4) were filled in by Philip [42] who first showed that for this proposition

$$P(n) \Rightarrow P(n-1).$$

Together with Walkup's result, this immediately implied that $P(n)$ cannot be true for *even* integers greater than or equal to 6. This left only the difficult case of $P(4)$ which Philip then settled.

Acknowledgments

I began collecting the examples of "quartic barriers" described in this paper more than thirty years ago. During this time I have discussed them with many people, all of whom have my gratitude for their willingness to listen and to contribute suggestions on things to include. I especially wish to thank Alexander Rinnooy Kan who recommended reference [25] used in Section 1.3, Timothy O'Meara and Daniel Bump who advised me on Section 2.4, Balder von Hohenbalken who pointed me to the phenomenon in Section 4.1 and Sheng-Pen Wang who carried out most of the calculations used there. I am grateful to Ingram Olkin, Bernhard von Stengel, and the referees for their many perceptive comments on earlier drafts of this paper. Finally, I wish to mention that the "blackboard bold" font MSBM10 came from $\mathcal{A}\mathcal{M}\mathcal{S}$-TEX.

Notes

1. That is, solvable through "rational operations and by extraction of roots, $\sqrt{\ }$, $\sqrt[3]{\ }$, ..." [52, p. 172].
2. van der Waerden [52, pp. 179–182] hints at the controversy by using quotation marks around attributions to Cardano.
3. See the interesting monograph [55] for example.
4. See also Dirac and Schuster [11].
5. A coloring of $G = (V, E)$ can be regarded as a function γ defined on V such that $\{v_1, v_2\} \in E$ implies $\gamma(v_1) \neq \gamma(v_2)$ if $v_1 \neq v_2$.
6. The reader may wish to substitute an appropriate gender-neutral term.
7. It will be convenient to call this a *segment*.
8. The exact moves are depicted in [25].
9. But no literature citation is given in [25] for this information. The same subject is addressed in [3, pp. 715–717] under the heading "Solitaire Army." I am grateful to J.H. Conway for some background on this.
10. There are exceptions to this practice. See [8, p. 4].
11. The *support* of a matrix A is defined as the set supp $A = \{(i, j) : a_{ij} \neq 0.\}$
12. The contributions of T.S. Motzkin to the theory of linear inequalities are well known from the publication of his doctoral dissertation [37] at the University of Basel.

13. This follows immediately from the *spectral decomposition* of A.
14. About this, Hall says "Professor A. Horn proved the existence of forms which did not have this property, and the author succeeded in constructing the following example."
15. We remark that the problems of interest belong to the family of diophantine equations.
16. If A were either positive or negative definite, the equation $x^{\mathrm{T}}Ax = 0$ would have only the *trivial* solution $x = 0$.
17. The reader should beware that elsewhere the symbols used for some of these classes have other meanings.
18. In other words, the cone is *pointed*.
19. There is a more unified expression for the volume of a hypersphere that involves the gamma function. See, for instance, [18, p. 118].

References

1. R.K. Ahuja, T.L. Magnanti, and J.B. Orlin, Network Flows: Theory, Algorithms and Applications, Prentice-Hall: Englewood Cliffs, NJ, 1993.
2. K. Appel and W. Haken, "Every planar map is four colorable," Bulletin of the American Mathematical Society, vol. 82, pp. 711–712, 1976.
3. E.R. Berlekamp, J.H. Conway, and R.K. Guy, Winning Ways: Games in Particular, vol. 2, Academic Press: London, 1982.
4. T. Bonnesen and W. Fenchel, Konvexe Körper, Chelsea Publishing Company: New York, 1948.
5. J.J.A.M. Brands and G. Laman, "Plane section of a tetrahedron," American Mathematical Monthly, vol. 70, pp. 338–339, 1963.
6. G. Chartrand and D.P. Geller, "On uniquely colorable planar graphs," Journal of Combinatorial Theory, vol. 6, pp. 271–278, 1969.
7. R.W. Cottle, "Minimal triangulation of the 4-cube," Discrete Mathematics, vol. 40, pp. 25–29, 1982.
8. R.W. Cottle, J.S. Pang, and R.E. Stone, The Linear Complementarity Problem, Academic Press: Boston, 1992.
9. H. Davenport, The Higher Arithmetic, 6th edition, Cambridge University Press: Cambridge, 1992.
10. P.H. Diananda, "On nonnegative forms in real variables some or all of which are nonnegative," Proceedings of the Cambridge Philosophical Society, vol. 58, pp. 17–25, 1962.
11. G.A. Dirac and S. Schuster, "A theorem of Kuratowski," Indagationes Mathematicae, vol. 16, pp. 343–348, 1954.
12. H.G. Eggleston, "Plane section of a tetrahedron," American Mathematical Monthly, vol. 70, p. 1108, 1963.
13. P. Erdös, A. Goodman, and L. Pósa, "The representation of a graph by set intersections," Canadian Journal of Mathematics, vol. 18, pp. 106–112, 1966.
14. R.E. Erickson, C.L. Monma, and A.F. Veinott, Jr., "Send-and-split method for minimum-concave-cost network flows," Mathematics of Operations Research, vol. 12, pp. 634–664, 1987.
15. H. Eves, An Introduction to the History of Mathematics, Rinehart: New York, 1953.
16. J. Faraut and A. Korányi, Analysis on Symmetric Cones, Oxford University Press: New York, 1994.
17. F. Granot and A.F. Veinott, Jr., "Substitutes, complements and ripples in network flows," Management Science, vol. 10, pp. 471–497, 1985.
18. H.W. Guggenheimer, Applicable Geometry, Robert E. Krieger Publishing Company, Inc.: Huntington, 1977.
19. M. Hall, "A problem in partitions," Bulletin of the American Mathematical Society, vol. 47, pp. 804–807, 1941.
20. M. Hall, Jr., "A survey of combinatorial analysis," in Some Aspects of Analysis and Probability, Surveys in Applied Mathematics IV, I. Kaplansky, M. Hall, Jr., E. Hewitt, and R. Fortet (Eds.), John Wiley & Sons: New York, 1958, pp. 35–104.
21. M. Hall, Jr., Combinatorial Theory, Blaisdell: Waltham, MA, 1967.
22. M. Hall, Jr. and M. Newman, "Copositive and completely positive quadratic forms," Proceedings of the Cambridge Philosophical Society, vol. 59, pp. 329–339, 1963.
23. F. Hausdorff, "Der Wertvorrat einer Bilinearform," Mathematische Zeitschrift, vol. 3, pp. 314–316, 1919.
24. P. Heawood, "Map color theorems," Quarterly Journal of Mathematics, vol. 24, pp. 332–338, 1890.

25. R. Honsberger, "A problem in checker-jumping," Mathematical Gems II, Mathematical Association of America: Washington, D.C., 1976, pp. 23–28.
26. R.A. Horn and C.R. Johnson, Matrix Analysis, Cambridge University Press: Cambridge, 1985.
27. R.A. Horn and C.R. Johnson, Topics in Matrix Analysis, Cambridge University Press: Cambridge, 1991.
28. C. Kuratowski, "Sur le probleme des courbes gauches en topologie," Fundamenta Mathematicae, vol. 15–16, pp. 271–283, 1930.
29. J.L. Lagrange, "Réflexions sur la résolution algébrique des équations," Oevres, vol. 3, pp. 207–421. (From a paper of 1771.)
30. A.-M. Legendre, Théorie des Nombres, F. Didot: Paris, 1830.
31. C.E. Lemke, "Bimatrix equilibrium points and mathematical programming," Management Science, vol. 11, pp. 681–689, 1965.
32. O. Ore, Theory of Graphs, American Mathematical Society: Providence, RI, 1962.
33. O. Ore, The Four-Color Problem, Academic Press: New York, 1967.
34. J.R. Maxfield and H. Minc, "On the matrix equation $X'X = A$," Proceedings of the Edinburgh Mathematical Society, vol. 13:2, pp. 125–129, 1962–1963.
35. A. Meyer, "Mathematische Mittheilungen," Vierteljahrschrift der Naturforschenden Gesellschaft in Zürich, vol. 29, pp. 209–222, 1884.
36. L.J. Mordell, Diophantine Equations, Academic Press: New York, 1969.
37. T.S. Motzkin, Beitrage zur Theorie der linearen Ungleichungen (Inaugural dissertation, Basel), Azriel: Jerusalem, 1936.
38. T.S. Motzkin, in National Bureau of Standards Report 1817, pp. 11–12, 1952.
39. B.N. Moyls and M.D. Marcus, "Field convexity of a square matrix," Proceedings of the American Mathematical Society, vol. 6, pp. 981–983, 1955.
40. T. Nagell, Introduction to Number Theory, John Wiley & Sons: New York; Almqvist & Wiksell: Stockholm, 1951.
41. O.T. O'Meara, Introduction to Quadratic Forms, Springer-Verlag: Berlin, 1963.
42. J. Philip, "Plane sections of simplices," Mathematical Programming, vol. 3, pp. 312–325, 1972.
43. K.W. Schmidt, "Problem 68-18, Binary matrices," SIAM Review, vol. 13, pp. 119–121, 1971.
44. D.E. Smith, History of Mathematics, vol. I, Dover: New York, 1951.
45. D.M.Y. Sommerville, An Introduction to the Geometry of N Dimensions, Dover: New York, 1958.
46. E. Szpilrajn-Marczewski, "Sur deux propriétés des classes d'ensembles," Fundamenta Mathematicae, vol. 33, pp. 303–307, 1945.
47. L. Toti Rigatelli, Evariste Galois, Birkhäuser: Basel, 1996.
48. A.F. Veinott, Jr., "Least d-majorized network flows with inventory and statistical applications," Management Science, vol. 17, pp. 547–567, 1971.
49. E.B. Vinberg, "Homogeneous cones," Soviet Mathematics Doklady, vol. 1, p. 787, 1960.
50. E.B. Vinberg, "The theory of convex homogeneous cones," Transactions of the Moscow Mathematical Society (1963), 1965, pp. 340–403.
51. E.B. Vinberg, private communication, May 5, 1968.
52. B.L. van der Waerden, Modern Algebra, vol. I, Ungar: New York, 1949.
53. D.W. Walkup, "A simplex with a large cross-section," American Mathematical Monthly, vol. 75, pp. 34–36, 1968.
54. R. Webster, Convexity, Oxford University Press: Oxford, 1994.
55. C. Zong, Strange Phenomena in Convex and Discrete Geometry, Springer-Verlag: New York, 1996.

Computational Optimization and Applications 12, 107–126 (1999)

A Partitioned ϵ-Relaxation Algorithm for Separable Convex Network Flow Problems*

RENATO DE LEONE
Dipartimento di Matematica e Fisica, Universitá di Camerino, Via Madonna delle Carceri, 62032 Camerino (MC), Italy

ROBERT R. MEYER rrm@cs.wisc.edu
ARMAND ZAKARIAN
Computer Sciences Department, University of Wisconsin-Madison, 1210 W. Dayton St., Madison, WI 53706

Received December 31, 1997; Accepted August 11, 1998

Abstract. A relaxation method for separable convex network flow problems is developed that is well-suited for problems with large variations in the magnitude of the nonlinear cost terms. The arcs are partitioned into two sets, one of which contains only arcs corresponding to strictly convex costs. The algorithm adjusts flows on the other arcs whenever possible, and terminates with primal-dual pairs that satisfy complementary slackness on the strictly convex arc set and ϵ-complementary slackness on the remaining arcs. An asynchronous parallel variant of the method is also developed. Computational results demonstrate that the method is significantly more efficient on ill-conditioned networks than existing methods, solving problems with several thousand nonlinear arcs in one second or less.

Keywords: network flows, relaxation methods, asynchronous

Dedication: The authors of this paper represent three of the many generations of mathematical programmers who developed their optimization skills at the University of Wisconsin-Madison in collaboration with Olvi Mangasarian. Robert Meyer joined the Faculty in 1973, Renato De Leone arrived with a visiting appointment in 1984, and Armand Zakarian came as a student in 1990. The topic of this paper—relaxation—represents just one of the many diverse (and ever-changing) areas to which Olvi has made fundamental contributions. Throughout the more than quarter-century that Olvi has been at Madison, his door has seemingly always been open, both literally and figuratively, for colleagues and students to enter for enlightening personal as well as professional discussions. We dedicate this paper to Olvi in appreciation of an open door symbolic of his generous willingness to share ideas.

1. Introduction

Single-commodity networks with separable convex objective functions arise in a variety of applications including electrical networks [6], pipeline networks [5], urban traffic flow [9] and communications networks [10]. Algebraically, such problems are defined via a connected directed graph $\mathcal{G}(\mathcal{N}, \mathcal{A})$ with n nodes and m arcs, a supply-demand vector b,

*This material is based on research supported by the Air Force Office of Scientific Research Grant F49620-94-1-0036, and by NSF Grants CDA-9024618 and CCR-9306807.

and an arcwise-defined objective, yielding the minimum-cost network flow problem:

$$\begin{aligned} \text{minimize} \quad & \sum_{(i,j)\in\mathcal{A}} f_{ij}(x_{ij}) \\ \text{subject to} \quad & \sum_{j\in\delta_+(i)} x_{ij} - \sum_{j\in\delta_-(i)} x_{ji} = b_i, \quad i \in \mathcal{N} \end{aligned} \tag{1.1}$$

where for each node i, $\delta_+(i)$ is the subset of nodes j such that $(i, j) \in \mathcal{A}$ and $\delta_-(i)$ is the subset of nodes j such that $(j, i) \in \mathcal{A}$. (Any upper and lower bounds on flows are taken into account via domain constraints on the objective function.)

We make the following assumptions with regard to the functions f_{ij}.

Assumption 1.1. *Each function f_{ij} is a closed proper convex function and the intersection of* dom$(\sum_{(i,j)\in\mathcal{A}} f_{ij})$ *with the set of vectors satisfying the constraints of* (1.1) *is nonempty.*

Assumption 1.2. *The conjugate functions*

$$f_{ij}^*(t_{ij}) := \sup_{x_{ij}}\{t_{ij}x_{ij} - f_{ij}(x_{ij})\}$$

are real-valued.

Assumption 1.1 guarantees that (1.1) is feasible. Assumption 1.2 implies that the objective function of (1.1) has bounded level sets and together with Assumption 1.1 guarantees that an optimal solution exists. A linear cost function must have a bounded domain in order to satisfy Assumption 1.2.

Often the functions f_{ij} imply flow capacity constraints:

$$f_{ij}(x_{ij}) := \begin{cases} \bar{f}_{ij}(x_{ij}) & \text{if } l_{ij} \le x_{ij} \le u_{ij} \\ +\infty & \text{otherwise.} \end{cases} \tag{1.2}$$

Here $\bar{f}_{ij}$ is a real-valued convex function and $l_{ij} \le u_{ij}$ are the (finite) lower and upper bounds on the flow on arc $(i, j) \in \mathcal{A}$. In this case Assumption 1.2 is trivially satisfied.

We denote the left and right derivative of f_{ij} at $x_{ij} \in \text{dom } f_{ij}$ by $f_{ij}^-(x_{ij})$ and $f_{ij}^+(x_{ij})$, respectively. For points x_{ij} to the left of dom f_{ij} we set $f_{ij}^-(x_{ij}) = f_{ij}^+(x_{ij}) = -\infty$ and for points x_{ij} to the right of dom f_{ij} we set $f_{ij}^-(x_{ij}) = f_{ij}^+(x_{ij}) = +\infty$. Theorem 24.1 in [11] states that f_{ij}^- and f_{ij}^+ are nondecreasing functions on **R**, finite on int dom f_{ij} and

$$f_{ij}^+(z_{ij}) \le f_{ij}^-(x_{ij}) \le f_{ij}^+(x_{ij}) \le f_{ij}^-(y_{ij})$$

whenever $z_{ij} < x_{ij} < y_{ij}$.

We make an additional assumption regarding the functions f_{ij}:

Assumption 1.3. $f_{ij}^-(x_{ij}) < +\infty$ *and* $f_{ij}^+(x_{ij}) > -\infty$ *for all* $x_{ij} \in \text{dom } f_{ij}$.

Assumption 1.3 guarantees that the feasible solutions of (1.1) are *regularly feasible* [12, Section 8D]. Together with the first two assumptions it also guarantees the existence of dual optimal solutions [12, Section 8L].

We say that a pair (x, π) satisfies complementary slackness (CS) on an arc $(i, j) \in \mathcal{A}$ if

$$f_{ij}^-(x_{ij}) \leq \pi_i - \pi_j \leq f_{ij}^+(x_{ij}). \tag{1.3}$$

For arbitrary vector π and f_{ij} strictly convex satisfying Assumption 1.2, there exists a unique x_{ij} satisfying (1.3). A vector x is optimal for (1.1) if it satisfies the flow conservation constraints and together with some π satisfies CS on every arc in $\mathcal{A}$.

A pair (x, π) satisfies ϵ-complementary slackness (ϵ-CS) on an arc $(i, j) \in \mathcal{A}$ if

$$f_{ij}^-(x_{ij}) - \epsilon \leq \pi_i - \pi_j \leq f_{ij}^+(x_{ij}) + \epsilon. \tag{1.4}$$

We will see below that ϵ-complementary slackness corresponds to near optimality for small ϵ.

Relaxation and auction-based methods can be used for solving (1.1). For the strictly convex cost case, convergence can be established for relaxation methods that are essentially coordinate ascent methods applied to the maximization of the dual function [2]. Auction algorithms based on ϵ-complementary slackness were originally developed for problems with linear cost functions [1]. The method that we develop here partitions the arcs into two sets and treats the arcs in one set by an ϵ-complementary slackness approach and handles the nonlinear arcs in the other set by preserving complementary slackness for them. The motivation for this partitioning is that an ϵ-complementary slackness viewpoint is useful for arcs with linear costs (and turns out to be good computationally for arcs with nearly linear costs), whereas complementary slackness is appropriate for arcs with highly nonlinear costs. This approach differs from that of [2] in that our method allows a mix of both linear and nonlinear arcs by partitioning the arcs into two sets and focuses on changing single node prices and flows on arcs incident to single nodes. Recently, Bertsekas et al. [3] have proposed a new algorithm that generalizes the ϵ-relaxation algorithm to convex cost functions and reduces to this method when *all* arcs are treated via an ϵ-complementary slackness approach.

The paper is organized as follows. A statement of the basics of the algorithm is given in the following section. Section 3 provides a convergence analysis and a quantitative interpretation of the significance of ϵ-complementary slackness. Some computational refinements of the basic method are presented in Section 4, together with a specialization of the algorithm to the mixed linear-quadratic case. Section 5 presents numerical results and a comparison with other approaches. Finally, parallel versions of the algorithm are discussed in Section 6.

2. The basic algorithm

We state the algorithm in its simplest form in this section. A brief discussion is given of the main ideas, followed by a formal statement of the method.

The algorithm converges to an approximate solution of problem (1.1) in the sense that it generates a sequence of iterates converging to a *feasible* solution satisfying ϵ-CS (as Lemmas 3.5–3.7 below demonstrate, this is a near-optimal solution for small ϵ). More precisely, we assume the arc set $\mathcal{A}$ has been partitioned into two subsets $\mathcal{A}_0$ and $\mathcal{A}_1$ such that only arcs with strictly convex cost functions are in $\mathcal{A}_1$, and that the algorithm starts with a primal-dual pair (x, π) that does not satisfy primal feasibility but does satisfy ϵ-CS on all arcs in $\mathcal{A}_0$ and CS on all arcs in $\mathcal{A}_1$. In the limit the primal-dual pair produced by the algorithm is feasible and still satisfies CS on each arc in $\mathcal{A}_1$ and ϵ-CS on each arc in $\mathcal{A}_0$.

In Step 1 below, we identify a node at which feasibility is violated by considering the node "surpluses" defined by

$$s_i := b_i - \sum_{j \in \delta_+(i)} x_{ij} + \sum_{j \in \delta_-(i)} x_{ji}.$$

Since it is easy to verify that the surpluses must sum to 0, feasibility is violated if and only if at least one node has a positive surplus. For such a node, Steps 2–4 are repeatedly applied to the node and its incident arcs until the surplus is driven to 0 (in a finite number of steps) at the node. In this process, the algorithm first tries to "push" this surplus to adjacent nodes by using only arcs in $\mathcal{A}_0$ (Steps 2 and 3) that violate CS by at least $\frac{\epsilon}{2}$ (such a push also reduces CS violation). If feasibility at the node cannot be accomplished by such flow pushes, Step 4 then raises the price (Step 4) by a positive amount that does not violate ϵ-CS on the arcs in $\mathcal{A}_0$ (which were close to satisfying CS as a result of Steps 2 and 3) but does allow surplus to then be "pushed away" via arcs in $\mathcal{A}_1$. We assume that ϵ and ϵ' are two given numbers such that $0 \le \epsilon' < \frac{\epsilon}{2}$.

Algorithm 1.

Step 0 [Initialization] Start with (x, π) satisfying ϵ-CS on all arcs in $\mathcal{A}_0$ and CS on all arcs in $\mathcal{A}_1$.

Step 1 [Choose node with positive surplus] Choose a node i such that

$$s_i > 0.$$

If no such node exists, STOP.

Step 2 [If possible, increase flow on an outgoing arc in $\mathcal{A}_0$] If there is no arc $(i, j) \in \mathcal{A}_0$ such that

$$f_{ij}^+(x_{ij}) - (\pi_i - \pi_j) \le -\frac{\epsilon}{2},$$

go to Step 3, else change the flow on such an arc (i, j):

$$x_{ij} \leftarrow x_{ij} + \min\{s_i, \delta\}$$

where $\delta > 0$ is the largest flow increase yielding ϵ'-CS for arc (i, j):

$$f_{ij}^-(x_{ij} + \delta) - \epsilon' \le \pi_i - \pi_j \le f_{ij}^+(x_{ij} + \delta) + \epsilon'$$

(The flows on the remaining arcs and the dual costs are not changed.) Update s_i and if its new value is 0 go to Step 1; else return to Step 2.

Step 3 [If possible, decrease flow on an incoming arc in $\mathcal{A}_0$] If there is no arc $(j, i) \in \mathcal{A}_0$ such that

$$f_{ji}^{-}(x_{ji}) - (\pi_j - \pi_i) \geq \frac{\epsilon}{2},$$

go to Step 4, else change the flow on such an arc (j, i):

$$x_{ji} \leftarrow x_{ji} - \min\{s_i, \delta\}$$

where $\delta > 0$ is the largest flow decrease yielding ϵ'-CS for arc (j, i):

$$f_{ji}^{-}(x_{ji} - \delta) - \epsilon' \leq \pi_j - \pi_i \leq f_{ji}^{+}(x_{ji} - \delta) + \epsilon'$$

(The flows on the remaining arcs and the dual costs are not changed.) Update s_i and if its new value is 0 go to Step 1; else return to Step 3.

Step 4 [Raise single node price, then change flows on incident arcs in $\mathcal{A}_1$] Compute

$$\Delta_1 := \min_{\{j \in \delta_+(i) | (i,j) \in \mathcal{A}_0\}} \{f_{ij}^{+}(x_{ij}) - (\pi_i - \pi_j)\} > -\frac{\epsilon}{2}$$

$$\Delta_2 := \min_{\{j \in \delta_-(i) | (j,i) \in \mathcal{A}_0\}} \{-(f_{ji}^{-}(x_{ji}) - (\pi_j - \pi_i))\} > -\frac{\epsilon}{2}.$$

For each γ and for each arc $(i, j) \in \mathcal{A}_1$ let $x_{ij}(\gamma)$ be the flow such that

$$f_{ij}^{-}(x_{ij}(\gamma)) \leq \pi_i + \gamma - \pi_j \leq f_{ij}^{+}(x_{ij}(\gamma)).$$

Similarly for each arc $(j, i) \in \mathcal{A}_1$ let $x_{ji}(\gamma)$ be the flow such that

$$f_{ji}^{-}(x_{ji}(\gamma)) \leq \pi_j - \pi_i - \gamma \leq f_{ji}^{+}(x_{ji}(\gamma))$$

and let $\Delta_3 > 0$ be the value of γ such that

$$b_i - \sum_{(i,j) \in \mathcal{A}_1} x_{ij}(\gamma) - \sum_{(i,j) \in \mathcal{A}_0} x_{ij} + \sum_{(j,i) \in \mathcal{A}_1} x_{ji}(\gamma) + \sum_{(j,i) \in \mathcal{A}_0} x_{ji} = 0$$

(where $\Delta_3 = +\infty$ if a solution does not exist).
Define:

$$\Delta := \min\{\Delta_1 + \epsilon, \Delta_2 + \epsilon, \Delta_3\}.$$

Change the price on node i:

$$\pi_i \leftarrow \pi_i + \Delta$$

and adjust the flow on the incident arcs in $\mathcal{A}_1$ so that CS is satisfied. Update s_i and if its new value is 0 go to Step 1; else return to Step 2.

We now verify that Step 2 maintains ϵ-CS on arc (i, j) (this can be done similarly for Step 3). Let x'_{ij} be the updated value of the flow on (i, j). We have

$$f_{ij}^{-}(x'_{ij}) - \epsilon \le f_{ij}^{-}(x_{ij} + \delta) - \epsilon \le f_{ij}^{-}(x_{ij} + \delta) - \epsilon' \le \pi_i - \pi_j$$

and

$$f_{ij}^{+}(x'_{ij}) + \epsilon \ge f_{ij}^{+}(x_{ij}) + \epsilon \ge \pi_i - \pi_j.$$

To see that $\Delta > 0$ in Step 4, note that $\Delta_1 > -\frac{\epsilon}{2}$ and $\Delta_2 > -\frac{\epsilon}{2}$ since no eligible arcs could be found in Step 2 and Step 3, respectively. Also, $\Delta_3 > 0$ since we require $x_{ij}(\Delta_3) > x_{ij}(0)$ for $(i, j) \in \mathcal{A}_1$ and $x_{ji}(\Delta_3) < x_{ji}(0)$ for $(j, i) \in \mathcal{A}_1$. Finally, it is easily seen that $\Delta_1 = \Delta_2 = \Delta_3 = +\infty$ cannot occur since this would imply that the surplus cannot be driven to 0 even if flows are set at (implicit) bounds.

We now verify that the price increase in Step 4 doesn't destroy ϵ-CS on the arcs in $\mathcal{A}_0$. Denote the new value of π_i by π'_i. We have

$$\pi'_i - \pi_j > \pi_i - \pi_j \ge f_{ij}^{-}(x_{ij}) - \epsilon$$

and

$$\pi'_i - \pi_j \le \pi_i + \Delta_1 + \epsilon - \pi_j = \min_{\{v \in \delta_+(i) \mid (i,v) \in \mathcal{A}_0\}} \{f_{iv}^{+}(x_{iv})\} + \epsilon \le f_{ij}^{+}(x_{ij}) + \epsilon.$$

Similarly, ϵ-CS is preserved on an arc $(j, i) \in \mathcal{A}_0$.

Finally, we show that Steps 2–4 eventually drive the surplus of node i to 0. Assume the contrary. Then $\Delta_3 = +\infty$ in every execution of Step 4 implying that this step increases π_i by at least $\frac{\epsilon}{2}$. Hence $\pi_i \to \infty$ which is impossible as will be shown in the proof of Lemma 3.1 in a more general context.

3. Convergence analysis

In this section we denote by (x^k, π^k) the distinct pairs generated by Algorithm 1, and we establish convergence of this sequence.

Lemma 3.1. *The sequence $\{\pi^k\}_{k=0}^{\infty}$ generated by Algorithm 1 converges to some point π^*.*

Proof: The sequence $\{\pi_i^k\}_{k=0}^{\infty}$ is nondecreasing for all $i \in \mathcal{N}$. Hence we only need to show it is bounded above. Assume the contrary, i.e., $\pi_i^k \to \infty$ for all $i \in \mathcal{N}_1$ while π_j^k stay bounded for $j \in \mathcal{N}_0$ as $k \to \infty$. Here $\{\mathcal{N}_0, \mathcal{N}_1\}$ is a partition of $\mathcal{N}$ with $\mathcal{N}_0 \ne \emptyset$ since there is always a node with negative surplus which is never chosen in Step 1. Pick an arc (i, j) with $i \in \mathcal{N}_1$ and $j \in \mathcal{N}_0$ (we assume that the network is connected, so either such an arc exists or an oppositely directed arc exists, and we consider the latter case below). We have $\pi_i^k - \pi_j^k \to \infty$ which together with the ϵ-CS condition

$$f_{ij}^{-}(x_{ij}^k) - \epsilon \le \pi_i^k - \pi_j^k \le f_{ij}^{+}(x_{ij}^k) + \epsilon$$

implies that $f_{ij}^+(x_{ij}^k) \to +\infty$. Hence for any $\bar{x}_{ij} \in \text{int dom } f_{ij}$ there exists an iteration number k_{ij} such that $x_{ij}^k \geq \bar{x}_{ij}$ whenever $k \geq k_{ij}$. Also, if $\bar{x}_{ij} \in \text{dom } f_{ij}$ is the right endpoint of $\text{dom } f_{ij}$ then $f_{ij}^-(\bar{x}_{ij}) < +\infty$ (Assumption 1.3) hence $f_{ij}^+(x)$ is uniformly bounded above on $\text{int dom } f_{ij}$ implying that $x_{ij}^k = \bar{x}_{ij}$ for $k \geq k_{ij}$.

Similarly, if (j, i) is an arc with $j \in \mathcal{N}_0$, $i \in \mathcal{N}_1$ and $\bar{x}_{ji} \in \text{dom } f_{ji}$ there exist an iteration number k_{ji} such that $x_{ji}^k \leq \bar{x}_{ji}$ whenever $k \geq k_{ji}$.

Now choose for each arc (i, j), $i \in \mathcal{N}_1$, $j \in \mathcal{N}_0$ a number $\bar{x}_{ij} \in \text{dom } f_{ij}$ and for each arc (j, i), $j \in \mathcal{N}_0$, $i \in \mathcal{N}_1$ a number $\bar{x}_{ji} \in \text{dom } f_{ji}$ such that

$$\sum_{i \in \mathcal{N}_1} b_i - \sum_{\{(i,j) | i \in \mathcal{N}_1, j \in \mathcal{N}_0\}} \bar{x}_{ij} + \sum_{\{(j,i) | i \in \mathcal{N}_1, j \in \mathcal{N}_0\}} \bar{x}_{ji} \leq 0.$$

(This is always possible since the problem is feasible.) For sufficiently large k we now have

$$\begin{aligned}
\sum_{i \in \mathcal{N}_1} s_i^k &= \sum_{i \in \mathcal{N}_1} b_i - \sum_{\{(i,j) | i \in \mathcal{N}_1, j \in \mathcal{N}_0\}} x_{ij}^k + \sum_{\{(j,i) | i \in \mathcal{N}_1, j \in \mathcal{N}_0\}} x_{ji}^k \\
&\leq \sum_{i \in \mathcal{N}_1} b_i - \sum_{\{(i,j) | i \in \mathcal{N}_1, j \in \mathcal{N}_0\}} \bar{x}_{ij} + \sum_{\{(j,i) | i \in \mathcal{N}_1, j \in \mathcal{N}_0\}} \bar{x}_{ji} \\
&\leq 0
\end{aligned}$$

which, given the observation that the surpluses of nodes considered in Step 1 are never driven below 0, contradicts the assumption that the nodes in $\mathcal{N}_1$ are chosen infinitely many times in Step 1. □

Lemma 3.2. *The sequence $\{x^k\}_{k=0}^\infty$ generated by Algorithm 1 converges to some $x^* \in \text{dom}(\sum_{(i,j) \in \mathcal{A}} f_{ij})$.*

Proof: We start by showing that the sequence $\{x^k\}_{k=0}^\infty$ is bounded. First note that the quantity $\sum_{i \in \mathcal{N}} |s_i^k|$ is a nonincreasing function of k. Assume that the flow on an arc $(i, j) \in \mathcal{A}$ is unbounded (i.e., $x_{ij}^{k_l} \to \infty$ where $\{k_l\}$ is an appropriately chosen index sequence). Hence, there is a cycle Y containing (i, j) such that the flows on the forward arcs of Y (denoted by Y^+) approach $+\infty$ while the flows on the backward arcs of Y (denoted by Y^-) approach $-\infty$. We assume that $\{k_l\}$ has been further thinned so that

$$x_{uw}^{k_l} \to +\infty, \quad (u, w) \in Y^+$$

$$x_{wu}^{k_l} \to -\infty, \quad (w, u) \in Y^-.$$

By summing the ϵ-CS condition

$$f_{uw}^-(x_{uw}^{k_l}) - \epsilon \leq \pi_u^{k_l} - \pi_w^{k_l} \leq f_{uw}^+(x_{uw}^{k_l}) + \epsilon$$

along Y we get

$$\sum_{(u,w) \in Y^+} f_{uw}^-(x_{uw}^{k_l}) - \sum_{(w,u) \in Y} f_{wu}^+(x_{wu}^{k_l}) \leq n\epsilon$$

which contradicts Assumption 1.2 since $x_{uw}^{k_l} \to +\infty$ implies $f_{uw}^{-}(x_{uw}^{k_l}) \to +\infty$, while $x_{wu}^{k_l} \to -\infty$ implies $f_{wu}^{+}(x_{wu}^{k_l}) \to -\infty$.

We now show that $\{x_{ij}^k\}_{k=0}^{\infty}$ converges for each $(i, j) \in \mathcal{A}_1$. Since f_{ij} is a proper closed strictly convex function, its conjugate f_{ij}^* is essentially smooth hence (using Assumption 1.2) (continuously) differentiable. We have [4, Lemma 5.1, p. 395]

$$x_{ij}^k = \text{grad}\, f_{ij}^*(\pi_i^k - \pi_j^k).$$

Lemma 3.1 shows that $\pi_i^k - \pi_j^k \to \pi_i^* - \pi_j^*$ hence $\{x_{ij}^k\}$ converges.

To see that $\{x_{ij}^k\}_{k=0}^{\infty}$ converges for $(i, j) \in \mathcal{A}_0$, note that for sufficiently large k (i, j) can be chosen only in Step 2 or only in Step 3 but not in both of them. Hence starting with some k, $\{x_{ij}^k\}$ is monotone, which together with its boundedness implies convergence.

Finally, $x_{ij}^* \in \text{dom}\, f_{ij}$ for each $(i, j) \in \mathcal{A}$ because (x^*, π^*) satisfy ϵ-CS implying that $f_{ij}^-(x_{ij}^*)$ and $f_{ij}^+(x_{ij}^*)$ cannot both be $+\infty$ or $-\infty$. □

The surplus of node $i \in \mathcal{N}$ for the flow x^k is denoted by

$$s_i^k := b_i - \sum_{j \in \delta_+(i)} x_{ij}^k + \sum_{j \in \delta_-(i)} x_{ji}^k. \tag{3.1}$$

The following assumption regarding the way nodes are chosen in Step 1 is needed in establishing convergence to a feasible point; its satisfaction may be guaranteed by periodically using cyclic priority for node selection in Step 1.

Assumption 3.1. *If $s_i^k > 0$ for infinitely many k then node i is chosen Step* 1 *infinitely often.*

Lemma 3.3. *The sequence $\{s^k\}_{k=0}^{\infty}$ generated by Algorithm* 1 *converges to* 0.

Proof: Let $i \in \mathcal{N}$ be a node that is chosen infinitely many times in Step 1 and let $\{k_l\}$ be the corresponding sequence of iteration numbers. Since the surplus at a node is nondecreasing between applications of Step 1 to the node, we have

$$\limsup_{k\to\infty} s_i^k = \limsup_{l\to\infty} s_i^{k_l} \le \limsup_{l\to\infty} \|x^{k_l} - x^{k_{l+1}}\|_1 = 0$$

(since the change in a node surplus doesn't exceed the sum of changes of arc flows). Hence $s_i^k \to 0$.

On the other hand, if $j \in \mathcal{N}$ is a node with $s_j^k \le 0$ for all sufficiently large k we have

$$s_j^k \ge \sum_{\{u \in \mathcal{N} | s_u^k < 0\}} s_u^k = - \sum_{\{u \in \mathcal{N} | s_u^k \ge 0\}} s_u^k \to 0$$

hence $s_j^k \to 0$. □

The preceding three lemmas constitute the proof of the first part of the main convergence theorem. The second part is concerned with the case when $\mathcal{A}_0 = \mathcal{A}$. In this case we define the *admissible network* w.r.t. some pair (x, π) to be $\mathcal{G}(\mathcal{N}, \hat{\mathcal{A}})$ where $(i, j) \in \hat{\mathcal{A}}$ iff

$$(i, j) \in \mathcal{A},\ f_{ij}^{+}(x_{ij}) - (\pi_i - \pi_j) \le -\frac{\epsilon}{2}$$

or

$$(j, i) \in \mathcal{A},\ f_{ji}^{-}(x_{ji}) - (\pi_j - \pi_i) \ge \frac{\epsilon}{2}.$$

Less formally, the admissible network contains an arc (i, j) if flow can be pushed from node i to node j according to the rules of the algorithm.

Assumption 3.2. *(x^0, π^0) are chosen so that the admissible network w.r.t. (x^0, π^0) is acyclic.*

The above assumption is trivially satisfied if (x^0, π^0) satisfy CS, for then the set of arcs of the admissible network is empty.

Theorem 3.4. *The sequence $\{(x^k, \pi^k)\}_{k=0}^{\infty}$ generated by Algorithm 1 converges to some (x^*, π^*) such that x^* is feasible for (1.1) and together with π^* satisfies ϵ-CS. Furthermore, if $\mathcal{A}_1 = \emptyset$ the algorithm terminates finitely with a primal-dual pair satisfying the same properties.*

Proof: The first part follows directly from Lemmas 3.1–3.3. For the second part we first show that if the admissible network is acyclic w.r.t. the initial choice of (x, π), it remains acyclic throughout the algorithm. Indeed, Steps 2 and 3 obviously cannot add new arcs to the admissible network. Step 4 increases π_i by more than $\frac{\epsilon}{2}$ and so may add some arcs originating at i to the admissible network. At the same time all arcs entering i leave the admissible network and so no cycle can be formed in it.

Now assume that the algorithm doesn't terminate. Since $\pi_i^k \to \pi_i^*$ and $\Delta > \frac{\epsilon}{2}$ in Step 4, we see that for sufficiently large k, $\pi_i^k = \pi_i^*$ for each $i \in \mathcal{N}$. Hence from some point on no new arcs are added to the admissible network. This together with its acyclicity guarantees that Algorithm 1 terminates finitely. □

The pair (x^*, π^*) computed by Algorithm 1 is nearly optimal for small ϵ. The following two lemmas (Corollary 3.1 and Proposition 3.6 from [2], respectively) give the precise statement. In them, $f(x)$ and $q(\pi)$ are used to denote the primal and dual objective values of (1.1):

$$f(x) := \sum_{(i,j)\in\mathcal{A}} f_{ij}(x_{ij}),$$

$$q(\pi) := \pi^T b - \sum_{(i,j)\in\mathcal{A}} f_{ij}^{*}(\pi_i - \pi_j).$$

Lemma 3.5. *Let $(x(\epsilon), \pi(\epsilon))$ satisfy ϵ-CS and let $x(\epsilon)$ satisfy the flow conservation constraints. Then $f(x(\epsilon)) - q(\pi(\epsilon)) \to 0$ as $\epsilon \to 0$.*

Lemma 3.6. *Let $(x(\epsilon), \pi(\epsilon))$ satisfy the assumptions of Lemma 3.5 and in addition assume that each f_{ij} is of the form (1.2) (i.e., finite capacity constraints). Then*

$$0 \leq f(x(\epsilon)) - q(\pi(\epsilon)) \leq \epsilon \sum_{(i,j)\in\mathcal{A}} (u_{ij} - l_{ij}).$$

In the case the f_{ij} satisfy for some $d_{ij} > 0$ the inverse Lipschitz condition

$$d_{ij}\left(x_{ij}^1 - x_{ij}^2\right) \leq f_{ij}^-\left(x_{ij}^1\right) - f_{ij}^+\left(x_{ij}^2\right) \quad \text{for } x_{ij}^1 > x_{ij}^2$$

(for example, when the f_{ij} are piecewise quadratic) we can derive a bound on the distance from the point computed by Algorithm 1 to the optimal solution set of (1.1).

Lemma 3.7. *Let (x^*, π^*) satisfy ϵ-CS and let x^* be feasible for (1.1). Let $\bar{x}$ be any optimal solution of (1.1). Then for every $(i, j) \in \mathcal{A}$*

$$d_{ij}|x_{ij}^* - \bar{x}_{ij}| \leq n\epsilon.$$

Proof: Fix $(i, j) \in \mathcal{A}$ and assume that $x_{ij}^* \neq \bar{x}_{ij}$. According to the Conformal Realization Theorem [1], $x^* - \bar{x}$ can be decomposed into a sum of conforming simple cycle flows. Let Y be any of the cycles containing (i, j). By summing the ϵ-CS condition (1.4) along Y (as in the proof of Lemma 3.2) we obtain

$$\sum_{(u,w)\in Y^+} f_{uw}^-(x_{uw}^*) - \sum_{(w,u)\in Y^-} f_{wu}^+(x_{wu}^*) \leq n\epsilon.$$

Similarly, by summing the CS condition (1.3) along Y we get

$$\sum_{(u,w)\in Y^+} f_{uw}^+(\bar{x}_{uw}) - \sum_{(w,u)\in Y^-} f_{wu}^-(\bar{x}_{wu}) \geq 0.$$

Now by subtracting the above two inequalities we get

$$\sum_{(u,w)\in Y^+} \{f_{uw}^-(x_{uw}^*) - f_{uw}^+(\bar{x}_{uw})\} - \sum_{(w,u)\in Y^-} \{f_{wu}^+(x_{wu}^*) - f_{wu}^-(\bar{x}_{wu})\} \leq n\epsilon. \tag{3.2}$$

For an arc $(u, w) \in Y^+$ we have $\bar{x}_{uw} < x_{uw}^*$ while for an arc $(w, u) \in Y^-$ we have $x_{wu}^* < \bar{x}_{wu}$ and so the conclusion of the Lemma follows immediately. □

4. Computational considerations

In this section, we discuss some issues related to the implementation of Algorithm 1 and some extensions to it.

One way of enforcing Assumption 3.1 in practice is to consider the nodes in Step 1 in some fixed order that includes all nodes in $\mathcal{N}$. A more economical solution is used in the linear code ϵ-RELAX [1]. There the nodes that have positive surplus are kept in a queue. Nodes are "selected" from the front of the queue; if during the course of an iteration the surplus of an adjacent node (not already in the queue) becomes positive, that node is added to the end of the queue.

Algorithm 1, as described, consists only of "up" iterations (i.e., a node with a positive surplus is selected and its price may be raised). It is possible to define in a symmetrical way a "down" iteration which starts by selecting a node with negative surplus and may decrease its price. Ref. [4] has an example of the ϵ-relaxation algorithm cycling if "up" and "down" iterations are mixed arbitrarily. If we assume, however, that either the number of "up" iterations or the number of "down" iterations started at any given node is finite, cycling cannot occur. Indeed, this assumption implies that from some point in the course of the algorithm on, either only "up" or only "down" iterations are executed at any given node and our convergence proof can be used with slight modifications. The computational results in Section 5 demonstrate that the performance of the algorithm for nonlinear problems is significantly improved by mixing both "up" and "down" iterations.

The above assumption can be computationally checked by using a device similar to the one used in the code ϵ-RELAX-N [1]. For a node i, let p_i^k be the number of "up" iterations executed at node i before iteration number k. Also let

$$S^k := \sum_{i \in \mathcal{N}} |s_i^k|.$$

We can now enforce the assumption by not allowing a "down" iteration for node i at iteration number k if

$$p_i^k > C_1 + C_2(S^0 - S^k)$$

where C_1 and C_2 are any two positive constants. The quantity $C_1 + C_2(S^0 - S^k)$ is nondecreasing and bounded above and is used as a measure of the progress made by the algorithm. As long as the algorithm makes sufficient progress, "up" and "down" iterations are allowed to be mixed. If, however, too many "up" iterations are executed at a given node without achieving progress, no more "down" iterations are allowed at this node.

Finally, we should note that the solution of problem (1.1) usually consists of more than one application of Algorithm 1. The algorithm is executed for some value of ϵ and terminated when the surplus of all nodes becomes relatively small. Then ϵ is reduced and the process is repeated. This is very similar to the way cost-scaling algorithms operate.

We now describe a specialization of Algorithm 1 for network flow problems with mixed linear and quadratic cost functions:

$$\begin{aligned} \text{minimize} \quad & \sum_{(i,j) \in \mathcal{A}} \bar{f}_{ij}(x_{ij}) \\ \text{subject to} \quad & \sum_{j \in \delta_+(i)} x_{ij} - \sum_{j \in \delta_-(i)} x_{ij} = b_i, \quad i \in \mathcal{N} \\ & 0 \le x_{ij} \le u_{ij} \end{aligned} \tag{4.1}$$

where $\bar{f}_{ij}(x_{ij})$ are convex continuously differentiable quadratic functions:

$$\bar{f}_{ij}(x_{ij}) = \frac{1}{2}d_{ij}[x_{ij}]^2 + c_{ij}x_{ij}$$

with $d_{ij} \geq 0$.

Again, let $\{\mathcal{A}_0, \mathcal{A}_1\}$ be a partition of $\mathcal{A}$ such that $\mathcal{A}_1$ contains only arcs (i, j) with $d_{ij} > 0$.

Algorithm 2.

Step 0 [Initialization] Start with (x, π) such that

$$x_{ij} = \begin{cases} 0 & \text{if } c_{ij} - (\pi_i - \pi_j) > \epsilon \\ u_{ij} & \text{if } d_{ij}u_{ij} + c_{ij} - (\pi_i - \pi_j) < -\epsilon. \end{cases}$$

For all the remaining arcs, the initial choice of x must satisfy the capacity constraints for the arcs with linear cost, and must satisfy

$$\max\left\{0, \min\left\{u_{ij}, -\frac{\epsilon}{d_{ij}} - \frac{c_{ij} - (\pi_i - \pi_j)}{d_{ij}}\right\}\right\} \leq x_{ij}$$
$$\leq \min\left\{u_{ij}, \max\left\{0, \frac{\epsilon}{d_{ij}} - \frac{c_{ij} - (\pi_i - \pi_j)}{d_{ij}}\right\}\right\}$$

for the remaining arcs in $\mathcal{A}_0$, as well as

$$x_{ij} = \max\left\{0, \min\left\{u_{ij}, -\frac{c_{ij} - (\pi_i - \pi_j)}{d_{ij}}\right\}\right\}$$

for the arcs in $\mathcal{A}_1$.

Step 1 [Choose node with positive surplus] Choose a node i such that

$$s_i := b_i - \sum_{j\in\delta_+(i)} x_{ij} + \sum_{j\in\delta_-(i)} x_{ji} > 0.$$

If no such node exists, STOP.

Step 2 [If possible, increase flow on an outgoing arc in $\mathcal{A}_0$] If there is no arc $(i, j) \in \mathcal{A}_0$ such that

$$r_{ij} := d_{ij}x_{ij} + c_{ij} - (\pi_i - \pi_j) \leq -\frac{\epsilon}{2}, \quad x_{ij} < u_{ij}$$

go to Step 3, else change the flow on arc (i, j):

$$x_{ij} \leftarrow x_{ij} + \min\left\{s_i, u_{ij} - x_{ij}, -\frac{r_{ij}}{d_{ij}}\right\}.$$

(The flows on the remaining arcs and the dual costs are not changed.) Update s_i and if its new value is 0 go to Step 1 else return to Step 2.

Step 3 [If possible, decrease flow on an incoming arc in $\mathcal{A}_0$] If there is no arc $(j, i) \in \mathcal{A}_0$ such that

$$r_{ji} := d_{ji}x_{ji} + c_{ji} - (\pi_j - \pi_i) \geq \frac{\epsilon}{2}, \quad x_{ji} > 0$$

go to Step 4, else change the flow on arc (j, i),

$$x_{ji} \leftarrow x_{ji} - \min\left\{s_i, x_{ji}, \frac{r_{ji}}{d_{ji}}\right\}.$$

(The flows on the remaining arcs and the dual costs are not changed.) Update s_i and if its new value is 0 go to Step 1 else return to Step 3.

Step 4 [Raise single node price and change flows on incident arcs in $\mathcal{A}_1$] Compute

$$\Delta_1 := \min_{\{j \in \delta_+(i) | (i,j) \in \mathcal{A}_0, x_{ij} < u_{ij}\}} \{c_{ij} - (\pi_i - \pi_j)\},$$

$$\Delta_2 := \min_{\{j \in \delta_-(i) | (j,i) \in \mathcal{A}_0, x_{ji} > 0\}} \{-(c_{ji} - (\pi_j - \pi_i))\}.$$

For each γ and for each arc $(i, j) \in \mathcal{A}_1$ let

$$x_{ij}(\gamma) = \max\left\{0, \min\left\{u_{ij}, -\frac{c_{ij} - (\pi_i + \gamma - \pi_j)}{d_{ij}}\right\}\right\}.$$

Similarly, for each arc $(j, i) \in \mathcal{A}_1$ let

$$x_{ji}(\gamma) = \max\left\{0, \min\left\{u_{ji}, -\frac{c_{ji} - (\pi_j - \pi_i - \gamma)}{d_{ji}}\right\}\right\}$$

and let Δ_3 be the value of γ such that

$$b_i - \sum_{(i,j) \in \mathcal{A}_1} x_{ij}(\gamma) - \sum_{(i,j) \in \mathcal{A}_0} x_{ij} + \sum_{(j,i) \in \mathcal{A}_1} x_{ji}(\gamma) + \sum_{(j,i) \in \mathcal{A}_0} x_{ji} = 0$$

(where $\Delta_3 = +\infty$ if a solution does not exist).
Define:

$$\Delta := \min\{\Delta_1 + \epsilon, \Delta_2 + \epsilon, \Delta_3\}.$$

Change the price on node i:

$$\pi_i \leftarrow \pi_i + \Delta$$

and adjust the flow on the incident arcs in $\mathcal{A}_1$ so that CS is satisfied. Update s_i and if its new value is 0 go to Step 1 else return to Step 2.

5. Numerical results

In this section we describe computational experience using Algorithm 2 to solve network flow test problems with mixed linear/quadratic and strictly quadratic cost functions. The code was written in C and run on a SPARCstation 20.

The quality of the computed primal-dual pair (x^*, π^*) can be measured in terms of the flow conservation constraints violation and relative accuracy of the computed objective value. For all runs reported below we had

$$\max_{i \in \mathcal{N}} \left| b_i - \sum_{j \in \delta_+(i)} x^*_{ij} + \sum_{j \in \delta_-(i)} x^*_{ji} \right| \leq 10^{-8}$$

which translates to relative feasibility violation given by

$$\max_{i \in \mathcal{N}} \left| b_i - \sum_{j \in \delta_+(i)} x^*_{ij} + \sum_{j \in \delta_-(i)} x^*_{ji} \right| \bigg/ \|b\|_\infty \leq 10^{-12}.$$

The computed objective value was accurate to nearly ten significant digits:

$$\left| \frac{f(x^*) - q(\pi^*)}{f(x^*)} \right| \leq 10^{-10}$$

We obtained our test problems by modifying the standard NETGEN [8] problems 1–10 and 16–25. The first ten problems are transportation problems while the rest are transshipment problems. The transshipment problems differ in number of transshipment sources and sinks, percentage of capacitated arcs and size of upper bounds. The last four problems in Table 1 were obtained by scaling the input data for the corresponding standard NETGEN problems by a factor of eight. By adding quadratic terms to the objective functions we generated the following three groups of test problems:

- "50% linear/50% quadratic." A quadratic term with coefficient 10.0 was added to the cost function of 50% of the arcs (randomly chosen). In the corresponding runs we chose $\mathcal{A}_0$ to be the set of linear arcs and $\mathcal{A}_1$ to be the set of quadratic arcs.
- "50% large quadratic/50% small quadratic." A quadratic term with coefficient 10.0 was added to half of the arc cost functions and a quadratic term with coefficient 0.001 was added to the other half. This corresponds to a strictly convex problem with an ill-conditioned Hessian. In the algorithm we chose $\mathcal{A}_1$ to be the set of arcs with large quadratic factors and $\mathcal{A}_0$ the set of arcs with small quadratic factors.
- "100% quadratic." A quadratic term with coefficient 10.0 was added to all arc cost functions. In the corresponding runs we chose $\mathcal{A}_0 = \emptyset$. $\mathcal{A}_1 = \mathcal{A}$. (With these choices the algorithm is practically equivalent to the relaxation method for strictly convex problems.)

Table 1 presents the sizes of the test problems and the execution times of Algorithm 2 in seconds for each of the three test groups. The results for the first and third groups

Table 1. Solution times for modified NETGEN problems using the ϵ-relaxation algorithm (figures in parentheses correspond to doing only "up iterations").

NETGEN number	Number of nodes	Number of arcs	50% linear 50% quadratic		50% large quadratic 50% small quadratic		100% quadratic	
1	200	1308	0.22	(0.99)	0.24	(0.83)	0.14	(4.72)
2	200	1511	0.23		0.25		0.14	
3	200	2000	0.39		0.34		0.15	
4	200	2200	0.29		0.30		0.17	
5	200	2900	0.44	(1.46)	0.40	(1.54)	0.19	(5.77)
6	300	3174	0.37		0.41		0.24	
7	300	4519	0.84		0.71		0.29	
8	300	5169	0.91		0.85		0.37	
9	300	6075	1.16		0.98		0.48	
10	300	6320	1.27		1.11		0.50	
16	400	1306	1.25	(1.77)	0.90	(1.67)	0.28	
17	400	2443	1.10		0.80		0.25	
18	400	1306	1.46		0.83		0.28	
19	400	2443	1.05		0.84		0.26	
20	400	1416	1.0		0.82		0.29	
21	400	2836	1.49	(3.0)	1.04	(2.17)	0.27	
22	400	1416	1.16		0.81		0.29	
23	400	2836	1.29		0.87		0.27	
24	400	1382	0.85		0.98		0.28	
25	400	2676	0.75		0.73		0.27	
6S	2400	25336	7.4		7.1		9.6	
10S	2400	50535	12.4		11.5		12.1	
24S	3200	11056	7.0		7.0		5.7	
25S	3200	21408	12.6		8.7		5.5	

are comparable (after taking into account the computer performance differential) to those reported in [2] for similarly modified NETGEN problems. However, the method of [2] had difficulties solving problems similar to "50% large quadratic/50% small quadratic," being more than five times slower for them compared to the corresponding "50% linear/50% quadratic" problems. Our method, on the other hand, solves problems in the second group in roughly the same time as corresponding problems in the first group. This shows that our algorithm is as efficient for ill-conditioned quadratic objectives as for mixed linear-quadratic objectives.

It is worth mentioning that mixing "up" and "down" iterations is very important for achieving good computational efficiency in practice. This is illustrate in Table 1 where numbers in parentheses correspond to runs where only "up" iterations were allowed (for some problems in the "100% quadratic" case executing only "up" iterations led to very large times not shown in the table). This should be contrasted to the observation that mixing "up" and "down" iterations does not lead to significant improvement in running time when solving linear min-cost problems using the ϵ-relaxation algorithm.

We also present the solution times for the smaller test problems using the piecewise linear approximation algorithm for separable convex network flow problems [7]. This algorithm constructs a sequence of problems having piecewise linear objective functions which converge to the original nonlinear objective and solves these problems using the network simplex method. Except for a few of the linear-quadratic problems, the times in Table 2 are larger than the ϵ-relaxation times. Some additional testing demonstrated that the piecewise-linear approach was competitive for problems in which most arcs had linear costs, but since such problem classes were not the main focus of this research, we do not present these results here.

Finally, our relaxation algorithm consistently outperformed by a significant factor the general nonlinear network optimization package LSNNO [13] on the test problems described above.

Table 2. Solution times for modified NETGEN problems using the piecewise-linear approximation method.

NETGEN number	Number of nodes	Number of arcs	50% linear 50% quadratic	50% large quadratic 50% small quadratic	100% quadratic
1	200	1308	0.74	1.69	2.17
2	200	1511	0.93	1.99	2.38
3	200	2000	1.18	2.84	3.36
4	200	2200	1.25	2.61	4.01
5	200	2900	1.46	3.49	5.26
6	300	3174	2.33	5.06	7.36
7	300	4519	2.65	8.31	10.3
8	300	5169	2.76	7.76	11.1
9	300	6075	3.65	9.06	14.3
10	300	6320	3.76	9.35	13.9
16	400	1306	1.31	8.26	5.44
17	400	2443	1.57	11.5	10.1
18	400	1306	0.97	6.27	4.04
19	400	2443	1.38	8.35	7.96
20	400	1416	1.86	17.2	7.36
21	400	2836	1.57	24.7	16.2
22	400	1416	1.49	7.57	5.14
23	400	2836	1.53	10.6	10.8
24	400	1382	1.98	11.0	6.18
25	400	2676	1.51	9.95	15.8

6. Parallel versions of the algorithm

An iteration of Algorithm 1 only involves a node in the given network and its adjacent arcs. Consequently, an obvious approach to parallelization would be to assign a node (or a group of nodes) to each processor in a parallel computing environment. Iterations at several different nodes would then be able to proceed simultaneously if only iterations at independent (i.e., no two are joined by an arc) nodes were allowed to proceed concurrently. A potential problem with this approach might be limited parallel efficiency due to insufficient number of independent nodes with nonzero surplus available at any given time.

Another approach (described below) removes the node independence requirement by allowing iterations at two adjacent nodes to be executed in parallel. In this case, the two nodes may set the flow on the arc joining them to two different values. As the algorithm progresses, the difference between the two nodes' estimates of the flow approaches zero. This algorithm may be viewed as a generalization of the totally asynchronous algorithm for linear network flow problems of [4], to nonlinear cost functions.

At time step t any node i has the following data:

- the price $\pi_i(t)$
- for any node $j \in \delta_+(i) \cup \delta_-(i)$, the price for node j at some previous time step (denoted by $\pi_j(i, t)$)
- for any arc $(i, j) \in \delta_+(i)$, its estimate of the flow $x_{ij}(i, t)$ and for any arc $(j, i) \in \delta_-(i)$, its estimate of the flow $x_{ji}(i, t)$

The surplus of node i at time t is defined to be

$$s_i(t) := b_i - \sum_{j\in\delta_+(i)} x_{ij}(i, t) + \sum_{j\in\delta_-(i)} x_{ji}(i, t).$$

At any time step t node i does one of the following:

- If $s_i(t) > 0$, executes an "up iteration" (with $\epsilon' = 0$) and updates the values of $\pi_i(t)$, $x_{ij}(i, t)$ for $(i, j) \in \delta_+(i)$ and $x_{ji}(i, t)$ for $(j, i) \in \delta_-(i)$.
- Receives messages from adjacent nodes j containing values $\pi_j(t')$ and $x_{ij}(j, t')$ for arcs $(i, j) \in \delta_+(i)$ or $x_{ji}(j, t')$ for arcs $(j, i) \in \delta_-(i)$ where $t' < t$. It uses these values to update $\pi_j(i, t)$ and $x_{ij}(i, t)$ for $(i, j) \in \delta_+(i)(x_{ji}(i, t)$ for $(j, i) \in \delta_-(i))$ according to the update rule outlined below.

Update rule: If $\pi_j(t') < \pi_j(i, t)$, do nothing. Else set $\pi_j(i, t) \leftarrow \pi_j(t')$. For an arc $(i, j) \in \delta_+(i)$, denote the largest flow that satisfies CS together with $\pi_i(t)$ and $\pi_j(i, t)$ by $y_{ij}(i, t)$. Now if $x_{ij}(i, t') < x_{ij}(i, t)$, set $x_{ij}(i, t)$ to the projection of $y_{ij}(i, t)$ on $[x_{ij}(j, t'), x_{ij}(i, t)]$. Similarly, for an arc $(j, i) \in \delta_-(i)$, denote the largest flow that satisfies CS together with $\pi_i(t)$ and $\pi_j(i, t)$ by $y_{ji}(i, t)$. If $x_{ji}(j, t') > x_{ji}(i, t)$, set $x_{ji}(i, t)$ to the projection of $y_{ji}(i, t)$ on $[x_{ji}(i, t), x_{ji}(j, t')]$.

The initial values of (x, π) should satisfy

$$\pi_i(0) \geq \pi_i(j, 0) \quad \forall j \neq i,$$
$$x_{ij}(i, 0) \geq x_{ij}(j, 0) \quad \forall (i, j)$$

together with the usual CS/ϵ-CS conditions.

The structure of the "up iteration", the update rule and the initial conditions imply the following properties of the iterates:

1. The sequence $\{\pi_i(t)\}_{t=0}^{\infty}$ is nondecreasing and

$$\pi_i(t) \geq \pi_i(j, t') \quad \forall t' \leq t.$$

2. ϵ-CS is satisfied "locally" at each node i:

$$f_{ij}^-(x_{ij}(i, t)) - \epsilon \leq \pi_i(t) - \pi_j(i, t) \leq f_{ij}^+(x_{ij}(i, t)) + \epsilon, \quad (i, j) \in \delta_+(i),$$
$$f_{ji}^-(x_{ji}(i, t)) - \epsilon \leq \pi_j(i, t) - \pi_i(t) \leq f_{ji}^+(x_{ji}(i, t)) + \epsilon, \quad (j, i) \in \delta_-(i).$$

3. The start node estimate of the flow on an arc dominates the end node estimate:

$$x_{ij}(i, t) \geq x_{ij}(j, t) \quad \forall t.$$

Proof: Let t be the first time step when the above fails. This can only happen if at time t node i received a message from j causing $x_{ij}(i, t)$ to decrease or j received a message from i causing $x_{ij}(j, t)$ to increase. Without loss of generality assume the former. Let t' be the time when the message in question was sent by j. Also let $s \leq t$ be the last time step when j received a message from i that increased $x_{ij}(j, t)$ and let s' be the time when this message was sent. We necessarily have $s > t'$ for otherwise

$$x_{ij}(i, t) \geq x_{ij}(j, t') \geq x_{ij}(j, t).$$

Now using $x_{ij}(j, t) > x_{ij}(i, t) \geq y_{ij}(i, t)$ and the definition of $y_{ij}(i, t)$ we get

$$f_{ij}^-(x_{ij}(j, t)) > \pi_i(t) - \pi_j(t') \geq \pi_i(s') - \pi_j(s)$$
$$\geq f_{ij}^-(y_{ij}(j, s)) \geq f_{ij}^-(x_{ij}(j, s)) \geq f_{ij}^-(x_{ij}(j, t)),$$

a contradiction.

4. There exists a node that never executes an "up iteration." This follows from the fact that once nonnegative, the surplus of a node remains nonnegative and from the following inequality

$$\sum_i s_i(t) = \sum_{(i,j)} \{x_{ij}(j, t) - x_{ij}(i, t)\} \leq 0 \quad \forall t.$$

In order to establish convergence of the algorithm, the usual assumption is needed that information in the system never becomes "too old." Using the above Property 4, it is easy to prove in a manner similar to Lemma 3.1 that the node prices converge.

We now show that the flow estimates of both end nodes of each arc converge to this same value:

$$x_{ij}(i,t) \to x_{ij}^*, \quad x_{ij}(j,t) \to x_{ij}^*.$$

Boundedness of both sequences can be established in the same way as for the sequential algorithm. Denote the largest flow value that satisfies CS together with π_i^* and π_j^* (where π_i^* and π_j^* denote the limit values for the prices of nodes i and j, respectively) by y_{ij}^*. Using the fact that $y_{ij}(i,t)$ is the largest element of $\partial f_{ij}(\pi_i(t) - \pi_j(i,t))$ and the continuity properties of the subdifferential mapping we get

$$y_{ij}(i,t) \to y_{ij}^*, \quad y_{ij}(j,t) \to y_{ij}^*.$$

Three cases need to be considered:

1. Both nodes i and j execute an infinite number of "up iterations" that affect their estimates of the flow on (i, j). From the logic of the "up iteration" and the update rule we get

   $$\limsup_t x_{ij}(i,t) \le y_{ij}^* \le \liminf_t x_{ij}(j,t)$$

   which together with Property 3 establishes the result.
2. Node i changes $x_{ij}(i,t)$ through "up iterations" only a finite number of times while node j changes $x_{ij}(j,t)$ through "up iterations" infinitely many times. Then $\{x_{ij}(i,t)\}$ is nonincreasing for sufficiently large t and hence converges to some x_{ij}^*. We have

   $$y_{ij}^* \le \liminf_t x_{ij}(j,t) \le x_{ij}^*$$

   so by the update rule

   $$\liminf_t x_{ij}(j,t) = x_{ij}^*.$$

   Hence $\{x_{ij}(j,t)\}$ also converges to x_{ij}^*.
3. Both nodes i and j change their respective estimates for the flow on (i, j) only finitely many times through "up iterations." Hence $\{x_{ij}(i,t)\}$ is nonincreasing while $\{x_{ij}(j,t)\}$ is nondecreasing for sufficiently large t. By the update rule they can only converge to the same value (projections of a converging sequence on a sequence of nested segments).

Finally, feasibility of the common limit points for the original problem can be established in the same way as for the serial algorithm.

7. Conclusions

We have developed a partitioned ϵ-relaxation method that is significantly more efficient on ill-conditioned networks than existing methods for nonlinear networks. Computational results with the procedure demonstrate that it solves networks with several thousand nonlinear arcs in one second or less. This method is also extended to an asynchronous parallel version that is amenable to parallel computation.

References

1. D. Bertsekas, Linear Network Optimization: Algorithms and Codes, The MIT Press: Cambridge, MA, 1991.
2. D.P. Bertsekas, P.A. Hosein, and P. Tseng, "Relaxation methods for network flow problems with convex arc costs," SIAM Journal on Control and Optimization, vol. 25, pp. 1219–1243, 1987.
3. D. Bertsekas, L. Polymenakos, and P. Tseng, "An ϵ-relaxation method for separable convex cost network flow problems," SIAM Journal on Optimization, vol. 7, pp. 953–970, 1997.
4. D. Bertsekas and J. Tsitsiklis, Parallel and Distributed Computation, Prentice Hall: Englewood Cliffs, NJ, 1989.
5. M. Collins, L. Cooper, R. Helgason, J. Kennington, and L. LeBlanc, "Solving the pipe network analysis problem using optimization techniques," Management Science, vol. 24, pp. 747–760, 1978.
6. T. Hu, "Minimum cost flows in convex cost networks," Naval Research Logistics Quarterly, vol. 13, pp. 1–9, 1966.
7. P. Kamesam and R. Meyer, "Multipoint methods for separable nonlinear networks," Mathematical Programming Study, vol. 22, pp. 185–205, 1984.
8. D. Klingman, A. Napier, and J. Stutz, "NETGEN—A program for generation of large-scale (un)capacitated assignment, transportation and minimum cost network problems," Management Science, vol. 20, pp. 814–822, 1974.
9. T. Magnanti, "Models and algorithms for predicting urban traffic equilibria," in Transportation Planning Models, M. Florian (Ed.), North-Holland: Amsterdam, 1984, pp. 153–186.
10. C. Monma and M. Segal, "A primal algorithm for finding minimum-cost flows in capacitated networks with applications," Bell System Technical Journal, vol. 61, pp. 449–468, 1982.
11. R. Rockafellar, Convex Analysis, Princeton University Press: Princeton, NJ, 1970.
12. R. Rockafellar, Network Flows and Monotropic Optimization, John Wiley: New York, 1984.
13. P. Toint and D. Tuyttens, "LSNNO: A Fortran subroutine for solving large-scale nonlinear network optimization problems," ACM Transactions on Mathematical Software, vol. 18, pp. 308–328, 1992.

Computational Optimization and Applications, 12, 127–155 (1999)

On a Primal-Dual Analytic Center Cutting Plane Method for Variational Inequalities*

M. DENAULT**
J.-L. GOFFIN† jlg@crt.umontreal.ca
GERAD, McGill University, Montréal
Québec, Canada

Received October 8, 1997; Accepted August 11, 1998

Abstract. We present an algorithm for variational inequalities $VI(\mathcal{F}, Y)$ that uses a primal-dual version of the Analytic Center Cutting Plane Method. The point-to-set mapping $\mathcal{F}$ is assumed to be monotone, or pseudomonotone. Each computation of a new analytic center requires at most four Newton iterations, in theory, and in practice one or sometimes two. Linear equalities that may be included in the definition of the set Y are taken explicitly into account.

We report numerical experiments on several well-known variational inequality problems as well as on one where the functional results from the solution of large subproblems. The method is robust and competitive with algorithms which use the same information as this one.

Keywords: variational inequalities; analytic center, cutting plane method, monotone mappings, interior points methods, Newton's method, primal-dual.

Dedication To Olvi Mangasarian, who always was an inspiration, as a researcher and as a human being. His now classic black and blue book on nonlinear programming was the cleanest book available on this topic when I (JLG) was a student. Its high and uncompromising standards had much to do with my training in this field.

Olvi's smile and personality have made many a meeting a more enjoyable experience than expected.

1. Introduction

The finite dimensional variational inequality problem (VI) is a very significant extension to the problem of optimizing a nonlinear function, as it allows for the modelling of systems involving multiple agents, each pursuing his own objective, while sharing joint resources. The computation of economic equilibria, Nash equilibria and discretized free boundary problems is often performed using a formulation as variational inequalities.

The variational inequality problem is a generalization of the linear and nonlinear complementarity problems and of the mixed nonlinear complementarity problem — to which it can often be reduced, but at the cost of an increase in the number of variables.

Excellent references on variational inequalities are the monograph by Harker [16], the book by Nagurney [35], the review article of Harker and Pang [17], the short course of Marcotte [31], and the proceedings book of Ferris and Pang [9].

* Corresponding author: J.-L. Goffin, GERAD, École des Hautes Études Commerciales, 3000 chemin de la Côte-Sainte-Catherine, Montréal (Québec), Canada, H3T 2A7.
** Research supported in part by McGill University Fellowships
† Research supported by NSERC grant OPG0004152 and by the FCAR

The development of algorithms for variational inequalities is an active and challenging research area. The variational inequality problem can be formulated as finding the zero of a generalized equation, as a fixed point problem or as a mixed nonlinear complementarity problem; these reformulations usually require additional assumptions.

The projection or fixed point methods may be viewed as extensions to VIs of subgradient optimization (Konnov [23]); more sophisticated versions include the extragradient method of Korpolevitch [24] and its variant by Solodov and Tseng [40]. Another class of approaches, pioneered by Mangasarian [29], in the case of nonlinear complementarity problems, uses a reformulation as a system of nonsmooth equations, which can also be viewed as a nonsmooth optimization problem; analogous smooth or nonsmooth reformulations can be found in the articles of Facchinei, Fischer, Kanzow [8] and Xiao, Harker [45, 46]. Optimization formulations involving the primal gap function, to which a proximal term is added, can be found in Fukushima [10], Kanzow, Qi [20] and Sun, Fukushima, Qi [42]. Ralph and Wright [38] describe an interior–point algorithm for VIs in a mixed nonlinear complementarity form.

Most of these algorithms need to make assumptions about the variational inequality, such as

1. the VI map is single-valued
2. it is differentiable, and the Jacobian can be used
3. the feasible set is given by a set of differentiable constraints.

Another approach, which extends nondifferentiable optimization to VIs, is to use a *cutting plane* description of the VI. This approach uses the fact that the VI functional separates the current point from the optimal set. Classical methods such as Kelley's cutting plane method, or by duality the Dantzig–Wolfe column generation, could be used. Lüthi [26] uses the ellipsoid method in this context, and Magnanti and Perakis [27] extend this approach to other centering methods. Lemaréchal, Nemirovskii, Nesterov [25] study the complexity of a level bundle method. The Analytic Center Cutting Plane (ACCPM), introduced in [11], uses the centering concept made familiar by interior point methods, i.e. *the analytic center*, in the context of column generation or cutting planes, and it has been shown, in practice, to be more effective than other centering schemes. The view of the VI as a convex feasibility problem was used by Goffin, Marcotte, Zhu [13] to describe and analyze a conceptual extension of ACCPM to the VI problem, using the results of [12]. A proximal analytic center cutting plane method, analyzed very nicely in a conic space, was introduced by Nesterov and Vial [37], and its complexity studied in the VI case.

The algorithm described here is a fully implementable version of [13], and we describe in detail a very effective method to compute the next analytic center from the previous one, using a primal-dual infeasible Newton method. The choice of an updating direction that involves a centering component leads, in practice, to a method that often needs no additional recentering Newton steps.

The algorithm applies to VIs defined by a monotone or pseudomonotone multivalued functional, possibly requiring the computation of subproblems, over a convex set Y that could be defined by a separation oracle, and linear inequalities or equalities. These features are useful if the Jacobian is expensive or impossible to evaluate, or if it does not exist.

The explicit accounting for linear equalities is useful for example in the case of network constraints.

As cutting plane methods do not necessarily generate a convergent sequence in the case of VIs, we introduce an auxiliary sequence of points, constructed by an averaging process, involving the primal variables computed by ACCPM, and which displays better convergence properties than the sequence of centers. Some intuitive justification can be found in Nesterov and Vial [37].

The article is organized as follows. We review in Section 2 some concepts of variational inequalities and analytic centers and we describe in Section 3 the algorithm in detail. In Section 4 we present some technical points concerning our MATLAB implementation of the method, and conclude in Section 5 with several numerical examples that range from typical economical equilibria to a pollution permits market model and to financial options pricing. The method is robust and competitive with algorithms which use the same information (mapping evaluations, but no Jacobian evaluations).

We use the following notation: $2^{\mathbf{R}^m}$ is the set of all subsets of $\mathbf{R}^m$; given a vector x, X is the diagonal matrix with x along the diagonal; e is a vector of ones whose dimension can be inferred from the context.

2. Basic Concepts

In this section we give the definitions and present the basic concepts of variational inequalities and analytic centers.

2.1. Variational Inequalities

Definition 1 *Let $\mathcal{F}$ be a point-to-set map from $\boldsymbol{R}^m$ into $2^{\boldsymbol{R}^m}$. Let Y be a nonempty convex subset of $\boldsymbol{R}^m$. Then a solution to the* variational inequality problem, *denoted as $VI(\mathcal{F}, Y)$, is a point $y^* \in Y$ such that*

$$F(y^*)^t(y - y^*) \geq 0 \text{ for some } F(y^*) \in \mathcal{F}(y^*) \text{ and } \forall y \in Y.$$

Geometrically, a solution y^* is a point of Y such that there exists a $F(y^*) \in \mathcal{F}(y^*)$ which makes an acute angle with all feasible directions, i.e. $-F(y^*)$ belongs to the *normal cone* to Y at y^*, denoted as $N_Y(y^*)$:

$$-F(y^*) \in N_Y(y^*) := \{z : (y - y^*)^t z \leq 0 \; \forall y \in Y\};$$

this can be written as:

$$0 \in \mathcal{F}(y^*) + N_Y(y^*),$$

which is the formulation as a generalized equation.

Definition 2 *A point-to-set mapping $\mathcal{F} : \boldsymbol{R}^m \to 2^{\boldsymbol{R}^m}$* is monotone *over a set Y if*

$$(F(y_1) - F(y_2))^t(y_1 - y_2) \geq 0$$
$$\forall F(y_1) \in \mathcal{F}(y_1), \forall F(y_2) \in \mathcal{F}(y_2), \forall y_1, y_2 \in Y.$$

It is maximal monotone *if the graph of $\mathcal{F}$, defined as*

$$G(\mathcal{F}) = \{(y,t) : y \in \operatorname{dom} Y, t \in \mathcal{F}(y)\}$$

is maximal, i.e. for any monotone map $\mathcal{F}'$, $G(\mathcal{F}) \subset G(\mathcal{F}')$ implies $G(\mathcal{F}) = G(\mathcal{F}')$.

A common example of a maximal monotone mapping is the subdifferential of a convex function.

Definition 3 *A point-to-set mapping $\mathcal{F} : \boldsymbol{R}^m \to 2^{\boldsymbol{R}^m}$ is* pseudomonotone *over a set Y if*

$$\begin{array}{l} F(y_2)^t(y_1 - y_2) \geq 0 \Rightarrow F(y_1)^t(y_1 - y_2) \geq 0 \\ \forall\, F(y_1) \in \mathcal{F}(y_1),\ \forall F(y_2) \in \mathcal{F}(y_2),\ \forall y_1, y_2 \in Y \end{array}$$

Definition 4 *Let $\mathcal{F}$ be a point-to-set map from $\boldsymbol{R}^m$ into $2^{\boldsymbol{R}^m}$. Let Y be a nonempty convex subset of $\boldsymbol{R}^m$. Then a* weak solution *to the variational inequality problem is a point $y^* \in Y$ such that*

$$F(y)^t(y - y^*) \geq 0 \ \ \forall y \in Y \text{ and } \forall F(y) \in \mathcal{F}(y).$$

The following theorem, from [36], relates weak solutions to solutions in the monotone case.

Theorem 1 *Let Y be a nonempty, closed, convex subset of $\boldsymbol{R}^m$, with nonempty interior, and let $\mathcal{F}$ be a monotone mapping with domain $dom(\mathcal{F})$, $int(Y) \subset dom(\mathcal{F}) \subset Y$, then*

1. *Every solution of the variational inequality is also a weak solution;*
2. *Conversely, provided either:*
   - *$Y \subset dom(\mathcal{F})$ and $\mathcal{F}$ is single-valued continuous, or*
   - *$\mathcal{F}$ is maximal monotone,*

   every weak solution solves the variational inequality.

In the case of pseudomonotone mappings, the following theorem was shown by Crouzeix [5] and Konnov [23]:

Theorem 2 *Let Y be a nonempty closed convex subset of $\boldsymbol{R}^m$ and let $\mathcal{F}$ be a pseudomonotone mapping from Y into $2^{\boldsymbol{R}^m}$, with $\mathcal{F}(y)$ assumed to be a nonempty convex set for all $y \in Y$, and $\mathcal{F}$ assumed to be upper semi–continuous on Y.*
Then every weak solution solves the variational inequality and conversely.

The key observation is that the solution set Y^* can be described as the intersection of an infinite number of half-spaces:

$$Y^* = \cap_{y \in Y} \cap_{F(y) \in \mathcal{F}(y)} \{y^* \in Y \mid F(y)^t(y - y^*) \geq 0\}$$

This formulation is the basis of the *cutting-plane* approach to variational inequalities.

It is thus possible to solve the $VI(\mathcal{F}, Y)$ as a convex feasibility problem, where $F(y)$ is the vector that separates y from Y^*. A generic algorithm to find a point $y^* \in Y^*$ would be as follows:

Step 0: Set $k = 0$, $Y_0 = Y$
Step 1: Choose a point $y_k \in Y_k$
Step 2: Select a vector $F(y_k) \in \mathcal{F}(y_k)$
Set $Y_{k+1} := Y_k \cap \{y : F(y_k)^t y_k \geq F(y_k)^t y\}$
Step 3: Check if $y_k \in Y^*$; if yes then stop.
Set $k := k + 1$
Return to step 1

Note that the set Y could be defined implicitly by a separation oracle, such that whenever $y_k \notin Y$, a *feasibility cut* separating y_k from Y is returned and added to the description of Y_k.

The selection of an appropriate stopping criterion is still an issue of some debate. We have used the primal gap function:

Definition 5 *The* primal gap function g *associated with* $VI(\mathcal{F}, Y)$ *is defined as*

$$g(y) = \min_{F(y) \in \mathcal{F}(y)} \min_{z \in Y} F(y)^t (z - y).$$

In general $g(y) \leq 0$ and in particular y^* is a solution of $VI(\mathcal{F}, Y)$ if and only if $g(y^*) = 0$. If furthermore Y is polyhedral and $\mathcal{F}$ single–valued, the gap can be evaluated by solving a linear optimization problem. An approximate solution can now be defined, in terms of the gap function:

Definition 6 *For the variational inequality VI(F,Y), the point* y^* *is an* ϵ–approximate solution *if* $y^* \in Y$ *and*

$$g(y^*) \geq -\epsilon.$$

If $\mathcal{F}$ is multivalued then a reasonable stopping criterion is:

$$\min_{z \in Y} F(y^*)^t (z - y^*) \geq -\epsilon,$$

for the $F(y^*) \in \mathcal{F}(y^*)$ selected in step 2 above.

The following assumption is common to most cutting plane methods, and will be made for the remainder of this paper:

Assumption 1 *The set Y is assumed to be convex and compact.*

2.2. *Analytic Centers*

The analytic center of a polyhedron defined by a set of linear inequalities was introduced by G. Sonnevend [41]. An explicit treatment of linear equalities in Y is introduced here, to avoid the traditional approach of rewriting unrestricted variables as the difference of two positive variables. Unrestricted variables arise through duality when equality contraints are present, and create instabilities in centering methods.

Consider the set

$$Y = \{\, y \mid A^t y \leq c,\ By = d \,\}$$

and the associated *dual potential function*

$$\varphi_D(y) = -\sum_{i=1}^{n} \ln(c_i - A_i^t y) = -\sum_{i=1}^{n} \ln(s_i)$$

where n is the number of components of c and the number of rows of A^t, while $s = c - A^t y$ are the dual slacks. In addition to the boundedness of Y, the following assumption is required for analytic centers to be well defined.

Assumption 2 *The set Y satisfies a Slater condition, that is:*

$$\{\, y \mid A^t y < c,\ By = d \,\}$$

is not empty.

Definition 7 *The* analytic center y^c *of Y is defined as the point minimizing the dual potential function over the set $\bar{Y} := \{\, y \mid A^t y < c,\ By = d \,\}$:*

$$y^c = \arg\min_{y \in \bar{Y}} \varphi_D(y).$$

It is well known that, under the above assumptions, the dual potential function is strictly convex on $\{\, y \mid A^t y < c,\ By = d \,\}$, and thus the center y^c is unique. The necessary and sufficient first-order optimality conditions to:

$$\begin{array}{ll} \text{minimize}_{y,s} & -\sum_{i=1}^{n} \ln s_i \\ \text{subject to} & A^t y + s = c \\ & By = d \\ & s > 0 \end{array}$$

are thus:

$$\begin{aligned} Ax + B^t \mu &= 0 & (1) \\ A^t y + s &= c & (2) \\ By &= d & (3) \\ Xs &= e & (4) \\ x, s &> 0 & (5) \end{aligned}$$

where x and μ are the variables associated with, respectively, the inequality and equality constraints. Equations (1) are referred to as the primal feasibility conditions, (2) and (3) are the dual feasibility conditions, and (4) the centrality conditions.

These optimality conditions are also those that minimize:

1. the primal potential function $\phi_P(x) = n\ln(c^t x + d^t\mu) - \sum_{i=1}^n \ln x_i$
2. the primal–dual potential fuction $\phi_{PD}(x,s) = n\ln x^t s - \sum_{i=1}^n \ln x_i s_i$; the minimum value of the primal- -dual potential function at the analytic center is the known value $n\ln n$. Note that for primal and dual feasible solutions $c^t x + d^t\mu = x^t s$ and thus $\phi_{PD} = \phi_D + \phi_P$.

An approximate analytic center is usually defined as a point (y, x, s, μ) which satisfies (1–3) and (5), and $\|e - Xs\| \leq \eta < 1$; the parameter η is thus used to control the quality of the approximation. This proximity measure could be replaced by one involving the primal–dual potential function.

3. An Analytic Center Cutting Plane Algorithm

We now describe our algorithm. We make the assumption of a *polyhedral* set Y; as mentioned earlier, a general convex Y can be treated by outer approximation (feasibility cuts).

Consider the problem VI($\mathcal{F}$,Y):

$$\text{Find } y^* \text{ and } F(y^*) \in \mathcal{F}(y^*) \text{ such that } F(y^*)^t(y - y^*) \geq 0 \; \forall y \in Y$$
$$\text{where } Y = \{ y \mid A^t y \leq c,\; By = d \}$$

where $\mathcal{F}$ is a pseudomonotone or monotone mapping, A is a $m \times n$ matrix and B is a $p \times m$ matrix with $p < m$. We assume that Y is bounded, possibly after the addition of a box $l \leq y \leq u$.

Then the following algorithm can be used to find an ϵ–approximate solution of VI($\mathcal{F}$,Y):

Step 0: **Initialization**
Set $k = 0$, $Y_0 = Y$

Step 1: **Analytic Center**
Find an approximate analytic center y_k of Y_k.

Step 2: **New Cut**
Compute the cut $F(y_k)^t y_k \geq F(y_k)^t y$
Set $Y_{k+1} := Y_k \cap \{y : F(y_k)^t y_k \geq F(y_k)^t y\}$

Step 3: **Termination Criterion**
Compute the primal gap $g(y_k)$
If $g(y_k) \geq -\epsilon$ then stop
Else $k := k+1$
Return to step 1

Step 1, computing an analytic center, for $k \geq 1$, is done with a primal-dual Newton method which uses the last center as a hot start.

Step 2 consists basically of a function evaluation; the relative time spent on this step varies with the application, as will become clear in the section on numerical experiments.

Although the gap computation in Step 3 involves solving a linear program, this can usually be done very quickly, as the size of most VIs is small relative to the size of LPs that can be solved efficiently. Furthermore, it is not crucial to evaluate the gap at every iteration. If $\mathcal{F}(y_k)$ is multivalued, and thus in practice not explicitly available, the gap is replaced by $\min_{y \in Y}\{F(y_k)^t(y - y_k)\}$.

The rest of this section is devoted to a method for finding the approximate analytic center y_{k+1} of Y_{k+1}, starting from y_k. The efficiency of this step is crucial to the whole algorithm.

3.1. *An Infeasible Primal-Dual Newton Step*

After k iterations the set of localization is:

$$Y_k = \{\, y \mid A_k^t y \le c_k,\ By = d \,\},$$

where $A_k^t y \le c_k$ includes both the n initial inequality constraints and the k generated inequalities (Step 2 of the algorithm). Let also $n_k = n + k$.

Assumption 3 *We assume that the original matrices A^t and B have maximum rank.*

Thus the current matrix A_k^t has also maximum rank.

At iteration k, the approximate analytic center $y_k \in \mathbf{R}^m$ of Y_k is a solution, together with $0 < x_k \in \mathbf{R}^{n_k}$, $\mu_k \in \mathbf{R}^p$ and $0 < s_k \in \mathbf{R}^{n_k}$, of the system of equations:

$$\begin{aligned} A_k x + B^t \mu &= 0 \\ A_k^t y + s &= c_k \\ By &= d \end{aligned}$$

while

$$\delta := e - Xs \quad \text{with} \quad \|\delta\| \le \eta < 1$$

The center y_{k+1} of Y_{k+1}, the next set of localization, solves:

$$\begin{aligned} A_k x + a\xi + B^t \mu &= 0 && (6) \\ A_k^t y + s &= c_k && (7) \\ a^t y + \sigma &= a^t y_k && (8) \\ By &= d && (9) \\ Xs &= e && (10) \\ \xi\sigma &= 1 && (11) \\ x, s, \xi, \sigma &> 0 && (12) \end{aligned}$$

where $a := F(y_k)$. Here ξ and σ are "new" unidimensional variables corresponding to the new constraint (8), and $n_{k+1} = n_k + 1 = n + k + 1$.

The system (6)–(12) is solved iteratively with Newton's method; the initial point we use is $y = y_k$, $x = x_k$, $s = s_k$ while the new variables σ and ξ are given positive values that must be carefully chosen to achieve the recovery of primal and dual feasibility and make the centering as efficient as possible. For values of $\xi = 0$ and $\sigma = 0$ this initial point is feasible in the sense of (6)–(9), although it will not be feasible with respect to strict positivity. For positive initial values of ξ and σ, which are required for interior point methods, the solution is infeasible to both (6) and (8). A full Newton step is needed to recover the feasibility conditions (6), (7) and (8).

The choice of a primal–dual method is motivated by the observation that it appears to be the most efficient interior point method available.

To improve the clarity of the notation, we now drop the index k, and replace it by a bar, so that the current center of $\bar{Y} := Y_k$ is denoted as $(\bar{y}, \bar{s} > 0, \bar{\mu}, \bar{x} > 0)$, and $\bar{A} := A_k$ and $\bar{c} := c_k$. The strictly positive values chosen for the new variables are denoted as $\bar{\xi}$ and $\bar{\sigma}$.

The Newton steps $dy, dx, ds, d\xi$ and $d\sigma$ from the initial point are given by the system:

$$\bar{A}dx + ad\xi + B^t d\mu = -\bar{A}\bar{x} - a\bar{\xi} - B^t\bar{\mu} = -a\bar{\xi} \tag{13}$$

$$\bar{A}^t dy + ds = -\bar{A}^t\bar{y} - \bar{s} + \bar{c} = 0 \tag{14}$$

$$a^t dy + d\sigma = -a^t\bar{y} - \bar{\sigma} + a^t\bar{y} = -\bar{\sigma} \tag{15}$$

$$Bdy = -B\bar{y} + d = 0 \tag{16}$$

$$\bar{S}dx + \bar{X}ds = e - \bar{X}\bar{s} =: \delta \tag{17}$$

$$\bar{\sigma}d\xi + \bar{\xi}d\sigma = 1 - \bar{\xi}\bar{\sigma} \tag{18}$$

From (14) and (17),

$$\begin{aligned} ds &= -\bar{A}^t dy \\ dx &= \bar{S}^{-1}(\delta - \bar{X}ds) \\ &= \bar{S}^{-1}(\delta + \bar{X}\bar{A}^t dy) \end{aligned}$$

Using (13) and the above we have

$$\bar{A}dx = \bar{A}\bar{S}^{-1}(\delta + \bar{X}\bar{A}^t dy) = -B^t d\mu - a(\bar{\xi} + d\xi)$$

from which, defining $\Delta := \bar{A}\bar{S}^{-1}\bar{X}\bar{A}^t$, Dikin's matrix, and $\xi_N := \bar{\xi} + d\xi$, $\sigma_N := \bar{\sigma} + d\sigma$, the values of the new variables after a full Newton step, we have:

$$dy = \Delta^{-1}(-\bar{A}\bar{S}^{-1}\delta - B^t d\mu - \xi_N a).$$

Combining (16) and the definition $H := B\Delta^{-1}B^t$, we obtain

$$Bdy = -B\Delta^{-1}B^t d\mu - B\Delta^{-1}(\bar{A}\bar{S}^{-1}\delta + \xi_N a) = 0$$

and

$$d\mu = -H^{-1}B\Delta^{-1}(\bar{A}\bar{S}^{-1}\delta + \xi_N a)$$

We also use the notation

$$P := I - \Delta^{-1/2}B^t H^{-1}B\Delta^{-1/2}$$

$$\bar{r}^2 := a^t\Delta^{-1}a - a^t\Delta^{-1}B^tH^{-1}B\Delta^{-1}a = a^t\Delta^{-1/2}P\Delta^{-1/2}a$$

where P is a projection matrix. Note that $\Delta^{-1/2}$ does not need to be computed; a LL^t Cholesky factorization of Δ is all that is needed in the computations. Clearly $a^t\Delta^{-1/2}P\Delta^{-1/2}a = \|P\Delta^{-1/2}a\|^2$ is nonnegative. The following lemma shows that $\bar{r} = 0$ if and only if the current iterate $\bar{y}$ is a solution of the VI.

Lemma 1 *If $\bar{r} = 0$, then $\bar{y}$ is a solution to the variational inequality.*

Proof:
$\bar{r} = 0$ implies that $P\Delta^{-1/2}a = 0$. Or $\Delta^{-1/2}a = \Delta^{-1/2}B^tH^{-1}B\Delta^{-1}a$.
Hence: $a = B^tH^{-1}B\Delta^{-1}a$, i.e. $a \in \text{Range}(B^t)$.
But as $\bar{s} > 0$, $A^t\bar{y} < c$, and thus Range(B^t) is both the normal cone of Y at $\bar{y}$ and its opposite. Q.E.D.

The expressions for dy and $d\mu$ above and (15) yield

$$\begin{aligned} -\sigma_N = a^tdy &= -a^t\Delta^{-1}(\bar{A}\bar{S}^{-1}\delta + B^td\mu + \xi_N a) \\ &= a^t\Delta^{-1}B^tH^{-1}B\Delta^{-1}a\xi_N - a^t\Delta^{-1}a\xi_N - a^t\Delta^{-1}\bar{A}\bar{S}^{-1}\delta \\ &\quad + a^t\Delta^{-1}B^tH^{-1}B\Delta^{-1}\bar{A}\bar{S}^{-1}\delta \\ &= -\bar{r}^2\xi_N - a^t\Delta^{-1}(I - B^tH^{-1}B\Delta^{-1})\bar{A}\bar{S}^{-1}\delta \end{aligned}$$

Introducing the notation

$$\omega := a^t\Delta^{-1}(I - B^tH^{-1}B\Delta^{-1})\bar{A}\bar{S}^{-1}\delta = a^t\Delta^{-1/2}P\Delta^{-1/2}\bar{A}\bar{S}^{-1}\delta,$$

we can then write

$$-\bar{r}^2\xi_N + \sigma_N = \omega.$$

If we also rewrite (18) as

$$\bar{\sigma}\xi_N + \bar{\xi}\sigma_N = 1 + \bar{\xi}\bar{\sigma},$$

we obtain the values for ξ_N and σ_N:

$$\xi_N = \frac{\bar{\xi} + \bar{\sigma}^{-1} - \bar{\xi}\bar{\sigma}^{-1}\omega}{1 + \bar{\sigma}^{-1}\bar{\xi}\bar{r}^2} \tag{19}$$

$$\sigma_N = \bar{r}^2\xi_N + \omega. \tag{20}$$

Replacing $d\mu$ in the expression for dy, and dy in the expressions for dx and ds one gets:

$$\begin{aligned} dy &= -\Delta^{-1/2}P\Delta^{-1/2}a\xi_N - \Delta^{-1/2}P\Delta^{-1/2}\bar{A}\bar{S}^{-1}\delta \\ ds &= \bar{A}^t\Delta^{-1/2}P\Delta^{-1/2}a\xi_N + \bar{A}^t\Delta^{-1/2}P\Delta^{-1/2}\bar{A}\bar{S}^{-1}\delta \\ dx &= \bar{S}^{-1}(\delta - \bar{X}ds). \end{aligned} \tag{21}$$

To complete the analysis, let:

$$K := \bar{S}^{-1/2}\bar{X}^{1/2}\bar{A}^t\Delta^{-1/2}P;$$

it can be checked that:

- $K^tK = P^2 = P$
- $KK^t = KK^tKK^t$, and thus KK^t is also a projection matrix, and we let $P_2 = I - KK^t$; KK^t is an oblique projection on the null space of ds.

Defining

$$\begin{aligned} p_x &:= P_2\bar{X}^{-1/2}\bar{S}^{-1/2}\delta \\ p_s &:= (I - P_2)\bar{X}^{-1/2}\bar{S}^{-1/2}\delta \\ \bar{p} &:= K\Delta^{-1/2}a, \end{aligned} \tag{22}$$

the Newton iterate for the old variables is then:

$$\begin{aligned} x_N &= \bar{X}(e + \bar{X}^{-1/2}\bar{S}^{-1/2}(p_x - \bar{p}\xi_N)) \\ s_N &= \bar{S}(e + \bar{X}^{-1/2}\bar{S}^{-1/2}(p_s + \bar{p}\xi_N)) \\ y_N &= \bar{y} - \Delta^{-1/2}P\Delta^{-1/2}a\xi_N - \Delta^{-1/2}P\Delta^{-1/2}\bar{A}\bar{S}^{-1}\delta \\ \mu_N &= \bar{\mu} - H^{-1}B\Delta^{-1}(\bar{A}\bar{S}^{-1}\delta + \xi_N a); \end{aligned} \tag{23}$$

The expression for y_N is composed of a centering term and an updating one:

$$-\Delta^{-1/2}P\Delta^{-1/2}a\xi_N;$$

this is a projection, in the metric given by Dikin's ellipsoid, of the functional $F(\bar{y})$ on the space of linear equality constraints. Thus the method may be viewed as a variable metric subgradient method, with a natural step–size (or trust region) given by Dikin's ellipsoid; this is the key reason for the efficiency of ACCPM. The scaled Newton direction for x and s is the sum of centering (p_x, p_s) components and an updating $(\bar{p})$ component.

The initial values $\bar{\xi}$ and $\bar{\sigma}$ to be given to the new variables ξ and σ are critical to the efficiency of a primal-dual infeasible Newton method. Four options are discussed below; for two of them, we give results on the number of Newton steps needed to recenter.

• Option 1 We note that the central cut analysis of [14] applies directly to our problem, even if [14] did not involve dual equalities $By = d$, and thus the algebra is different. That is, the following results hold:

1. $\bar{p}^T p_x = 0$
2. $\|\bar{p}\| = \bar{r}$
3. $p_x^t p_s = 0$
4. $p_x + p_s = \bar{X}^{-1/2}\bar{S}^{-1/2}\delta$
5. $\|p_x \circ p_s\| \le \frac{\sqrt{2}}{4}\frac{\eta^2}{1-\eta}$: this is the key lemma on the convergence of the primal–dual method, due to Mizuno, see [34]
6. $\|p_x\| \le \frac{\eta}{\sqrt{1-\eta}}$ and $\|p_s\| \le \frac{\eta}{\sqrt{1-\eta}}$
7. $|\omega| = |\bar{p}^T p_s| \le \|\bar{p}\|\|p_s\| \le \bar{r}\|\bar{X}^{-1/2}\bar{S}^{-1/2}\delta\| \le \bar{r}\frac{\eta}{\sqrt{1-\eta}}$,

where $\circ$ denotes the Hadamard or componentwise product. Denoting $\bar{\omega} := \omega/\bar{r}$, $x^+ := [x_N; \xi_N]$ and $s^+ := [s_N; \sigma_N]$, the following theorem can be proved (see [14]):

Theorem 3 *If the infeasibilities are chosen as*

$$\bar{\xi} = \bar{\sigma}^{-1} = \frac{1 - \sqrt{1 - \beta(\beta + \bar{\omega})}}{\bar{r}(\beta + \bar{\omega})},$$

then

$$\bar{r}\xi_N = \beta,$$

$$\sigma_N/\bar{r} = \beta + \bar{\omega},$$

and

$$\|X^+ s^+ - e\| \le \left(\beta^4 + (1 - \beta(\beta + \bar{\omega}))^2\right)^{1/2} + \frac{\sqrt{2}}{4}\frac{\eta^2}{1-\eta} + \beta\frac{\eta}{\sqrt{1-\eta}}. \tag{24}$$

Thus for $\eta = .04$ and $\beta = 1/\sqrt{2}$, x^+ and s^+ are primal and dual feasible, and approximately centered, i.e.:

$$\|X^+ s^+ - e\| \le .757,$$

and thus 3 centering steps (1 dual, 2 primal-dual)[1] *will suffice to recenter within $\eta = .04$.*

Experimental results indicate clearly that the very strong centering ($\eta = .04$) required in the proof of the above theorem is not best in practice. By loosening the recentering criterion (say to $\eta = .9$ or $\eta = .95$), centering steps are often not needed at all.

If a loose centering criterion is used it is clearly not possible to guarantee the approximate centering of the next iterate (x^+, s^+); this can be managed as the primal– dual (or the dual or the primal) potential function can be used to guide the algorithm to the next center. However, in a primal–dual method, the strict positivity of x^+, s^+ is absolutely essential (as the strict positivity of x^+ or the strict positivity of s^+ are required for a primal or a dual method, respectively).

The conditions for the positivity of the next iterate are:

1. $x_N > 0$ if and only if $\bar{p}_j\xi_N < (\bar{x}_j\bar{s}_j)^{1/2} + (p_x)_j\ \forall j$
2. $s_N > 0$ if and only if $-\bar{p}_j\xi_N < (\bar{x}_j\bar{s}_j)^{1/2} + (p_s)_j\ \forall j$
3. $\sigma_N > 0$ if and only if $\bar{r}^2\xi_N > -\omega$.
4. $\xi_N > 0$.

These four conditions amount to a ratio test that may not have, if η is too large, any positive solution in ξ_N. Note that, from item 3, $\omega < 0$ makes matters more difficult than $\omega \ge 0$. In fact it is easy to see that $\eta < 1/3$ implies that the ratio test will be successful, but this may lead, at least in theory, to poor centering of (x^+, s^+). This statement follows (see [14]) as: $x_N > 0$ and $s_N > 0$ is implied by $\bar{r}\xi_N \le \frac{1-2\eta}{\sqrt{1-\eta}}$, and $|\bar{\omega}| \le \frac{\eta}{\sqrt{1-\eta}}$, so that a feasible value of $\bar{r}\xi_N$ exists if

$$\frac{\eta}{\sqrt{1-\eta}} < \frac{1-2\eta}{\sqrt{1-\eta}};$$

that is $\eta < 1/3$.

• Option 2 Another option would be to use instead of the linearized centering equation corresponding to the new variables

$$\bar{\sigma} d\xi + \bar{\xi} d\sigma = 1 - \bar{\xi}\bar{\sigma}$$

the actual quadratic relation

$$(\bar{\xi} + d\xi)(\bar{\sigma} + d\sigma) = 1,$$

which can be written as

$$\xi_N \sigma_N = 1.$$

This amounts to enforcing the exact centering for the new variables. This also maximizes the values of σ_N, and thus of ξ_N in (19) and (20). This is achieved if $\bar{r}\bar{\xi} = \bar{r}/\bar{\sigma} = \frac{\sqrt{\bar{\omega}^2+4}-\bar{\omega}}{2}$, leading to $\bar{r}\xi_N = \frac{\sqrt{\bar{\omega}^2+4}-\bar{\omega}}{2}$ and $\sigma_N/\bar{r} = \frac{\sqrt{\bar{\omega}^2+4}+\bar{\omega}}{2}$, which are both positive for any value of $\bar{\omega}$. This can also be seen as setting $\beta = \frac{\sqrt{\omega^2+4}-\bar{\omega}}{2}$ in Theorem 3.

This option cannot guarantee that $(x_N, s_N, \xi_N, \sigma_N) > 0$, as it is identical to option 1 if a linesearch on ξ_N is used. In practice however, both seem to work, and this option seems more effective than option 1, if no linesearch is used.

• Option 3 If the above fails, then an alternative approach would be to assume in equation (17) that $\delta = 0$, thus ignoring the centering component of the Newton updating step; this is a pure predictor step, which ignores the centering component of the previous options. The formulas derived above are all still valid, provided one sets $\omega = 0$ and $p_x = p_s = 0$.

The ratio test then becomes:

1. $x_N > 0$ if and only if $\bar{p}_j \xi_N < (\bar{x}_j \bar{s}_j)^{1/2}$
2. $s_N > 0$ if and only if $-\bar{p}_j \xi_N < (\bar{x}_j \bar{s}_j)^{1/2}$
3. $\xi_N > 0$, which implies $\sigma_N > 0$,

which always has a positive solution in ξ_N. A sufficient condition for this ratio test to pass is that $\bar{\xi}$ and $\bar{\sigma}$ be set as in the theorem below, with the parameter β satisfying $\beta < \sqrt{1-\eta}$.

Theorem 4 *If the infeasibilities are chosen as*

$$\bar{\xi} = \bar{\sigma}^{-1} = \frac{\beta}{\bar{r}(1+\sqrt{1-\beta^2})} = \frac{1-\sqrt{1-\beta^2}}{\bar{r}\beta},$$

then

$$\bar{r}\xi_N = \beta,$$

$$\sigma_N/\bar{r} = \beta,$$

and,

$$\|X^+ s^+ - e\| \le \left(\beta^4 + (1-\beta^2)^2\right)^{1/2} + \eta. \tag{25}$$

Thus for $\eta = .08$ and $\beta = 1/\sqrt{2}$, x^+ and s^+ are primal and dual feasible, and approximately centered. i.e.:

$$\|X^+s^+ - e\| \leq .7872,$$

and thus 3 centering steps (1 dual, 2 primal-dual) will suffice to recenter within $\eta = .08$.

Proof: The primal-dual proximity measure after the updating step is given by, denoting $\bar{p} \circ \bar{p}$ by $\bar{p}^2$,

$$\begin{aligned} (\, X^+s^+ - e \,) &= \begin{pmatrix} X_N s_N - e \\ \xi_N \sigma_N - 1 \end{pmatrix} \\ &= \begin{pmatrix} -\bar{p}^2 \xi_N^2 \\ \xi_N \sigma_N - 1 \end{pmatrix} + \begin{pmatrix} -\delta \\ 0 \end{pmatrix}. \end{aligned}$$

But we have

$$\|\bar{p}^2\|^{1/2} \leq \|\bar{p}\| = \bar{r}.$$

Hence

$$\begin{aligned} \|X^+s^+ - e\| &\leq \|\delta\| + \sqrt{(\|\bar{p}^2\|\xi_N^2)^2 + (1 - \sigma_N \xi_N)^2} \\ &\leq \eta + \sqrt{(\bar{r}\xi_N)^4 + (1 - \sigma_N \xi_N)^2}, \end{aligned}$$

and the theorem follows. □

In this case if η is chosen rather loosely, say 0.9, a value of $\beta < \sqrt{1-\eta}$ can be viewed as giving an initial value $\beta/\bar{r}$ to ξ_N, and then a linesearch can be used that minimizes the primal–dual potential.

• Option 4 A fourth approach could consist in using $\alpha\delta$, with $0 < \alpha < 1$, in equation (17), thus limiting but not ignoring the centering component of the Newton updating step. We did not test this last option in practice.

3.2. *Centering Newton Steps*

Once a primal and dual feasible point has been computed, Newton's method is used for computing an approximate analytic center of the updated set Y_{k+1}. ACCPM [11] used the primal projective method due to Karmarkar. One could also use a dual Newton method or a primal–dual Newton method. We describe the primal–dual Newton method first.

Let the bar notation now account for the $(k+1)^{th}$ generated cut (8), $\bar{A} := A_{k+1}$ and $\bar{c} := c_{k+1}$. Then the new analytic center must solve

$$\begin{aligned} \bar{A}x + B^t\mu &= 0 \\ \bar{A}^t y + s &= \bar{c} \\ By &= d \\ Xs &= e \\ x, s &> 0. \end{aligned}$$

After the update Newton step of section 3.1, the current point $(\bar{y}, \bar{s}, \bar{x}, \bar{\mu})$ is assumed to satisfy:

$$\begin{aligned} \bar{A}\bar{x} + B^t\bar{\mu} &= 0 \\ \bar{A}^t\bar{y} + \bar{s} &= \bar{c} \\ B\bar{y} &= d \\ \|\bar{X}\bar{s} - e\| &\leq \eta < 1 \\ \bar{x}, \bar{s} &> 0. \end{aligned}$$

The Newton iterate (y^+, s^+, x^+, μ^+) can be computed exactly as in the previous section, just dropping the updating component, and thus:

$$\begin{aligned} x^+ &= \bar{X}(e + \bar{X}^{-1/2}\bar{S}^{-1/2}p_x) \\ s^+ &= \bar{S}(e + \bar{X}^{-1/2}\bar{S}^{-1/2}p_s) \\ y^+ &= \bar{y} - \Delta^{-1/2}P\Delta^{-1/2}\bar{A}\bar{S}^{-1}\delta \\ \mu^+ &= \bar{\mu} - H^{-1}B\Delta^{-1}\bar{A}\bar{S}^{-1}\delta; \end{aligned} \tag{26}$$

Mizuno's [34] Lemma 1 applies and:

$$\|X^+s^+ - e\| = p_x \circ p_s \leq \frac{\sqrt{2}}{4}\frac{\eta^2}{1-\eta}$$

This shows that the recentering steps will improve $\|X^+s^+ - e\|$ quadratically as long as $\eta < \frac{4}{4+\sqrt{2}} < 0.739$, since this implies $\frac{\sqrt{2}}{4}\frac{\eta^2}{1-\eta} < \eta$.

On the other hand, a dual Newton method would take a Newton step for the dual potential, constrained by $\bar{A}^t y + s = \bar{c}$, $By = d$, starting at $(\bar{y}, \bar{s})$. This gives the next dual iterate (y^+, s^+). The primal iterate (x^+, μ^+) is then computed by

$$\min_{(x,\mu)}\{\|S^+x - e\| : Ax + B^t\mu = 0\};$$

the formulas are not given here, but are similar to the ones given above, with $\bar{X}$ replaced by $\bar{S}^{-1}$, and thus using a different factorization. The standard quadratic convergence lemma is then:

$$\|X^+s^+ - e\| \leq \eta^2.$$

If $\frac{4-\sqrt{2}}{4} \leq \eta < 1$, a dual recentering can be used, as it is, in theory, stronger than a primal-dual centering. If $\eta > 1$ then a linesearch involving the primal-dual potential along the primal–dual Newton direction should be used; see the book by Ye [47, pp. 102–106] for details.

3.3. *Convergence and Averaging*

The convergence proofs of centering schemes that do not include a regularization term are very delicate. Lüthi [26] and Magnanti and Perakis [27] prove for a variety of volume reduction schemes that the volume of Y_k converges to zero; convergence and complexity results require that $\mathcal{F}$ be single–valued and strongly monotone or f-monotone (this is also known as co–coercive).

The analytic center cutting plane method is not a volume reduction algorithm, and the convergence proof shows that the size of the largest sphere included in Y_k converges to zero [12], and this implies convergence and complexity only under additional assumptions on $\mathcal{F}$.

Convergence and complexity were shown in [13] under the assumption that $\mathcal{F}$ is single–valued pseudo–co–coercive[2]. The following theorem extends the results of [13] to the case involving linear equalities.

Theorem 5 *The cutting plane algorithm which uses approximate analytic centers, and the primal-dual updating scheme, constructs a set Y_k that does not contain a sphere of radius $\bar{\epsilon}$ after $O^*(\frac{(m-p)^2}{\bar{\epsilon}^2})$ iterations.*[3] *Under the additional condition on $\mathcal{F}$ that it is single-valued and pseudo-co-coercive, the algorithm terminates with a primal gap $g(y_k) \geq -\epsilon$ in $O^*(\frac{(m-p)^2}{\epsilon^2})$ iterations.*[4]

The sequence of analytic centers $\{y_0, y_1, y_2, \ldots, y_k, \ldots\}$ produced by cutting plane methods does not need to converge to a solution without these strong assumptions. Nesterov and Vial [37] exhibit a two-dimensional monotone variational inequality for which the sequence of analytic centers converges to a point that is *not* a solution. They also show, for the proximal analytic center cutting plane method, that a sequence of certain convex combinations of previous centers will converge to a solution for any monotone VI problem.

Drawing an analogy between the Nesterov-Vial algorithm and ours, we construct, corresponding to each analytic center y_k, a point $\tilde{y}_k$ that is a convex combination of the k previous analytic centers (recall that the initial analytic center is labeled y_0). The weights of the combination that we use are the last k elements of the primal variable x, which are approximately the inverses of the slacks for the cuts generated at each of the previous centers. These weights are normalized by their sum; $\tilde{y}_k$ is then:

$$\tilde{y}_k = \sum_{i=1}^{k} \frac{x_{(n+i)}}{\sum_{j=1}^{k} x_{(n+j)}} y_i \tag{27}$$

Here $x_{(n+j)}$ refers to the $(n+j)^{\text{th}}$ component of x_k; also, recall that n is the number of initial linear inequality constraints.

The sequence given by the $\tilde{y}_k$ points converges in practice to the solution for the Nesterov-Vial problem. Also, the speed of convergence improves significantly for all problems tested. A full explanation and proof of the behaviour of this sequence is not yet available; but this averaging approach is traditional in many algorithms for fixed point problems and VIs ([27], [28]).

Note that if the gap is computed at each iteration, two function evaluations are necessary per iteration: $F(y_k)$ to generate the cut and $F(\tilde{y}_k)$ to obtain the gap value. In step 3 of the algorithm in section 3, the evaluation of the gap $g(\tilde{y}_k)$ thus replaces the evaluation of the gap $g(y_k)$.

4. Implementation of the algorithm

The algorithm was implemented using the MATLAB language and environment, version 5.1. MATLAB is a convivial setting for matrix computations and algorithm development. With the computing intensive tasks devoted to compiled subroutines, it can also be competitive with C or Fortran codes, see for example the LIPSOL solver of Zhang [48].

• Factorization The update Newton step described in Section 3.1 implies the solution of a linear system, as well as the solution of a second smaller system when equality constraints are present. The same thing is true for the recentering Newton step of Section 3.2.

The solution of linear systems is performed by dense Cholesky factorizations. Because of the density of the introduced cuts, the matrices to be factored are dense; however, they have the dimension $m \times m$ of the problem's size, independent of the potentially large number of cuts generated.

When equality constraints are present, one additional $p \times p$ matrix is factored at each step. We apply, when necessary, the simple technique of overweighing the diagonal elements if the matrix appears to become singular.

• Sparsity and Products The sparsity of the initial constraints, box and others, is exploited whenever possible in the matrix products. This is made easy by MATLAB's sparsity handling capability.

• Initial center An initial analytic center is needed to start the algorithm. When only box constraints are involved, this is a trivial task. When more complex constraints are present, the recentering subroutine is applied to a point that is feasible by construction.

• Stopping criterion The gap function $g(y)$ is used as a stopping criterion. Its evaluation at the point $\tilde{y}_k$ (see (27)) corresponds to solving the linear optimization problem

$$\min_{z \in Y} F(\tilde{y}_k)^t (z - \tilde{y}_k).$$

We use the CPLEX software to solve this linear problem; given the dimension of the Y's under consideration —up to a few hundred variables—, this step is very rapid.

• Cuts removal After the addition of several dozens or hundreds of cuts, it is possible to remove many of them without hindering the convergence of the algorithm while improving its speed. The approach of removing cuts with the greatest slacks can slightly improve the speed of the algorithm although convergence cannot be guaranteed. In the experiments reported here, we keep all cuts.

5. Numerical Results

We report here on our computational experience with the MATLAB implementation of the algorithm. We include examples both with and without equality constraints. Most problems, except the larger ones, are solved to the precision level $g(y) > -10^{-4}$, evaluated at the convex combination points (27); this is a stringent criterion which yields very good accuracies $\|y_k - y^*\|$. In most examples, initial box constraints must be set before the first iterations. The variables of all the problems happen to be bounded below by 0; the upper bounds in the problems without equalities were set at 1×10^3, and at 1×10^2 for those with equality constraints. This choice has to do with the magnitude of the solution, not with the presence of equality constraints. Some problems are inherently bounded upwards.

Concerning the four options of section 3.1, we proceed as follows. We tested the algorithm using option 2, both with a tight ($\eta < 0.08$) and relaxed ($\eta < 0.9$) centering. We also tested option 3: with a relaxed centering ($\eta < 0.9$) and a linesearch on the update step, and with a tight centering ($\eta < 0.08$) but no linesearch. Among these four methods, the safest is option 3 with tight centering, which is guaranteed to converge. The results with option 1 were always very similar but slightly worse than to those with option 2; we therefore do not report on those tests. We did not try option 4. Finally, on the large and difficult mmmr problem, section 5.1, we only tested option 2 (for both tight and relaxed centerings).

Results are reported as follows: *Center precision* refers to the value of $\eta = \|Xs - e\|$ at which a point is considered centered. The *# iterations* is the number of iterations required to meet the primal gap stopping criterion; it is therefore also the number of analytic centers, and of cutting planes, generated for the problem. The *# of centering steps* is the total number of centering steps that were required; this excludes the first step at each iteration, i.e. those steps that we called update steps. As a rule, either one or no centering step at all is required at each iteration. The *cpu* column indicates the seconds or minutes of cpu time that were required for the completion of the algorithm. All runs except one were performed on an UltraSparc 1 workstation, model 200E, 128 Mb of memory. The CO_2 environmental model was run on the multiprocessor HP machine of the E.T.H. in Zürich.

5.1. An environmental model for pricing CO_2 emission permits

This recent extension of the well-known MARKAL model was developed by a group of researchers at the ETH and the Paul Scherrer Institute in Zürich (see [1] and [3]) Their model, called MMMR for Markal Macro Multi-Regional, integrates the energy models of three countries to allow the possibility of the trade of pollution emission permits; a country can thus buy a right to pollute, while the selling country must reduce its pollution by the same amount.

The MMMR model has two levels. At the international level, the model is a $VI(\mathcal{F}, Y)$ in ten variables and one equality constraint; this VI represents economic equilibrium conditions. The mapping $\mathcal{F}$ is the (international) excess supply of different goods, including pollution certificates, and is a function of y, the prices of the goods. At the national level, a large MARKAL-MACRO model (nonlinear objective, linear constraints) is solved for each country. The national level NLPs take the prices y as parameters, and return the (national) excess supply amounts. The link between the two levels is that the international excess supply is the sum of the national excess supplies. That is, $\mathcal{F}(y)$ is evaluated by solving each of the national level NLPs and adding the national excess supplies achieved by the national solutions. The equality constraint represents a scaling of the prices y. One cannot expect the NLPs to have unique solutions, and therefore $\mathcal{F}$ may be a point-to-set mapping, ruling out the existence of the Jacobian of $\mathcal{F}$. Also, the experiments of [1] show that the

Table 1. Example `mmmr` dimension = 10, with option 2

Center precision	# iterations	# of centering steps	gap
0.1	100	101	-1.2e-2
0.9	106	0	-1.0e-2

mapping F is "rather" monotone on Y, and "almost" pseudo-monotone (in the sense that pseudo-monotonicity holds between most pairs of points and monotonicity between many of them). A further challenge in solving the variational inequality comes from the fact that simply evaluating $\mathcal{F}$ at a certain point takes several minutes on a fast computer, because of the large size of the national level NLPs.

We do not report cpu times for this example: 100 iterations can take 10 hours and more, but spent almost exclusively on the national level NLPs. The 10-dimensional VI itself, excluding the function evaluation, is solved in 0.01 second or less per iteration. Results are shown in Table 1.

5.2. *A Walrasian Equilibrium problem:* `scarf`

A *Walrasian* or *General Equilibrium* is attained when there are no goods whose demand exceeds the supply. Scarf ([39]) has described an economy with a number of goods, consumers and producers, that are related through prices variables and production activities variables.

With the definitions

p	prices of the goods (m by 1)
y	activities levels (n by 1)
$d(p)$	consumers demand functions (m by 1)
A	constant technology or activity matrix; translates activity levels into outputs (m by n)
b	initial endowment of goods (m by 1)

the equilibrium is determined by the Nonlinear Complementarity Problem (see [39] and [32] for example)

$$\begin{array}{lll} A^t p \leq 0 & y \geq 0 & y^t A^t p = 0 \\ b - d(p) + Ay \geq 0 & p \geq 0 & p^t(b - d(p) + Ay) = 0 \end{array}$$

This formulation, which includes both primal and dual conditions on top of the complementarity, is necessary to apply NCP algorithms. It also hides the equivalent $VI(F, Y)$ with

$$F(p) = b - d(p), \quad Y = \{p \mid Ap \leq 0,\ p \geq 0\}.$$

This primal-only formulation involves m variables instead of the NCP's $m + n$.

In [39], two Walrasian equilibrium examples are given, the largest one with 14 goods and 26 production activities. The resulting NCP has 40 variables, but the equivalent VI only 14; the mapping $d(p)$ is of the form $d_i(p) = (v_i^t p)/p_i$, where v_i is a constant vector.

Table 2. Example `scarf` dimension = 14

Option	Center precision	# iterations	# of centering steps	cpu secs
2	0.9	104	2	2.9
2	0.08	101	101	4.0
3, with search	0.9	102	78	5.3
3, no search	0.08	100	200	4.65

Table 3. Progression in `scarf` (option 2, center precision 0.9)

# iterations	gap	y_1	y_2	y_{12}	y_{13}	y_{14}
1	-308	0.0217	0.0239	0.5665	0.0363	0.0304
50	-0.1429	0.0617	0.0583	0.0629	0.0361	0.0930
103	-1.0792e-04	0.062144	0.058334	0.062012	0.036515	0.092786
104	-9.0878e-05	0.062145	0.058334	0.062012	0.036515	0.092786

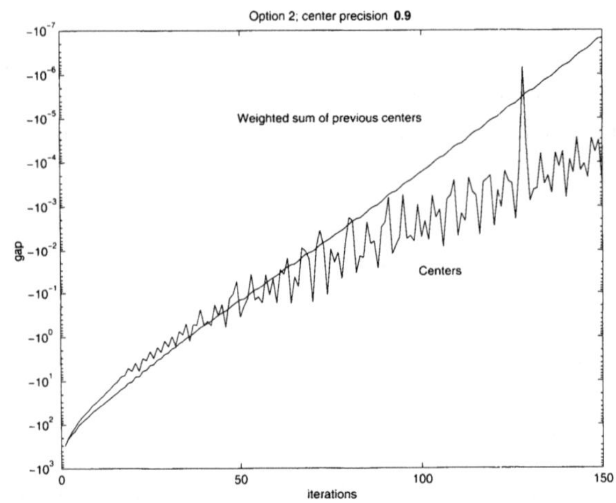

Figure 1. Example `scarf`: Gap vs iterations

Numerical results are given in Tables 2 and 3. Figure 1 shows the evolution of the gap with respect to the iterations. We display both the gap associated to the analytic centers, and the gap at the weighted average of past centers.

5.3. *Problems from the projection-method literature*

Recently, Solodov and Tseng [40] presented a projection-type method for $VI(F, Y)$ which shares two features with the extragradient method ([24]): the mapping needs not be differentiable, as no Jacobian is used, and in fact needs only be monotone and continuous. These two projection-type methods have a special interest to us, because they tackle the same class

Table 4. Example `qHPHard` dimension = 20

Option	Center precision	# iterations	# of centering steps	`etime` secs
2	0.9	220	0	6.6
2	0.08	214	218	8.3
3 with search	0.9	214	49	13
3 no search	0.08	213	426	10

Table 5. Example `Nash 10` dimension = 10

Option	Center precision	# iterations	# of centering steps	`etime` secs
2	0.9	76	0	1.9
2	0.08	69	74	2.2
3 with search	0.9	70	22	2.6
3 no search	0.08	70	138	2.4

of problems for which our algorithm is devised: low-monotonicity, no Jacobian, general Y. We tested two problems reported in [40], that are the two largest ones without special structure (the authors also solved linear programs and linear complementarity problems with their method). The first problem, called `qHPhard`, is custom-made and its mapping is the sum of a linear term and a term $\max\{0, y_i\}^2$; it has 20 variables. The second problem, called `Nash10`, is a 10-variable oligopoly problem with a nonlinear mapping, from [15]. In both cases, the set Y used is $\{y \in \mathbf{R}^m_+ \mid y_1 + y_2 + \ldots + y_m = m\}$. We report our results in Tables 4 and 5. For the sake of comparison with the results of [40], we report here the elapsed time (MATLAB function `etime`), and we used the stopping criterion from [40], which is based on projections. With their method and for, respectively, `qHPHard` and `Nash10`, Solodov and Tseng report 555 function evaluations and 251.6 seconds, and 192 function evaluations and 10.6 seconds. They also tested the extragradient method; on these problems, it is 25% to 50% slower than their own method, but requires 10% to 15% less function evaluations.

We report our results in Tables 4 and 5, and in Figure 2, below.

5.4. *An option pricing model in finance*

In the area of finance, the pricing of stock options and other derivative products has become of paramount importance in the last two decades.

In [19], the authors reformulated the Black-Scholes model for American type options as an infinite-dimensional VI which, through discretization by finite-differences, is approximated by a finite-dimensional VI. For a simple American option with one underlying asset, the discretization takes place along two axes, the current price of the asset and the time to expire. A time-stepping approach allows the problem to be solved as a sequence of VIs, one for each time step, all with the same dimension. We solved a problem with the data:

- Strike price: $K = \$10$
- Time span: $T = 6$ intervals of 2 weeks
- Price span: $[\$0, \$20]$ in intervals of $0.05.

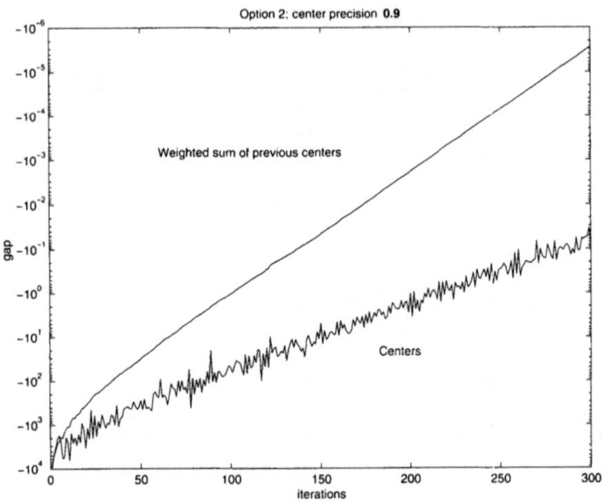

Figure 2. Example qHPHard: Gap vs iterations

Table 6. Example option dimension = 400

Option	Center precision	# iterations	# of centering steps	cpu mins
2	0.9	1304	0	170
2	0.08	1318	1339	340
3 with search	0.9	1272	225	305
3 no search	0.08	1317	2634	542

- Risk-free interest rate: $r = 3\%$
- Diffusion term: $\sigma = 0.2$

The resulting sequence of problems $VI(F, Y)$ have 400 variables, and have no other initial constraints than boxes. The mapping F is linear with a non-symmetric Jacobian. This problem is highly structured and sparse while its size grows very rapidly with finer discretizations; in a sense, it is very simple and very large. Our purpose here was not to compete with more specifically-adapted algorithms, but to show that our algorithm can solve problems with a few hundred variables. We were here only interested in solving a "real-world" large VI, so that we report only on the first of the six 400-variable problems. For more details on the model, see [44] or the appendix of the thesis [6]. Huang and Pang [18] discuss the pricing of options with LCP (Linear Complementarity Problem) algorithms which take advantage of the inherent linearity and sparsity of this application.

Numerical results are shown in Tables 6 and 7. Given the size of the problem, we stopped the algorithm when it reached a gap $> -10^{-3}$. Note that because of boundary conditions, the values at either end of the solution vector in $\mathbf{R}^{400}$, i.e. $y_1^*, y_2^*, \ldots$ and $\ldots, y_{399}^*, y_{400}^*$ are known in advance with rather good precision. Also, the units in Table 7 are dollars, so one observes that already at iteration 1000, the accuracy is within a tenth of a cent.

Table 7. Progression in `option`

# iterations	gap	y_1	y_2	y_{199}	y_{200}	y_{399}	y_{400}
1	-885	10.0500	10.0125	2.6250	2.5875	0.0750	0.0625
500	-0.999	10.0316	9.9844	0.2387	0.2229	0.0223	0.0170
1000	-0.0088	10.0018	9.9518	0.2163	0.2018	0.0021	0.0016
1303	-0.0010	10.0003	9.9503	0.2159	0.2014	0.0006	0.0005
1304	-0.00099	10.0003	9.9503	0.2159	0.2014	0.0006	0.0005

Table 8. Example `choi` dimension = 13

Option	Center precision	# iterations	# of centering steps	cpu secs
2	0.9	67	0	13
2	0.08	66	70	13
3 with search	0.9	62	18	13.2
3 no search	0.08	63	124	13.6

5.5. *Other problems without linear equality constraints*

The problems in this section do not use the capability of the algorithm to handle equality constraints.

The first example is a Nash equilibrium, the second one a spatial price equilibrium problem, the third is a traffic assignment problem and the fourth is an energy model. They are available from MCPLIB [7], and more information on them is available in the specific references as well as in [7].

• Example `choi`: Choi, DeSarbo and Harker [4] have used a Nash equilibrium in a model for the pricing of analgesics. With p the vector of prices of 13 analgesics, the Nash equilibrium of this application is a solution of the $VI(F, Y)$; here $F(p)$ is a vector with components $-d\Pi_j(p)/dp_j$ and $\Pi_j(p)$ is the profit function of the jth firm. The profit function $\Pi_j(p)$ is considered to be a function of p_j only, the other prices remaining constant.

In the instance described in [4], the mapping F is nonlinear and is not a gradient mapping, i.e. its Jacobian is not symmetric. The results are shown in Table 8 and in Figure 3.

• Example `tobin`: The second problem is a well-known spatial price equilibrium model from Tobin [43]. It is a so-called *price* formulation: supply and demand at the different markets are functions of the prices. In opposition to the quantity formulation, this one involves no equality constraints. This model has two interacting commodities on five markets or regions. The transportation cost, demand and supply functions are nonlinear, and the problem is asymmetric, as well as non-monotone at the solution. See Tables 9 and 10 below. Figure 4 shows, for two levels of recentering precision, the evolution of the gap with respect to the iterations. We display both the gap associated to the analytic centers, and the gap at the weighted average of past centers. No satisfactory explanation could be found for the kink around 720 iterations.

• Example `gafni` and `bertsekas`: This interesting application to traffic assignment was described in [2]. There are two strictly equivalent formulations of the same problem

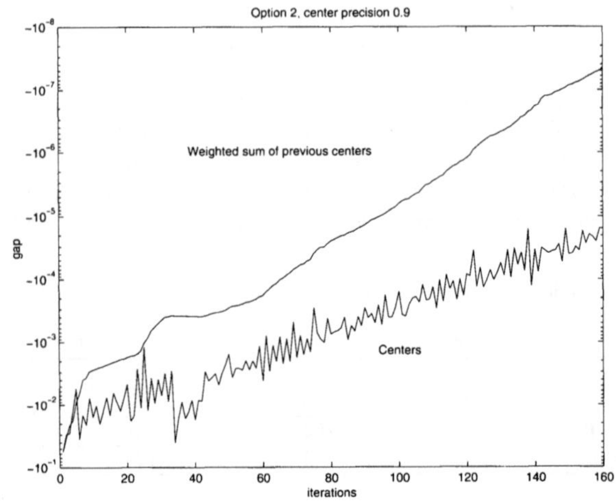

Figure 3. Example `choi`: Gap vs iterations

Table 9. Example `tobin` dimension = 42

Option	Center precision	# iterations	# of centering steps	cpu secs
2	0.9	703	0	47
2	0.08	688	692	75
3 with search	0.9	697	125	218
3 no search	0.08	687	1368	163

Table 10. Progression in `tobin` (option 2, center precision 0.9)

# iterations	gap	y_1	y_9	y_{17}	y_{40}	y_{41}	y_{42}
1	-1e12	500	500	500	500	500	500
250	-113.1288	0.0689	3.9642	12.7263	22.7554	18.7534	29.1753
500	-0.0994	3.7882e-05	3.8827	12.7678	22.7644	18.7604	29.1615
702	-0.00011	7.1856e-08	3.8827	12.7679	22.7644	18.7604	29.1615
703	-0.81e-05	6.99e-08	3.8827	12.7679	22.7644	18.7604	29.1615

Table 11. Example `bertsekas` dimension = 15

Option	Center precision	# iterations	# of centering steps	cpu secs
2	0.9	197	0	4.9
2	0.08	194	193	7.3
3 with search	0.9	191	36	11.8
3 no search	0.08	195	338	11.5

(see the results in Tables 11 and 12). The first formulation is a nonlinear complementarity problem (NCP) in 15 variables while the second is a more general mixed complementarity problem (MCP) in 5 variables. The delay functions are quadratic, and the nonsymmetry of the problem is controlled through a parameter γ. While both formulations are just specialized variational inequality problems and can thus be solved as VIs, the lower dimension of the MCP is very favorable to the speed of the algorithm.

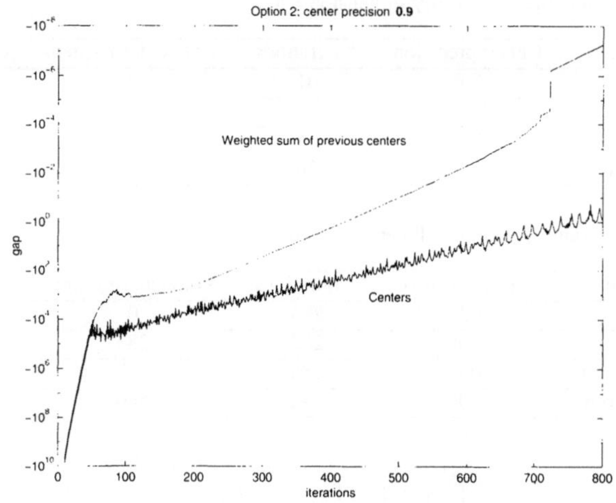

Figure 4. Example `tobin`: Gap vs iterations

Table 12. Example `gafni` dimension = 5

Option	Center precision	# iterations	# of centering steps	cpu secs
2	0.9	43	0	0.95
2	0.08	41	41	1.4
3 with search	0.9	41	13	1.8
3 no search	0.08	41	79	1.9

Table 13. Example `etamge` dimension = 114

Option	Center precision	# iterations	# of centering steps	cpu minutes
2	0.9	2980	7	82
2	0.08	3487	3137	199
3 with search	0.9	3255	348	399
3 no search	0.08	3186	6370	376

• Example `etamge`: This is a general equilibrium model of A. Manne [30]. The model represents the interactions between the energy sector and the economy of the United States and comprises 114 variables. See the numerical results in Table 13 below.

5.6. *Other problems with linear equality constraints*

We present computational experience with two spatial price equilibrium problems that are readily available in the literature. The so-called *quantity formulation* of such problems uses inverse demand and supply functions, i.e. price functions in terms of the quantities. With this formulation the equilibrium is defined as a variational inequality problem with (linear) conservation of flow constraints; the general format is:

Table 14. Example `nagurney` dimension = 8

Option	Center precision	# iterations	# of centering steps	cpu secs
2	0.9	61	0	1.2
2	0.08	55	59	1.7
3 with search	0.9	55	19	1.9
3 no search	0.08	56	111	1.9

Table 15. Example `harker` dimension = 26 $\gamma = 1$

Option	Center precision	# iterations	# of centering steps	cpu secs
2	0.9	293	0	15
2	0.08	282	283	19
3 with search	0.9	288	67	30
3 no search	0.08	283	566	24

$$[\pi(s^*), c(T^*), -\rho(d^*)] \cdot \begin{bmatrix} s - s^* \\ T - T^* \\ d - d^* \end{bmatrix} \geq 0, \ \ \forall\, (s, T, d) \in K$$

where s, T and d are respectively the supply, shipments, and demand quantities, π, c and ρ are supply, demand, and shipment prices, and $K = \{(s, T, d) | s_i = \sum_{j=1}^{n} T_{ij}, d_j = \sum_{i=1}^{m} T_{ij}\}$ ensures flow conservation.

For a general reference, see [35]. Our algorithm allows us to work directly with this original formulation.

• Example `nagurney`: Our first example is a small problem from A. Nagurney's book [35]. It has 2 supply markets and 2 demand markets; with the shipment quantities on the 4 arcs, it has dimension 8. The transportation costs, supply prices and demand prices are all linear functions. The Jacobian of F in the $VI(F, Y)$ formulation is asymmetric, so that no equivalent convex optimization formulation exists.

The results are given in Table 14.

• Example `harker`: The second equality-constrained example (see [15]) has 5 markets all acting as demand and supply markets for the traded commodity. There are 16 connecting arcs for a total of 26 variables. All functions are polynomial and nonlinear. This problem includes a parameter γ which controls the level of asymmetry of the Jacobian of F, with $\gamma = 0$ implying a symmetric matrix. Results are shown for two values of γ, 1 and 15; see Tables 15 and 16. It is especially interesting to note the influence of asymmetry. Symmetrized methods such as the ones discussed in [15] are, quite naturally, sensitive to deviations from symmetry: there the $\gamma = 15$ case takes twice the time and the iterations of the $\gamma = 1$ case (4 minutes vs 2 minutes, on a pre-1988 computer). Our algorithm seems on the other hand to be unaffected by asymmetry.

6. Conclusion

We have presented an algorithm for pseudomonotone or monotone variational inequalities $VI(\mathcal{F}, Y)$ over convex sets. At each iteration, it requires one function evaluation to generate

Table 16. Example `harker` dimension = 26 $\gamma = 15$

Option	Center precision	# iterations	# of centering steps	cpu secs
2	0.9	265	0	13
2	0.08	261	261	17
3 with search	0.9	272	70	27
3 no search	0.08	261	522	22

a cut, and a second one if the gap at the weighted average point is to be computed. Beyond that, each iteration basically consists of solving from one to four linear systems, depending on the desired recentering precision (which controls the number of centering steps) and the presence or not of equality constraints. The mapping $\mathcal{F}$ need not be differentiable and may be multivalued. The set Y can be defined by linear equalities and inequalities, or implicitly by a separation oracle.

Our approach may not be competitive with specialized methods for problems with special features: LCP's, differentiable NCP's and MCP's, strongly monotone and differentiable VI's, etc. However, when the Jacobian is difficult to obtain, or simply not defined (e.g. in multi-valued cases), it is our opinion that there are few other efficient alternatives. Furthermore, the low level of monotonicity required in practice (pseudo-monotonicity or plain monotonicity) makes the ACCPM-VI method with linear cuts attractive both in terms of robustness and speed. This is well illustrated in the MMMR application above.

The update from one center to the next is done very efficiently using option 2: recentering steps are almost never needed if a loose centering condition is used, and this justifies the use of a centering term in the updating step. This shows that a primal-dual approach is a little bit more efficient than a pure primal or a pure dual one.

Acknowledgments

We want to express our gratitude to B. Büeler and H.-J. Lüthi of the E.T.H. Zürich for making both their MMMR model and their computer available to us for testing. We also thank J. Detemple, P. Marcotte, Y. Nesterov and J.-Ph. Vial for stimulating discussions and helpful comments. The very careful analysis by two anonymous referees significantly improved the presentation of this paper.

Notes

1. centering steps are discussed later
2. Pseudo–co–coercive with modulus α is defined as $F(y_1)^t(y_2 - y_1) \Rightarrow F(y_2)^t(y_2 - y_1) \geq \alpha\|F(y_1) - F(y_2)\|^2$ for all $y_1, y_2 \in Y$.
3. The notation O^* indicates that lower order terms are ignored
4. A formal proof of this would require a very significant and lengthy rewriting of the proofs of [12] and [13].

References

1. O. Bahn, B. Büeler, S. Kypreos and H.-J. Luethi, "Modeling an international market of CO2 emission permits," International Journal Global Energy Issues, submitted 1997.

2. D. Bertsekas and E. Gafni, "Projection Methods for variational inequalities with applications to the traffic assignment problem," Mathematical Programming, vol. 17, pp. 139–159, 1982.
3. B. Büeler, "Computing Economic Equilibria and its Application to International Trade of CO_2 Permits: An Agent-Based Approach," Ph.D. thesis, Swiss Federal Institute of Technology, Zürich, Switzerland, December 1997.
4. S. Choi, W. DeSarbo and P. Harke, "Product positioning under price competitition," Management Science, vol. 36-2, pp. 175-199, 1990.
5. J.-P. Crouzeix, "Pseudomontone variational inequality problems: Existence of solutions," Mathematical Programming, vol. 78, pp. 305–314, 1997.
6. M. Denault, "Variational Inequalities with Analytic Center Cutting-Plane Methods," Ph.D. thesis, McGill University, Montréal, Canada, 1998.
7. S. Dirkse and M. Ferris, "MCPLIB: A collection of nonlinear mixed complementarity problems," Optimization Methods and Software, vol. 5, pp. 319–345, 1995.
8. F. Facchinei, A. Fischer and C. Kanzow, "A semismooth Newton method for variational inequalities: the case of box constraints," in Complementarity and variational problems: State of the art, Society for Industrial and Applied Mathematics (SIAM), Philadelphia, 1997.
9. M. Ferris and J.-S. Pang, "Complementarity and variational problems: State of the art," Society for Industrial and Applied Mathematics (SIAM), Philadelphia, 1997.
10. M. Fukushima, "Equivalent differentiable optimization problems and descent methods for asymmetric variational inequality problems," Mathematical Programming, vol. 53, pp. 99–110, 1992.
11. J.-L. Goffin, A. Haurie and J.-P. Vial, "Decomposition and nondifferentiable optimization with the projective algorithm," Management Science, vol. 38, pp. 284–302, 1992.
12. J.-L. Goffin, Z.-Q. Luo and Y. Ye, "Complexity Analysis of an Interior Cutting Plane Method for Convex Feasibility Problems," SIAM Journal on Optimization, vol 6, pp. 638–652, 1996.
13. J.-L. Goffin, P. Marcotte and D. Zhu, "An analytic center cutting plane method for pseudomonotone variational inequalities," Operations Research Letters, vol. 20, pp. 1–6, 1997.
14. J.-L. Goffin and F. Mokhtarian, "Using the primal-dual infeasible Newton method in the analytic center method for problems defined by deep cutting planes," Journal of Optimization Theory and Applications, vol. 37, pp. 181-210, 1998.
15. P. Harker, "Accelerating the convergence of the diagonalization and projection algorithms for finite-dimensional variational inequalities," Mathematical Programming, vol. 41, pp. 29–59, 1988.
16. P. Harker, "Lectures on Computation of Equilibria with Equation-Based Methods," , CORE Lecture Series, CORE Foundation, Louvain-la-Neuve, 1993.
17. P. Harker and J.-S. Pang, "Finite-dimensional variational inequality and nonlinear complementarity problems: a survey of theory, algorithms and applications," Mathematical Programming, vol. 48 (1990), 161–220, 1990.
18. J. Huang and J.-S. Pang, "Option Pricing and Linear Complementarity," Journal of Computational Finance, submitted 1997.
19. P. Jaillet, D. Lamberton, and B. Lapeyre, "Variation Inequalities and the Pricing of American Options," Acta Applicandae Mathematicae, vol. 21, pp. 263–289, 1990.
20. C. Kanzow and H.-D. Qi, "A QP-free constrained Newton-type method for variational inequality problems," Preprint 121, Institute of Applied Mathematics, University of Hamburg, Hamburg, Germany, 1997.
21. S. Karamardian, "Complementarity problems over cones with monotone and pseudomonotone maps," Journal of Optimization Theory and Applications, vol. 18, pp. 445–454, 1976.
22. D. Kinderlehrer and G. Stampacchia, "An Introduction to Variational Inequalities and their applications," Academic Press, New York, 1980.
23. I. V. Konnov, "A combined relaxation method for variational inequalties with nonlinear constraints," Mathematical Programming, vol. 80, pp. 239–252, 1998.
24. G. M. Korpelevich, "The extragradient method for finding saddle points and other problems," Matecon, vol. 12, pp. 747–756, 1976.
25. C. Lemaréchal, A. Nemirovskii and Y. Nesterov, "New Variants of Bundle Methods, " Mathematical Programming, vol. 69, pp.111–147, 1995.
26. H.-J. Lüthi, "On the solution of variational inequalities by the ellipsoid method," Mathematics of Operations Research, vol. 10-3, pp. 515–522, 1985.
27. T. Magnanti and G. Perakis, "A unifying geometric solution framework and complexity analysis for variational inequalities," Mathematical Programming, vol. 71, pp. 327-351, 1995.

28. T. Magnanti and G. Perakis, "Averaging schemes for variational inequalities and systems of equations," Mathematics of Operations Research, pp. 3-22, pp. 568–587, 1997.
29. O. Mangasarian, "Equivalence of the complementarity problem to a system of nonlinear equations," SIAM Journal on Applied Mathematics, vol. 31, pp. 89–92, 1976.
30. A.S. Manne, "ETA-MACRO: A Model of Energy- Economy Interactions," in Modeling Energy-Economy Interactions, Resources for the Future, Washington, D.C, 1977.
31. P. Marcotte, "Inéquations variationnelles: Motivation, algorithmes de résolution et quelques applications," Centre de Recherche sur les Transports, Publication CRT-97-02, 1997.
32. L. Mathiesen, "An algorithm based on a sequence of linear complementarity problems applied to a wlarasian equilibrium model: an example," Mathematical Programming, vol 37, pp. 1–18, 1997.
33. G. Minty, "Monotone (nonlinear) operators in Hilber space," Duke Mathematics Journal, vol. 29, pp. 341–346, 1962.
34. S. Mizuno, "A new polynomial time method for a linear complementarity problem," Mathematical Programming, vol. 56, pp. 31–43, 1992.
35. A. Nagurney, "Network Economics: a variational inequality approach," Kluwer Academic Publishers: Boston, MA, 1993.
36. Y. Nesterov and A. Nemirovskii, "Interior-Point Polynomial Algorithms in Convex Programming," SIAM Studies in Applied Mathematics, SIAM: Philadelphia, 1994.
37. Y. Nesterov and J.-Ph. Vial, "Homogeneous analytic center cutting plane methods for convex problems and variational inequalities," Logilab Technical Report, 1997.4, 1997.
38. D. Ralph and S. Wright, "Superlinear convergence of an interior-point method for monotone variational inequalities," in Complementarity and variational problems: State of the art, Society for Industrial and Applied Mathematics (SIAM): Philadelphia, 1997.
39. H. Scarf (with the collaboration of T. Hansen), "The Computation of Economic Equilibria," Yale University Press, 1973.
40. M. Solodov and P. Tseng, "Modified projection-type methods for monotone variational inequalities," SIAM Journal on Control and Optimization, vol. 34-5, pp. 1814–1830, 1996.
41. Gy. Sonnevend, "New algorithms in convex programming based on a notion of "centre" (for systems of analytic inequalities) and on rational extrapolation," in Trends in Mathematical Optimization, Birkhäuser Verlag, pp. 311–326, 1988.
42. D. Sun, M. Fukushima and L. Qi, "A computable generalized hessian of the D-gap function and Newton-type methods for variational inequality problems," in Complementarity and variational problems: State of the art, Society for Industrial and Applied Mathematics (SIAM): Philadelphia, 1997.
43. R. Tobin, "A variable dimension solution approach for the general spatial price equilibrium problem, "Mathematical Programming, vol. 40, pp. 33–51, 1988.
44. P. Wilmott, J.N. Dewynne and S.D. Howison, "Option Pricing: Mathematical Models and Computation," Oxford Financial Press, 1993.
45. B. Xiao and P. Harker, "A nonsmooth Newton method for variational inequalities, I: theory", Mathematical Programming, vol. 65, pp. 151–194, 1994.
46. B. Xiao and P. Harker, "A nonsmooth Newton method for variational inequalities, II: numerical results," Mathematical Programming, vol. 65, 195–216, 1994.
47. Y. Ye, "Interior Point Algorithms: Theory and Analysis," John Wiley and Sons, Inc., 1997.
48. Y. Zhang, "Solving Large-Scale Linear Programs by Interior-Point Methods Under the MATLAB Environment," Technical Report TR96-01, University of Maryland Baltimore County, 1996.

Computational Optimization and Applications, 12, 157–188 (1999)

A Shifted-Barrier Primal-Dual Algorithm Model for Linearly Constrained Optimization Problems*

GIANNI DI PILLO dipillo@dis.uniroma1.it
STEFANO LUCIDI lucidi@dis.uniroma1.it
LAURA PALAGI palagi@dis.uniroma1.it
Dipartimento di Informatica e Sistemistica, Università di Roma "La Sapienza", via Buonarroti 12, 00185 Roma, Italy.

Received January 29, 1998; Accepted July 18, 1998

Abstract. In this paper we describe a Newton-type algorithm model for solving smooth constrained optimization problems with nonlinear objective function, general linear constraints and bounded variables. The algorithm model is based on the definition of a continuously differentiable exact merit function that follows an exact penalty approach for the box constraints and an exact augmented Lagrangian approach for the general linear constraints. Under very mild assumptions and without requiring the strict complementarity assumption, the algorithm model produces a sequence of pairs $\{x^k, \lambda^k\}$ converging quadratically to a pair $(\bar{x}, \bar{\lambda})$ where $\bar{x}$ satisfies the first order necessary conditions and $\bar{\lambda}$ is a KKT multipliers vector associated to the linear constraints. As regards the behaviour of the sequence $\{x^k\}$ alone, it is guaranteed that it converges at least superlinearly. At each iteration, the algorithm requires only the solution of a linear system that can be performed by means of conjugate gradient methods. Numerical experiments and comparison are reported.

Keywords: linearly constrained optimization, primal-dual algorithm, Penalty-Lagrangian merit function

Dedication:
The fundamental and important Olvi's research contribution speaks on its own. Olvi is one of the most active and prolific researchers in the field of Mathematical Programming and his fundamental papers make of him a father (although a young one !) of this area. But Olvi is not only an outstanding researcher, he is also a special man in the fact that he always puts a lot of joyful enthusiasm in his research work. Enthusiasm is an essential ingredient in our job as well as in many other everyday life human activities and Olvi is able to transfer it to every one who has the luck of spending a period with him. Besides Mathematical Programming, he appreciates also music and art, good cooking and wine, sports (particularly swimming). All these things make Olvi a very pleasant and interesting man. We are all very happy to have met him and we take this opportunity to thank him so much and to wish him all the best.

1. Introduction

In this paper we describe a new class of efficient algorithms for solving linearly constrained optimization problems characterized by the fact that the constraints can be divided into *general linear* constraints $Ax \le b$ and *box* constraints $l \le x \le u$, that is optimization problems of the type

$$\min \{f(x) : Ax \le b, \ l \le x \le u\}.$$

* This work was partially supported by MURST National Research Program "Metodi per l'Ottimizzazione di Sistemi e Tecnologie (MOST)".

Problems with such a structure have been widely studied in the literature and many algorithms have been proposed for tackling them by following both a linesearch approach (see, for example [20, 19, 21, 27] and the references therein) and a trust region approach (see, for example [1, 5]).

In this paper we exploit the possibility to cope with a constrained optimization problem by defining an *exact merit function*, namely a function whose unconstrained minimization provides a solution of the original constrained problem. In particular, we are interested in merit functions which are *continuously differentiable*, since the unconstrained minimization of a continuously differentiable function has been widely studied in the literature and many efficient algorithms can be used or adapted. As discussed in [9] and [13], a continuously differentiable exact merit function can be constructed by combining the original objective function of the constrained problem with one or more terms which *penalize* the violation of the feasibility and of the KKT conditions. In order to construct such penalty terms we need some information on the KKT multipliers. A usual way to obtain this information is to employ *multiplier functions*, namely functions which yield an estimate of the KKT multipliers as a function of the variable x. The continuously differentiable merit functions obtained by using multiplier functions are called *exact penalty functions*. Unfortunately, in general, the computation of a multiplier function turns out to be too expensive when the number of constraints is large since it requires the solution of a linear system that involves the Jacobian of the constraints. A different approach to obtain the information on the KKT multipliers is adopted in *exact augmented Lagrangian functions*. These functions are defined on an extended space of variables made up of the original variables and of *multiplier (or dual) variables*, which are as many as the constraints. Therefore by an exact augmented Lagrangian functions approach we avoid the need of using multiplier functions but, on the other hand, we pay the price of working on a larger space of variables.

When a constrained minimization problem presents some peculiarity in its structure, it could be very convenient from the theoretical and computational point of view to define particular merit functions which exploit as much as possible the structure of the problem. In this paper, taking into account that the constraints are divided into general linear constraints $Ax \leq b$ and simple box constraints $l \leq x \leq u$, we show that it is possible to construct a new merit function by "mixing" the exact penalty functions approach with the exact augmented Lagrangian functions approach. More in particular, the penalty functions approach is used for dealing with the box constraints, since their simplicity allows us to define very simple multiplier functions, while the augmented Lagrangian functions approach is used to tackle the general linear constraints. In this way, we have the advantage over the exact penalty functions approach of using very simple multiplier functions whose computation does not require the solution of any linear system, and over the augmented Lagrangian functions approach we have the advantage of working on the extended space of the primal variables and of the dual variables associated only to the general linear constraints. Making use of the Penalty-Lagrangian merit function and under assumptions milder than the usual ones, we define a globally and superlinearly convergent primal-dual algorithm that requires at each iteration only the solution of a linear system of dimension equal *at most* to the number of primal variables plus the number of dual variables associated only to the general linear constraints.

The paper is organized as follows. In Section 2 we formalise the problem and the assumptions used. In Section 3 we introduce the Penalty-Lagrangian merit function and in Section 4 we analyse its exactness properties. In Section 5 we introduce a class of primal-dual algorithms based on the Penalty- Lagrangian merit function. Finally, in Section 6 we report a numerical experimentation of the approach.

Before going into further developments, we introduce some notation.

Given a vector $v \in \mathbb{R}^p$, we indicate by the uppercase V the diagonal matrix $V = \text{diag}_{1\leq i \leq p}\{v_i\}$. We denote by e the vector of all ones, so that E denotes the identity matrix.

For sake of simplicity, given a vector $v \in \mathbb{R}^p$ and a positive scalar q, we make use of the non standard notation v^{-q} to denote the vector with components $1/v_i^q$.

Given two vectors $v, w \in \mathbb{R}^p$, the operation $\max\{v, w\}$ is intended componentwise, namely $\max\{v, w\}$ denotes the vector with components $\max\{v_i, w_i\}$ for $i = 1, \dots, p$. We denote by $\circ$ the Hadamard product between two vectors, that is $v \circ w$ is a vector with components $v_i w_i$. For short notation, the Hadamard product $w \circ w$ of a vector w by itself is denoted by w^2.

We denote by $\|\cdot\|_p$ the ℓ_p norm, and when p is not specified we intend $p = 2$. Given a positive definite matrix Q, we denote $\|y\|_Q^2 = y^TQy$. Finally, given a real and symmetric matrix Q, we denote by $\mu_{\min}(Q)$ and $\mu_{\max}(Q)$ respectively the smallest and largest eigenvalue of the matrix Q.

Given an $m \times n$ matrix Q we denote by q_i^T the i-th row.

Let $\mathcal{K} \subseteq \{1, \dots, m\}$ be an index subset, v be an m vector and Q be a $m \times n$ matrix. We denote by $v_\mathcal{K}$ the subvector of v with components v_i such that $i \in \mathcal{K}$ and by $Q_\mathcal{K}$ the submatrix of Q made up of the rows q_i^T with $i \in \mathcal{K}$. For the transpose of $Q_\mathcal{K}$, $(Q_\mathcal{K})^T$, we adopt the short notation $Q_\mathcal{K}^T$. Moreover, let $\mathcal{H} \subseteq \{1, \dots, n\}$ be an index subset, we denote by $Q_{\mathcal{K},\mathcal{H}}$ the submatrix made up of the elements q_{ij} with $i \in \mathcal{K}$ and $j \in \mathcal{H}$.

If E is the $n \times n$ identity matrix and $\mathcal{H}_1, \mathcal{H}_2$ and $\mathcal{H}_3$ constitute a partition of the index set $\{1, \dots, n\}$ we can write

$$\begin{aligned} v &= E_{\mathcal{H}_1}^T v_{\mathcal{H}_1} + E_{\mathcal{H}_2}^T v_{\mathcal{H}_2} + E_{\mathcal{H}_3}^T v_{\mathcal{H}_3}, \\ v_{\mathcal{H}_i} &= E_{\mathcal{H}_i} v, \quad i = 1, 2, 3. \end{aligned}$$

2. Problem definition and assumptions

As already said we consider an optimization problem of the type

$$\begin{aligned} \min_x \quad & f(x) \\ & Ax - b \leq 0, \\ & l \leq x \leq u. \end{aligned} \tag{1}$$

Here $x \in \mathbb{R}^n$, $f : \mathbb{R}^n \to \mathbb{R}$, is a twice continuously differentiable function, A is an $m \times n$ matrix, and $l, u \in \mathbb{R}^n$ with $-\infty < l < u < \infty$. We denote the feasible set of Problem (1) by

$$\mathcal{F} = \{x \in \mathbb{R}^n : Ax - b \leq 0, \ l \leq x \leq u\}.$$

We remark that, due to the bound constraints, the feasible set $\mathcal{F}$ is bounded. Naturally, we rely on the fact that this assumption is always satisfied in real application.

The Lagrangian function for Problem (1) is given by:

$$L(x, \lambda, \rho, \sigma) = f(x) + \lambda^T(Ax - b) + \rho^T(l - x) + \sigma^T(x - u),$$

where $\lambda \in \mathbb{R}^m$ and $\rho, \sigma \in \mathbb{R}^n$ are the Karush-Kuhn-Tucker (KKT) multipliers.

A KKT point for Problem (1) is a point $(\bar{x}, \bar{\lambda}, \bar{\rho}, \bar{\sigma}) \in \mathbb{R}^{3n+m}$ such that $\bar{x} \in \mathcal{F}$ and

$$\nabla_x f(\bar{x}) + A^T\bar{\lambda} - \bar{\rho} + \bar{\sigma} = 0 \tag{2}$$

$$\begin{array}{ll} \bar{\rho}^T(l - \bar{x}) = 0, & \bar{\rho} \geq 0, \\ \bar{\sigma}^T(\bar{x} - u) = 0, & \bar{\sigma} \geq 0 \\ \bar{\lambda}^T(A\bar{x} - b) = 0, & \bar{\lambda} \geq 0. \end{array} \tag{3}$$

Since the feasible set is bounded, a KKT point always exists.

We suppose that the Assumptions A1 and A2 below are satisfied, so that we do not recall them any more in the sequel.

Assumption A1. A feasible point $\tilde{x}$ is available.

Assumption A2 [LICQ]. The gradients of the active constraints are linearly independent on $\mathcal{F}$.

The following Assumption A3 is recalled explicitly when needed.

Assumption A3 [Strong Second-Order Sufficient Condition (SSOSC)]. At a KKT point $(\bar{x}, \bar{\lambda}, \bar{\rho}, \bar{\sigma})$ it results that

$$z^T\nabla^2 f(\bar{x})z > 0,$$

for all z such that

$$z_i = 0, \quad \text{for } i \in \{i = 1, \dots, n : \ \bar{x}_i = l_i, \ \bar{\rho}_i > 0, \ \text{ or } \ \bar{x}_i = u_i, \ \bar{\sigma}_i > 0\}$$

and

$$a_j^T z = 0, \quad \text{for } \ j \in \{j = 1, \dots, m : \ a_j^T\bar{x} - b = 0, \ \bar{\lambda}_j > 0\}.$$

We point out that Assumption A1 is not restrictive due to the linearity of the constraints. Assumption A2 is a standard one in nonlinear programming. As regards Assumption A3, it is stronger than the usual sufficient second order condition; however, if the strict complementarity condition holds, namely if the multiplier of the active constraints are strictly positive, then (SSOSC) reduces to the usual one.

3. The Penalty-Lagrangian merit function

In this section, we introduce a merit function that exploits the structure of the constraints of Problem (1). As already said, merit functions can be constructed by adding to the objective function penalty terms which weigh the violation of the feasibility and the violation of the KKT condition (2) (3).

To this aim, Hestenes-Powell-Rockafeller proposed the augmented Lagrangian function that can be written in the form:

$$L_a(x, \lambda, \rho, \sigma; \varepsilon) = f(x) + \frac{1}{2\varepsilon}\Big[\phi(Ax - b, \varepsilon\lambda) + \phi(l - x, \varepsilon\rho) + \phi(x - u, \varepsilon\sigma)\Big],$$

with $\varepsilon \in \mathrm{IR}_+$ and where the function $\phi : \mathrm{IR}^p \times \mathrm{IR}^p \to \mathrm{IR}$ is given by:

$$\phi(v, z) = 2z^T \max\{v, -z\} + \| \max\{v, -z\}\|^2, \tag{4}$$

with $v, z \in \mathrm{IR}^p$. It is well known that the function L_a does not possess exactness properties; this is due to the fact that, roughly speaking, the function L_a does not convey enough information on the KKT multipliers (see [13] for more detailed discussion).

In order to get information on the KKT multipliers, there are two main approaches employed respectively in continuously differentiable exact penalty functions and exact augmented Lagrangian functions. In the exact penalty approach, the multiplier variables (λ, ρ, σ) in L_a are substituted by multiplier functions $(\lambda(x), \rho(x), \sigma(x))$, namely functions which yield an estimate of the KKT multipliers as a function of the variable x. On the other hand, the exact augmented Lagrangian function derives from L_a by adding an additional term that penalizes the difference between the variables (λ, ρ, σ) and the KKT multipliers.

In both cases a main role is played by a function that weighs the errors in the violation of the KKT condition, namely:

$$\Psi(\lambda, \rho, \sigma; x) = \|\nabla_x L(x, \lambda, \rho, \sigma)\|^2 + \|(l - x)\circ\rho\|^2 + \|(x - u)\circ\sigma\|^2 + \|(Ax - b)\circ\lambda\|^2 + r(x)\|\lambda\|^2$$

where

$$r(x) = \| \max\{0, l - x\}\|_p^p + \| \max\{0, x - u\}\|_p^p + \| \max\{0, Ax - b\}\|_p^p, \quad p > 2.$$

If the LICQ assumption holds, the function Ψ is a strictly convex function of (λ, ρ, σ). Moreover, if $(\bar{x}, \bar{\lambda}, \bar{\rho}, \bar{\sigma})$ is a KKT point, we have:

$$\begin{pmatrix} \bar{\lambda} \\ \bar{\rho} \\ \bar{\sigma} \end{pmatrix} = \arg\min \Psi(\lambda, \rho, \sigma; \bar{x});$$

or equivalently, the point $(\bar{\lambda}, \bar{\rho}, \bar{\sigma})$ is the unique solution of the system

$$\nabla_{(\lambda,\rho,\sigma)} \Psi(\lambda, \rho, \sigma; \bar{x}) = 0.$$

Then, in the exact penalty approach [26], the multiplier functions are defined as

$$\begin{pmatrix} \lambda(x) \\ \rho(x) \\ \sigma(x) \end{pmatrix} = \arg\min \Psi(\lambda, \rho, \sigma; x),$$

whereas in the exact augmented Lagrangian approach a term of the type

$$\|\nabla_{(\lambda,\rho,\sigma)} \Psi(\lambda, \rho, \sigma; x)\|^2$$

is added to L_a.

Unfortunately, as it has been already remarked, both approaches have some drawbacks. In particular, the computation of the multiplier functions can be too expensive because it requires, for each penalty function evaluation, the solution of a linear system of dimension $2n+m$. When n and m are large, this limits the applicability of the exact penalty functions approach. On the other hand, the exact augmented Lagrangian function approach pays the price of working in the enlarged space of dimension $3n+m$.

In order to overcome these difficulties we follow here a hybrid approach, that is we adopt an exact penalty approach for the bound constraints and an augmented Lagrangian approach for the general linear constraints. In other words, we work in the extended space of the primal variables x and of the dual variables λ associated to the constraints $Ax - b \leq 0$. The rationale behind this idea is that, taking λ as a variable, it is possible to define two continuously differentiable multiplier functions $\rho(x,\lambda)$ and $\sigma(x,\lambda)$ for the box constraints that are very easy to evaluate, not requiring any matrix inversion.

In fact, we define $\rho(x,\lambda)$ and $\sigma(x,\lambda)$ as the unique solution of the linear system

$$\begin{aligned} &(l-x)^2 \circ \rho + (x-u)^2 \circ \sigma = 0, \\ &\nabla f(x) + A^T\lambda - \rho + \sigma = 0. \end{aligned} \tag{5}$$

We easily obtain the following expressions for the multiplier functions:

$$\begin{aligned} \rho(x,\lambda) &= (x-u)^2 \circ d(x) \circ \left(\nabla f(x) + A^T\lambda\right), \\ \sigma(x,\lambda) &= -(l-x)^2 \circ d(x) \circ \left(\nabla f(x) + A^T\lambda\right), \end{aligned} \tag{6}$$

where $d(x)$ is a vector with components given by

$$d_i(x) = \frac{1}{(l_i - x_i)^2 + (x_i - u_i)^2} \qquad i = 1, \dots, n.$$

We can verify that:

PROPOSITION 1 *If $(\bar{x}, \bar{\lambda}, \bar{\rho}, \bar{\sigma})$ is a KKT point for Problem (1) it results $\rho(\bar{x}, \bar{\lambda}) = \bar{\rho}$ and $\sigma(\bar{x}, \bar{\lambda}) = \bar{\sigma}$.*

As regards the dual variable λ, we follow an exact augmented Lagrangian approach and we need a term that penalizes the difference between λ and the corresponding KKT multiplier. We use the continuously differentiable term

$$\varphi(x,\lambda) = \|\nabla_\lambda \Psi(\lambda, \rho(x,\lambda), \sigma(x,\lambda); x)\|^2,$$

that is

$$\varphi(x,\lambda) = \left\|M(x)\lambda + Ad(x) \circ (l-x)^2 \circ (x-u)^2 \circ \nabla f(x)\right\|^2$$

with

$$M(x) = \operatorname{diag}^2(Ax-b) + r(x)E + A(L-X)^2 D(x)(X-U)^2 A^T. \tag{7}$$

The term $\varphi(x,\lambda)$ is a strictly convex function w.r.t. λ. Indeed, in Appendix A.1 we prove that:

PROPOSITION 2 *For every $x \in \mathbb{R}^n$, the matrix $M(x)$ given by (7) is positive definite.*

Therefore, we obtain a merit function of the type:

$$P(x, \lambda; \varepsilon) = f(x) + \frac{1}{2\varepsilon}\Big[\phi(Ax - b, \varepsilon\lambda) + \phi(l - x, \varepsilon\rho(x, \lambda)) \\ + \phi(x - u, \varepsilon\sigma(x, \lambda))\Big] + \varphi(x, \lambda). \tag{8}$$

This merit function inherits the features of the original constrained problem in the sense that, for sufficiently small values of the penalty parameter, there is a one to one correspondence between stationary points of the merit function and KKT points of the Problem (1). However, in order to guarantee that the unconstrained minimization of the function $P(x, \lambda; \varepsilon)$ is well defined, we need to enforce additional properties on its level sets. This can be obtained by including shifted barrier terms in the expression of the merit function in such a way that it goes to infinity on the boundary of some open perturbation of the feasible set.
Here we can include barrier terms related only to the box constraints. In particular, we consider the following open perturbation of the feasible set

$$\mathcal{S} = \{x \in \mathbb{R}^n : \; l - \alpha < x < u + \beta\},$$

where α and β are given positive vectors in $\mathbb{R}^n$. Obviously we have $\mathcal{F} \subset \mathcal{S}$. Then we consider the vector functions

$$q(x) = \alpha - l + x, \qquad p(x) = \beta + u - x.$$

For every $x \in \mathcal{S}$ it results $q(x) > 0$ and $p(x) > 0$.

Moreover, given the diagonal non singular matrices $Q(x)$ and $P(x)$ we define the functions $\phi_{Q(x)} : \mathbb{R}^n \times \mathbb{R}^n \to \mathbb{R}$ and $\phi_{P(x)} : \mathbb{R}^n \times \mathbb{R}^n \to \mathbb{R}$ by modifying the function ϕ given in (4) in the following way:

$$\phi_{Q(x)}(v, z) = 2z^T Q(x)^{-1} \max\{v, -z\} + \|\max\{v, -z\}\|^2_{Q(x)^{-1}}, \\ \phi_{P(x)}(v, z) = 2z^T P(x)^{-1} \max\{v, -z\} + \|\max\{v, -z\}\|^2_{P(x)^{-1}}. \tag{9}$$

By substituting in (8) the functions $\phi_{Q(x)}$, $\phi_{P(x)}$ for the functions ϕ related to the bound constraints, we finally get the following expression of the merit function:

$$P(x, \lambda; \varepsilon) = f(x) + \frac{1}{2\varepsilon}\Big[\phi(Ax - b, \varepsilon\lambda) + \phi_{Q(x)}(l - x, \varepsilon q(x) \circ \rho(x, \lambda)) \\ + \phi_{P(x)}(x - u, \varepsilon p(x) \circ \sigma(x, \lambda))\Big] + \varphi(x, \lambda), \tag{10}$$

where we assume, without loss of generality, that $\varepsilon \in (0, \varepsilon_M]$ and ε_M is any given positive scalar. It is easily seen that the introduction of the functions ϕ_Q, ϕ_P guarantees that $P \to \infty$ as x approaches the boundary of $\mathcal{S}$.

The merit function can be written explicitly as follows:

$$
\begin{aligned}
P(x,\lambda;\varepsilon) = \ & f(x) + \lambda^T \max\{Ax-b, -\varepsilon\lambda\} + \frac{1}{2\varepsilon}\|\max\{Ax-b,-\varepsilon\lambda\}\|^2 \\
& +\rho(x,\lambda)^T \max\{l-x, -\varepsilon q(x)\circ\rho(x,\lambda)\} \\
& +\sigma(x,\lambda)^T \max\{x-u, -\varepsilon p(x)\circ\sigma(x,\lambda)\} \\
& +\frac{1}{2\varepsilon}\Big[\|\max\{l-x,-\varepsilon q(x)\circ\rho(x,\lambda)\}\|^2_{Q(x)^{-1}} \\
& +\|\max\{x-u,-\varepsilon p(x)\circ\sigma(x,\lambda)\}\|^2_{P(x)^{-1}}\Big] \\
& +\left\|M(x)\lambda + Ad(x)\circ(l-x)^2\circ(x-u)^2\circ\nabla f(x)\right\|^2 .
\end{aligned}
$$

The merit function P is continuously differentiable with gradient ∇P given by:

$$
\begin{aligned}
\nabla_x P(x,\lambda;\varepsilon) = \ & \frac{1}{\varepsilon}A^T\max\{Ax-b,-\varepsilon\lambda\} \\
& +\left[\nabla_x\rho(x,\lambda) - \frac{1}{\varepsilon}Q(x)^{-1}\right]\max\{l-x,-\varepsilon q(x)\circ\rho(x,\lambda)\} \\
& +\left[\nabla_x\sigma(x,\lambda) + \frac{1}{\varepsilon}P(x)^{-1}\right]\max\{x-u,-\varepsilon p(x)\circ\sigma(x,\lambda)\} \\
& -\frac{1}{2\varepsilon}q(x)^{-2}\circ\max\{l-x,-\varepsilon q(x)\circ\rho(x,\lambda)\}^2 \\
& +\frac{1}{2\varepsilon}p(x)^{-2}\circ\max\{x-u,-\varepsilon p(x)\circ\sigma(x,\lambda)\}^2 \\
& +2\Big[2A^T\mathrm{diag}(Ax-b)\Lambda + \nabla^2 f(x)(L-X)^2D(x)(X-U)^2A^T \\
& +\nabla r(x)\lambda^T + \left[(L-X)^3-(X-U)^3\right]D(x)\Phi(x,\lambda)A^T\Big] \\
& \times\left[M(x)\lambda + Ad(x)\circ(l-x)^2\circ(x-u)^2\circ\nabla f(x)\right]
\end{aligned} \tag{11}
$$

$$
\begin{aligned}
\nabla_\lambda P(x,\lambda;\varepsilon) = \ & \max\{Ax-b,-\varepsilon\lambda\} \\
& +Ad(x)\circ(x-u)^2\circ\max\{l-x,-\varepsilon q(x)\circ\rho(x,\lambda)\} \\
& -Ad(x)\circ(l-x)^2\circ\max\{x-u,-\varepsilon p(x)\circ\sigma(x,\lambda)\} \\
& +2M(x)\left[M(x)\lambda + Ad(x)\circ(l-x)^2\circ(x-u)^2\circ\nabla f(x)\right],
\end{aligned} \tag{12}
$$

where

$$
\begin{aligned}
&\nabla_x\rho(x,\lambda) = \nabla^2 f(x)D(x)(X-U)^2 + (L-U)D(x)\Phi(x,\lambda) \\
&\nabla_x\sigma(x,\lambda) = -\nabla^2 f(x)D(x)(L-X)^2 + (L-U)D(x)\Phi(x,\lambda) \\
&\nabla r(x) = p\left[\max\{0,x-u\}^{p-1} - \max\{0,l-x\}^{p-1}\right] + pA^T\max\{0,Ax-b\}^{p-1} \\
&\Phi(x,\lambda) = 2(L-X)(X-U)D(x)\mathrm{diag}\{\nabla f(x) + A^T\lambda\}.
\end{aligned}
$$

4. Exactness properties of the function P

4.1. *Preliminaries*

In this section we describe the exactness properties of the merit function $P(x, \lambda; \varepsilon)$ given by (10).

First of all we note that, similarly to many merit functions previously proposed in the literature, some of its properties follow directly from the structure of the terms ϕ, $\phi_{P(x)}$, $\phi_{Q(x)}$, $\varphi(x, \lambda)$. Indeed, repeating with minor modifications the arguments in [13], it is possible to prove the following proposition.

PROPOSITION 3

(i) *For any $\varepsilon > 0$, if $(\bar{x}, \bar{\lambda}, \bar{\rho}, \bar{\sigma})$ is a KKT point of Problem (1), then $(\bar{x}, \bar{\lambda})$ is a stationary point of $P(x, \lambda; \varepsilon)$; moreover it results $P(\bar{x}, \bar{\lambda}; \varepsilon) = f(\bar{x})$, $\rho(\bar{x}, \bar{\lambda}) = \bar{\rho}$ and $\sigma(\bar{x}, \bar{\lambda}) = \bar{\sigma}$.*

(ii) *Assume that, for some value of ε, every stationary point of $P(x, \lambda; \varepsilon)$ yields a KKT point of Problem (1); then,*

(iia) *if $(\bar{x}, \bar{\lambda})$ is a local minimum point of $P(x, \lambda; \varepsilon)$, then $\bar{x}$ is a local minimum point for Problem (1) and $(\bar{\lambda}, \rho(\bar{x}, \bar{\lambda}), \sigma(\bar{x}, \bar{\lambda}))$ are the corresponding multipliers;*

(iib) *if $(\bar{x}, \bar{\lambda})$ is a global minimum point of $P(x, \lambda; \varepsilon)$, then $\bar{x}$ is a global minimum point for Problem (1); conversely, if $\bar{x}$ is a global minimum point for Problem (1) and $\bar{\lambda}$ is the multiplier associated to the general linear constraints, then $(\bar{x}, \bar{\lambda})$ is a global minimum point for $P(x, \lambda; \varepsilon)$.*

Points (i) and (ii) of Proposition 3 clearly point out that Problem (1) is equivalent to an unconstrained minimization of the merit function $P(x, \lambda; \varepsilon)$, if it is possible to prove that there exists a nonempty set of values of the penalty parameter ε such that

- the function $P(x, \lambda; \varepsilon)$ has an unconstrained global minimum point over $\mathcal{S} \times \mathbb{R}^m$;
- every stationary point of the merit function $P(x, \lambda; \varepsilon)$ is a KKT point for Problem (1).

In the following we show that the merit function $P(x, \lambda; \varepsilon)$ enjoys these two properties.

4.2. *Existence of an unconstrained global minimizer of the function $\boldsymbol{P(x, \lambda; \varepsilon)}$ over $\boldsymbol{\mathcal{S}} \times \mathbb{R}^m$*

In order to investigate this point, let us introduce the level set:

$$\Omega(x^0, \lambda^0; \varepsilon) = \{(x, \lambda) : x \in \mathcal{S}, \lambda \in \mathbb{R}^m, P(x, \lambda; \varepsilon) \le P(x^0, \lambda^0; \varepsilon)\}, \tag{13}$$

where $(x^0, \lambda^0) \in \mathcal{S} \times \mathbb{R}^m$ is given.

The following proposition states that the level set defined by (13) is compact in the extended space of the variables (x, λ), for $x^0 \in \mathcal{F}$ and for sufficiently small values of ε.

PROPOSITION 4 *Let* (x^0, λ^0) *be a given point in* $\mathcal{F} \times \mathbb{R}^m$*, then there exists a value* $\bar{\varepsilon}$ *such that, for every* $\varepsilon \in (0, \bar{\varepsilon}]$*, the level set* $\Omega(x^0, \lambda^0; \varepsilon)$ *is compact. In particular, there exists a positive constant* $c_1(x^0, \lambda^0)$ *such that*

$$\Omega(x^0, \lambda^0; \varepsilon) \subseteq \left\{(x, \lambda) : \ x \in \mathcal{S}, \ \|\lambda\| \leq c_1(x^0, \lambda^0)\right\}.$$

The proof is reported in Appendix A.2.

As it is well known, in unconstrained minimization the compactness of the level sets is of main importance. In fact, it guarantees that the merit function $P(x, \lambda; \varepsilon)$ admits a global minimum point (and hence a stationary point) in $\mathcal{S} \times \mathbb{R}^m$. Furthermore, from the computational point of view, the compactness of the set $\Omega(x^0, \lambda^0; \varepsilon)$ is a standard assumption employed to prove global convergence towards stationary points.

4.3. *Existence of a threshold value for* ε *such that stationary points of* $\boldsymbol{P(x, \lambda; \varepsilon)}$ *are KKT points*

The following result provides the basis to prove that a threshold value $\bar{\varepsilon}$ exists, such that, for $\varepsilon \in (0, \bar{\varepsilon}]$ every stationary point of the function $P(x, \lambda; \varepsilon)$ gives a KKT point of Problem (1). Indeed, from the definition of $P(x, \lambda; \varepsilon)$, we can easily verify that the following proposition holds.

PROPOSITION 5 *For all* $\varepsilon > 0$*, if* (x, λ) *is such that*

$$\begin{aligned} &\max\{x - u, -\varepsilon p(x) \circ \sigma(x, \lambda)\} = 0, \\ &\max\{l - x, -\varepsilon q(x) \circ \rho(x, \lambda)\} = 0, \\ &\max\{Ax - b, -\varepsilon \lambda\} = 0, \end{aligned} \tag{14}$$

then $(x, \lambda, \rho(x, \lambda), \sigma(x, \lambda))$ *is a KKT point of Problem (1).*

Hence, it is sufficient to show that there exists a threshold value of the penalty parameter ε such that every stationary point of $P(x, \lambda; \varepsilon)$ satisfies system (14). This point follows from the property, stated in the following proposition, that the norm of the gradient of P provides an upper bound on the violation of system (14) for sufficiently small ε. The proof is quite technical and is reported in Appendix A.3.

PROPOSITION 6 *Let* $(x, \lambda) \in \mathcal{S} \times \mathbb{R}^m$*. Then, for every* $\theta > 0$*, there exists a value* $\bar{\varepsilon}$ *such that for all* $\varepsilon \in (0, \bar{\varepsilon}]$*, the following inequality holds:*

$$\begin{aligned} \|\nabla P(x, \lambda; \varepsilon)\|^2 \geq \theta \ \Big(&\|\max\{Ax - b, -\varepsilon \lambda\}\|^2 + \|\max\{l - x, -\varepsilon q(x) \circ \rho(x, \lambda)\}\|^2 \\ &+ \|\max\{x - u, -\varepsilon p(x) \circ \sigma(x, \lambda)\}\|^2 \Big). \end{aligned}$$

The previous results imply that the correspondence between stationary points of P and KKT points of Problem (1) can be proved for sufficiently small values of the penalty parameter ε; in particular, as a consequence of Propositions 5 and 6 it is possible to state the following proposition.

PROPOSITION 7 *A threshold value $\bar{\varepsilon} > 0$ exists such that, for any $\varepsilon \in (0, \bar{\varepsilon}]$, if $(\bar{x}, \bar{\lambda}) \in \mathcal{S} \times \mathbb{R}^m$ is a stationary point for $P(x, \lambda; \varepsilon)$, then the point $(\bar{x}, \bar{\lambda}, \rho(\bar{x}, \bar{\lambda}), \sigma(\bar{x}, \bar{\lambda}))$ satisfies the KKT conditions for Problem (1).*

Finally, by collecting the previous results, we can summarize the exactness properties of the merit function $P(x, \lambda; \varepsilon)$ in the following proposition.

PROPOSITION 8

(i) *Let $(\bar{x}, \bar{\lambda}, \bar{\rho}, \bar{\sigma})$ be a KKT point of Problem (1), then, for any $\varepsilon > 0$, $(\bar{x}, \bar{\lambda})$ is a stationary point of $P(x, \lambda; \varepsilon)$; moreover it results $P(\bar{x}, \bar{\lambda}; \varepsilon) = f(\bar{x})$, $\rho(\bar{x}, \bar{\lambda}) = \bar{\rho}$ and $\sigma(\bar{x}, \bar{\lambda}) = \bar{\sigma}$.*

(ii) *A threshold value $\bar{\varepsilon}$ exists such that, for any $\varepsilon \in (0, \bar{\varepsilon}]$:*

(iia) *if $(\bar{x}, \bar{\lambda}) \in \mathcal{S} \times \mathbb{R}^m$ is a stationary point for $P(x, \lambda; \varepsilon)$, then the point $(\bar{x}, \bar{\lambda}, \rho(\bar{x}, \bar{\lambda}), \sigma(\bar{x}, \bar{\lambda}))$ satisfies the KKT conditions for Problem (1);*

(iib) *if $(\bar{x}, \bar{\lambda})$ is a local minimum point of $P(x, \lambda; \varepsilon)$, then $\bar{x}$ is a local minimum point for Problem (1) and $(\bar{\lambda}, \rho(\bar{x}, \bar{\lambda}), \sigma(\bar{x}, \bar{\lambda}))$ are the corresponding multipliers;*

(iic) *if $(\bar{x}, \bar{\lambda})$ is a global minimum point of $P(x, \lambda; \varepsilon)$, then $\bar{x}$ is a global minimum point for Problem (1); conversely, if $\bar{x}$ is a global minimum point for Problem (1) and $\bar{\lambda}$ is the multiplier associated to the general linear constraints, then $(\bar{x}, \bar{\lambda})$ is a global minimum point for $P(x, \lambda; \varepsilon)$.*

4.4. *Second order analysis*

In this subsection we assume that the function $f(x)$ is three times continuously differentiable. From the definition of $P(x, \lambda; \varepsilon)$ and the differentiability of f, it follows that the function P is an SC^1 function for all $(x, \lambda) \in \mathcal{S} \times \mathbb{R}^m$, that is a continuously differentiable function with a semismooth gradient (see [28]). Hence, its generalised Hessian $\partial^2 P(x, \lambda; \varepsilon)$, in Clarke's sense [3], exists and is given by:

$$\begin{aligned}\partial^2 P(x, \lambda; \varepsilon) = \; & \mathrm{co}\{H \in \mathbb{R}^{(n+m)\times(n+m)} : \exists\{(x^k, \lambda^k)\} \to (x, \lambda) \text{ with} \\ & \nabla P \text{ differentiable at } (x^k, \lambda^k) \text{ and } \{\nabla^2 P(x^k, \lambda^k; \varepsilon)\} \to H\}.\end{aligned}$$

The generalised Hessian $\partial^2 P$ is a nonempty, convex, compact set of symmetric matrices; furthermore, the point-to-set map $(x, \lambda) \mapsto \partial^2 P(x, \lambda; \varepsilon)$ is bounded on bounded sets [24]. In order to perform an analysis of the second order properties of the merit function P we introduce a partition of the index set $\{1, \dots, m\}$ given by the subsets $\mathcal{I}_a$, $\mathcal{I}_n$, and a partition of the index set $\{1, \dots, n\}$ given by the subsets $\mathcal{L}$, $\mathcal{U}$, $\mathcal{N}$ and we denote $\mathcal{W} = \{\mathcal{I}_a, \mathcal{I}_n, \mathcal{L}, \mathcal{U}, \mathcal{N}\}$. Let us define the $(n+m) \times (n+m)$ symmetric matrix $H(x, \lambda; \varepsilon, \mathcal{W})$, which depends on these partitions and which is given block-wise by (for simplicity, we omit the arguments in the r.h.s.):

$$
\begin{aligned}
H_{xx}(x,\lambda;\varepsilon,\mathcal{W}) &= \nabla^2 f - E_{\mathcal{L}}^T \nabla_x \rho_{\mathcal{L}}^T - \nabla_x \rho_{\mathcal{L}} E_{\mathcal{L}} + E_{\mathcal{U}}^T \nabla_x \sigma_{\mathcal{U}}^T + \nabla_x \sigma_{\mathcal{U}} E_{\mathcal{U}} \\
&\quad + \frac{1}{\varepsilon}\left(E_{\mathcal{L}}^T Q_{\mathcal{L},\mathcal{L}}^{-1} E_{\mathcal{L}} + E_{\mathcal{U}}^T P_{\mathcal{U},\mathcal{U}}^{-1} E_{\mathcal{U}} + A_{\mathcal{I}_a}^T A_{\mathcal{I}_a}\right) \\
&\quad -\varepsilon\left[\nabla_x \rho_{\mathcal{N}} Q_{\mathcal{N},\mathcal{N}} \nabla_x \rho_{\mathcal{N}}^T + \nabla_x \sigma_{\mathcal{N}} P_{\mathcal{N},\mathcal{N}} \nabla_x \sigma_{\mathcal{N}}^T\right] \\
&\quad + 2\nabla^2 f (L-X)^2 D (X-U)^2 A^T A (X-U)^2 D (L-X)^2 \nabla^2 f \\
H_{\lambda\lambda}(x,\lambda;\varepsilon,\mathcal{W}) &= -\varepsilon\begin{bmatrix} 0 & E_{\mathcal{I}_n}^T \end{bmatrix} - \varepsilon \nabla_\lambda \rho_{\mathcal{N}} Q_{\mathcal{N},\mathcal{N}} \nabla_\lambda \rho_{\mathcal{N}}^T \\
&\quad -\varepsilon \nabla_\lambda \sigma_{\mathcal{N}} P_{\mathcal{N},\mathcal{N}} \nabla_\lambda \sigma_{\mathcal{N}}^T + 2\tilde{M}^2 \\
H_{x\lambda}(x,\lambda;\varepsilon,\mathcal{W}) &= \begin{bmatrix} A_{\mathcal{I}_a}^T & 0 \end{bmatrix} - \varepsilon \nabla_x \rho_{\mathcal{N}} Q_{\mathcal{N},\mathcal{N}} \nabla_\lambda \rho_{\mathcal{N}}^T - \varepsilon \nabla_x \sigma_{\mathcal{N}} P_{\mathcal{N},\mathcal{N}} \nabla_\lambda \sigma_{\mathcal{N}}^T \\
&\quad - E_{\mathcal{L}}^T \nabla_\lambda \rho_{\mathcal{L}}^T + E_{\mathcal{U}}^T \nabla_\lambda \sigma_{\mathcal{U}}^T + 2\nabla^2 f (L-X)^2 (X-U)^2 D A^T \tilde{M}.
\end{aligned}
\tag{15}
$$

where

$$
\tilde{M}(x) = A(X-U)^2 D(x)(L-X)^2 A^T + \begin{bmatrix} 0 & 0 \\ 0 & \text{diag}\,(Ax-b)^2_{\mathcal{I}_n,\mathcal{I}_n} \end{bmatrix},
$$

and

$$
\begin{aligned}
\nabla_\lambda \rho(x,\lambda) &= AD(x)(X-U)^2, \\
\nabla_\lambda \sigma(x,\lambda) &= -AD(x)(L-X)^2.
\end{aligned}
$$

Let us consider the index sets

$$
\begin{array}{ll}
\mathcal{I}_0(x) = \{i:\ a_i^T x = b_i\}, & \mathcal{I}_+(x,\lambda) = \{i:\ a_i^T x = b_i,\ \ \lambda_i > 0\}, \\
\mathcal{L}_0(x) = \{j:\ x_j = l_j\}, & \mathcal{L}_+(x,\rho) = \{j:\ x_j = l_j,\ \rho_j > 0\}, \\
\mathcal{U}_0(x) = \{j:\ x_j = u_j\}, & \mathcal{U}_+(x,\sigma) = \{j:\ x_j = u_j,\ \sigma_j > 0\}.
\end{array}
\tag{16}
$$

Given a KKT point $(\bar{x},\bar{\lambda},\bar{\rho},\bar{\sigma})$ of Problem (1) let us define the set of partitions

$$
\begin{aligned}
\mathcal{P} = \{\mathcal{W} = \{\mathcal{I}_a,\mathcal{I}_n,\mathcal{L},\mathcal{U},\mathcal{N}\} :\ & \mathcal{I}_+(\bar{x},\bar{\lambda}) \subseteq \mathcal{I}_a \subseteq \mathcal{I}_0(\bar{x}),\ \mathcal{I}_n = \{1,\ldots,m\}\setminus\mathcal{I}_a, \\
& \mathcal{L}_+(\bar{x},\bar{\rho}) \subseteq \mathcal{L} \subseteq \mathcal{L}_0(\bar{x}),\ \mathcal{U}_+(\bar{x},\bar{\sigma}) \subseteq \mathcal{U} \subseteq \mathcal{U}_0(\bar{x}),\ \mathcal{N} = \{1,\ldots,n\}\setminus \mathcal{L}\cup\mathcal{U}\}.
\end{aligned}
$$

The following proposition holds.

PROPOSITION 9 *Let $(\bar{x},\bar{\lambda},\bar{\rho},\bar{\sigma})$ be a KKT point of Problem (1) and ε be given. Then there exists a neighborhood $\mathcal{B}$ of $(\bar{x},\bar{\lambda})$ such that for all $(x,\lambda)\in\mathcal{B}$, we have:*

$$
\partial^2 P(x,\lambda;\varepsilon) \subseteq \tilde{\partial}^2 P(x,\lambda;\varepsilon),
$$

with

$$
\tilde{\partial}^2 P(x,\lambda;\varepsilon) = \text{co}\{H(x,\lambda;\varepsilon,\mathcal{W}) + K(x,\lambda;\mathcal{W}) \text{ for all } \mathcal{W}\in\mathcal{P}\},
$$

where $H(x,\lambda;\varepsilon,\mathcal{W})$ is given by (15) and $K(x,\lambda;\mathcal{W})$ is a $(n+m)\times(n+m)$ matrix such that $\|K(x,\lambda;\mathcal{W})\| \le \gamma(x,\lambda)$, with $\gamma(x,\lambda)$ a nonnegative function which vanishes at $(\bar{x},\bar{\lambda})$.

The proof easily follows from first part of the proof of Proposition 6.1 in [14].

Under (SSOSC), the next proposition provides a converse of (iib) of Proposition 8.

PROPOSITION 10 *Let $(\bar{x}, \bar{\lambda}, \bar{\rho}, \bar{\sigma})$ be a KKT point of Problem (1) which satisfies the (SSOSC). Then there exists an $\bar{\varepsilon}$ such that for all $\varepsilon \in (0, \bar{\varepsilon}]$, $(\bar{x}, \bar{\lambda})$ is an isolated minimum point of $P(x, \lambda; \varepsilon)$ and all the matrices in $\tilde{\partial}^2 P(\bar{x}, \bar{\lambda}; \varepsilon)$ are positive definite.*

The proof can be found in the Appendix A.4.

At a KKT point where the strict complementarity holds, it results $\mathcal{I}_+(\bar{x}, \bar{\lambda}) = \mathcal{I}_0(\bar{x})$, $\mathcal{L}_+(\bar{x}, \bar{\rho}) = \mathcal{L}_0(\bar{x}), \mathcal{U}_+(\bar{x}, \bar{\sigma}) = \mathcal{U}_0(\bar{x})$ and $\mathcal{P} = \mathcal{W}_0 = \{\mathcal{I}_0, \{1, \ldots, m\} \backslash \mathcal{I}_0, \mathcal{L}_0, \mathcal{U}_0, \{1, \ldots, n\} \setminus \mathcal{L}_0 \cup \mathcal{U}_0\}$. In this case $\partial^2 P(\bar{x}, \bar{\lambda}; \varepsilon)$ reduces to a singleton, and in particular, in a neighbourhood of the KKT point we have that $P(x, \lambda; \varepsilon)$ is twice continuously differentiable, with Hessian matrix given by:

$$\nabla^2 P(x, \lambda; \varepsilon) = H(x, \lambda; \varepsilon, \mathcal{W}_0) + K(x, \lambda; \mathcal{W}_0),$$

as it follows by specialising the proof of Proposition 10 to the case $\mathcal{I}_+(\bar{x}, \bar{\lambda}) = \mathcal{I}_0(\bar{x})$, $\mathcal{L}_+(\bar{x}, \bar{\rho}) = \mathcal{L}_0(\bar{x}), \mathcal{U}_+(\bar{x}, \bar{\sigma}) = \mathcal{U}_0(\bar{x})$.

5. The primal-dual algorithm

In this section we propose an iterative scheme for the solution of Problem (1) that turns out to be globally convergent with superlinear convergence rate. As it has been pointed out, if the strict complementarity condition holds at a KKT point, the penalty function $P(x, \lambda; \varepsilon)$ is twice continuously differentiable in a neighbourhood of that point, and therefore, under this condition, any unconstrained Newton algorithm can be easily adapted in order to define globally convergent methods which show a superlinear rate of convergence.

However, it is possible to define globally and superlinearly convergent algorithms that require neither the evaluation of the second derivatives of P nor the strict complementarity assumption. In particular, in [8], we propose a Newton-type algorithm with respect to the merit function $P(x, \lambda; \varepsilon)$ that is based on a powerful active set strategy that allows rapid changes in the working set.

First, we describe a linesearch based algorithm model for the unconstrained minimization of the merit function P. Since the working space is that of the primal variables x and of the dual variables λ, the algorithm can be considered a primal-dual one.

Algorithm PLA (Penalty-Lagrangian Algorithm)

Initialization. It is given a point $(\tilde{x}, \tilde{\lambda}) \in \mathcal{F} \times \mathbb{R}^m$, a parameter $\varepsilon^0 > 0$ and a scalar $0 < \theta < 1$.

Set $k = 0$, $j = 0$, and $(x^k, \lambda^k) = (\tilde{x}, \tilde{\lambda})$

Step 1. If $(x^k, \lambda^k, \rho(x^k, \lambda^k), \sigma(x^k, \lambda^k))$ is a KKT point for Problem (1), stop.

Step 2. If $\|\lambda^k\| > 1/\varepsilon^j$ go to Step 6.

Step 3. Evaluate a direction $d^k = (d_x^k, d_\lambda^k)$.

Step 4. If

$$\begin{aligned}\nabla P(x^k, \lambda^k; \varepsilon^j)^T d^k > & -\|\max\{Ax^k - b, -\varepsilon^j \lambda^k\}\|^2 \\ & -\|\max\{l - x^k, -\varepsilon^j q(x^k) \circ \rho(x^k, \lambda^k)\}\|^2 \\ & -\|\max\{x^k - u, -\varepsilon^j p(x^k) \circ \sigma(x^k, \lambda^k)\}\|^2\end{aligned}$$

go to Step 6.

Step 5. Evaluate a steplength δ^k along d^k; then compute

$$\begin{pmatrix} x^{k+1} \\ \lambda^{k+1} \end{pmatrix} = \begin{pmatrix} x^k \\ \lambda^k \end{pmatrix} + \delta^k \begin{pmatrix} d_x^k \\ d_\lambda^k \end{pmatrix}; \tag{17}$$

set k=k+1 and go to Step 1.

Step 6. Set $\varepsilon^{j+1} = \theta\varepsilon^j$.

If $P(x^k, \lambda^k; \varepsilon^{j+1}) < P(\tilde{x}, \tilde{\lambda}; \varepsilon^{j+1})$ then $(x^0, \lambda^0) = (x^k, \lambda^k)$

else $(x^0, \lambda^0) = (\tilde{x}, \tilde{\lambda})$; set $k = 0$, $j = j + 1$ and go to Step 2.

The algorithm PLA performs an outer iteration, indexed by j, and an inner iteration, indexed by k. For a fixed j the inner iteration generates a sequence $\{(x^k, \lambda^k)\} \in \Omega(x^0, \lambda^0)$. We note that, even if x^0 is feasible, the sequence $\{x^k\}$ is not forced to stay in the feasible region. The outer iteration generates a sequence $\{\varepsilon^j\}$; in particular, the outer iteration performs an automatic updating of the penalty parameter in order to enforce the conditions of Propositions 4 and 6.

The convergence properties of the algorithm are established in the next proposition.

PROPOSITION 11 *Let* $(x^0, \lambda^0) \in \mathcal{F} \times \mathbb{R}^m$ *be a given starting point. Assume that for all* ε

(i) *the direction* (d_x^k, d_λ^k) *satisfies the conditions*

$$\begin{aligned} &\nabla P(x^k, \lambda^k; \varepsilon)^T d^k \le -c\|\nabla P(x^k, \lambda^k; \varepsilon)\|^2 \\ &c\|d^k\| \le \|\nabla P(x^k, \lambda^k; \varepsilon)\| \end{aligned} \tag{18}$$

where c is a given positive constant;

(ii) *the linesearch procedure computes* δ^k *such that*

$$\begin{aligned} &x^{k+1} \in \mathcal{S} \\ &P(x^{k+1}, \lambda^{k+1}; \varepsilon) \le P(x^0, \lambda^0; \varepsilon) \\ &\lim_{k\to\infty} \nabla P(x^k, \lambda^k; \varepsilon)^T d^k = 0. \end{aligned}$$

Then either the algorithm terminates at some $(x^p, \lambda^p) \in \mathcal{S} \times \mathbb{R}^m$ *and* $(x^p, \lambda^p, \rho(x^p, \lambda^p), \sigma(x^p, \lambda^p))$ *is a KKT point for Problem* (1), *or the algorithm produces an infinite sequence* $\{(x^k, \lambda^k)\} \in \mathcal{S} \times \mathbb{R}^m$ *such that every limit point* $(\bar{x}, \bar{\lambda})$ *yields a KKT point* $(\bar{x}, \bar{\lambda}, \rho(\bar{x}, \bar{\lambda}), \sigma(\bar{x}, \bar{\lambda}))$ *for Problem* (1).

The proof of the preceding proposition follows from the fact that, for ε sufficiently small Propositions 4 and 6 hold and, since the direction satisfies the conditions (18), the test at Step 2 and the test at Step 4 of Algorithm PLA are eventually violated, so that the penalty parameter is no longer updated. Then the assertion follows from (i), (ii), the test at Step 4 and Proposition 5.

Then, we briefly describe how we compute a search direction in order to guarantee locally a superlinear rate of convergence.

The definition of the direction $d^k = (d_x^k, d_\lambda^k)$ is based on a powerful active set identification technique. In particular, at each point (x^k, λ^k) we define the following index sets:

$$\begin{aligned}
&\mathcal{I}_a^k = \{i : a_i^T x^k - b_i \geq -\varepsilon \lambda_i^k\}, && \mathcal{I}_n^k = \{1, \dots, m\} \setminus \mathcal{I}_a^k, \\
&\mathcal{L}^k = \{j : l_j - x_j^k \geq -\varepsilon q_j(x^k) \rho_j(x^k, \lambda^k)\}, && \\
&\mathcal{U}^k = \{j : x_j^k - u_j \geq -\varepsilon p_j(x^k) \sigma_j(x^k, \lambda^k)\}, && \mathcal{N}^k = \{1, \dots, n\} \setminus (\mathcal{L}^k \cup \mathcal{U}^k).
\end{aligned}$$

At each iteration k, we can write

$$d_x^k = E_{\mathcal{L}^k}^T d_{x_{\mathcal{L}^k}} + E_{\mathcal{U}^k}^T d_{x_{\mathcal{U}^k}} + E_{\mathcal{N}^k}^T d_{x_{\mathcal{N}^k}}. \tag{19}$$

In the sequel, where this does not produce confusion, we do not report explicitly the dependence on the iteration k.

The direction is obtained by the following rules: first solve with respect to $d_\mathcal{N}$ and z the linear system of dimension $|\mathcal{N}| + |\mathcal{I}_a|$

$$\begin{bmatrix} E_\mathcal{N} \nabla^2 f E_\mathcal{N}^T & E_\mathcal{N} A_{\mathcal{I}_a}^T \\ A_{\mathcal{I}_a} E_\mathcal{N}^T & 0 \end{bmatrix} \begin{pmatrix} d_{x_\mathcal{N}} \\ z \end{pmatrix} = - \begin{pmatrix} E_\mathcal{N}(\nabla f + \nabla^2 f (E_\mathcal{L} d_\mathcal{L} + E_\mathcal{U} d_\mathcal{U})) \\ (Ax - b)_{\mathcal{I}_a} + A_{\mathcal{I}_a}(E_\mathcal{L} d_\mathcal{L} + E_\mathcal{U} d_\mathcal{U}) \end{pmatrix}; \tag{20}$$

then the component d_x is obtained according to (19) by taking $d_{x_{\mathcal{N}^k}} = d_{x_\mathcal{N}}$ and

$$\begin{aligned} d_{x_{\mathcal{L}^k}} &= (l - x)_{\mathcal{L}^k} \\ d_{x_{\mathcal{U}^k}} &= -(x - u)_{\mathcal{U}^k}; \end{aligned} \tag{21}$$

the component d_λ^k is obtained by taking:

$$(d_\lambda^k)_i = \begin{cases} z_i - \lambda_i^k, & i \in \mathcal{I}_a^k \\ -\lambda_i^k, & i \in \mathcal{I}_n^k. \end{cases} \tag{22}$$

A first property of this direction is reported in the following proposition, whose proof can be found in [8], by observing that the multiplier functions that we have proposed satisfies the Assumption B of [8].

PROPOSITION 12 *If the direction (d_x^k, d_λ^k) obtained by (19)-(22) is zero then $(x^k, \lambda^k, \rho(x^k, \lambda^k), \sigma(x^k, \lambda^k))$ is a KKT point for Problem* (1).

In order to state additional properties of the direction, in the remaining of this section we assume that f is three times continuously differentiable.

The following proposition characterises the properties of the direction defined by (19)-(22) and is proved in [8].

PROPOSITION 13

(i) *Let* $(\bar{x}, \bar{\lambda}, \bar{\rho}, \bar{\sigma})$ *be a KKT point for Problem (1) which satisfies the (SSOSC). Then there exist a neighbourhood* $\mathcal{B}(\bar{x})$ *of* $\bar{x}$ *and a neighbourhood* $\mathcal{B}(\bar{\lambda})$ *of* $\bar{\lambda}$ *such that, if* $(x^0, \lambda^0) \in \mathcal{B}(\bar{x}) \times \mathcal{B}(\bar{\lambda})$*, the system (20) is well defined for all k, the sequence* $\{x^k\}$ *produced by the iteration (17) with* $\delta_k = 1$ *satisfies* $x^k \in \mathcal{B}(\bar{x})$ *for all k and it converges to* $\bar{x}$ *with superlinear rate of convergence. Furthermore, the sequence* $\{\lambda^k\}$ *remains in* $\mathcal{B}(\bar{\lambda})$*, converges to* $\bar{\lambda}$*, and the sequence* $\{(x^k, \lambda^k)\}$ *converges quadratically to* $(\bar{x}, \bar{\lambda})$*.*

(ii) *The direction* d^k *obtained by (19)-(22) satisfies a system of the kind:*

$$\tilde{H}(x^k, \lambda^k; \varepsilon)d = -\nabla P(x^k, \lambda^k; \varepsilon),$$

where the matrix $\tilde{H}(x^k, \lambda^k; \varepsilon)$ *has the property that, if the sequence* $\{(x^k, \lambda^k)\}$ *converges to a* KKT *pair* $(\bar{x}, \bar{\lambda})$ *for Problem (1), then*

$$\lim_{k\to\infty} \operatorname{dist}\left[\tilde{H}(x^k, \lambda^k; \varepsilon) \middle| \tilde{\partial}^2 P(x^k, \lambda^k; \varepsilon)\right] = 0.$$

Point (i) characterises the local rate of convergence of the sequence generated by the algorithm, whereas Point (ii) shows that the direction is obtained by solving a system that tends to a non-smooth version of the Newton system.

For ε sufficiently small, in a neighbourhood of a KKT point where SSOSC holds, the direction d^k defined by (19)-(22) satisfies the conditions required at point (i) of Proposition 11 and it is able to avoid the Maratos effect. In particular we have the following result, whose proof follows easily from (ii) of Proposition 13, from Proposition 10 and from the results of [18].

PROPOSITION 14 *Let* $(\bar{x}, \bar{\lambda})$ *be a KKT pair which satisfies the (SSOSC). If the direction is obtained by (19)-(22), then there exists an* $\bar{\varepsilon}$ *such that for all* $\varepsilon \in (0, \bar{\varepsilon}]$*, there exists a neighbourhood* $\mathcal{B}(\bar{x}, \bar{\lambda})$ *of* $(\bar{x}, \bar{\lambda})$ *such that, for all* $(x^k, \lambda^k) \in \mathcal{B}(\bar{x}, \bar{\lambda})$*:*

- *the search direction* d^k *satisfies the conditions:*

$$\begin{aligned} &\nabla P(x^k, \lambda^k; \varepsilon)' d^k \leq -c\|\nabla P(x^k, \lambda^k; \varepsilon)\|^2 \\ &c\|d^k\| \leq \|\nabla P(x^k, \lambda^k; \varepsilon)\|, \end{aligned}$$

where c is a positive constant;

- *an Armijo-type linesearch accepts the unit stepsize.*

The preceding proposition shows that locally the direction defined by (19)-(22) can be used in Algorithm PLA, in order to conciliate global convergence with superlinear rate of convergence. In particular, in order to tackle also large scale problems, we use a conjugate gradient based method for solving approximatively the linear system (20).

However, far from the solution, the direction d^k may not satisfy the conditions needed to ensure the convergence. In this case, we need some alternative direction. In particular, in the algorithm implemented we obtain a gradient related direction by applying a Truncated-Newton scheme [7, 22] to the system:

$$H(x^k, \lambda^k; \varepsilon, \mathcal{W}^k)d^k = -\nabla P(x^k, \lambda^k; \varepsilon),$$

where H is the matrix defined in (15) and $\mathcal{W}^k = (\mathcal{I}_a^k, \mathcal{I}_n^k, \mathcal{L}^k, \mathcal{U}^k, \mathcal{N}^k)$.

Finally, as concerns the linesearch procedure, it must guarantee the satisfaction of the conditions (ii) of Proposition 11 and must not deteriorate the rate of convergence. We choose to use a nonmonotone stabilization scheme. In particular we have easily adapted the scheme proposed in [23].

6. Numerical results

The approach proposed in this paper has been developed in order to cope with a real world application, namely the Yield Management (YM) problem in the railway transportation. This problem consists in determining how many seats to assign for reservation on each possible origin-destination served by a train so as to maximise the overall revenue. A detailed description of the YM problem can be found in [2]. An essential feature of the railway YM problem dealt with is that, when the whole railway network is considered, a large number of large dimensional instances of the problem have to be solved in a limited time, so that a specific optimization code was required. We operated in this way: we performed preliminary a numerical experimentation by using a rough Fortran90 implementation of Algorithm PLA on a set of academic problems. Since the results were acceptable, we developed a more sophisticated code that better exploits the structure of the constraints and applied it to the solution of the YM problem. In this second case the code was implemented in C, in order to interface with a C code that simulates the railway YM system using real data. In the following we comment on the two experimentations and on the results obtained.

The first test problems considered are taken from the CUTE collection by selecting most of the linearly inequality constrained test problems. Since there are few problems that are formulated with a complete box, we selected also those problems that have incomplete bound constraints. In the case that some $l_i = -\infty$ and/or some $u_i = +\infty$ we set l_i and/or u_i respectively to a small and a large value ($l_i = -10^5$, $u_i = 10^5$). If these fictitious constraints are binding at the optimal solution, we enlarged the box.

As regards to the starting point (x^0, λ^0) we have chosen heuristically as x^0 the point supplied by CUTE even if it is not feasible, and this has not affected the convergence of the algorithm. In fact we observed the occurence of a feasible point in the sequence generated by the algorithm. The initial value λ^0 has been chosen in such a way that $\varphi(x^0, \lambda^0) = 0$; this choice is motivated by the fact that the value obtained is the same that would be obtained by the evaluation of a multiplier function. The values of the barrier terms α_i and β_i are chosen in such a way to account for the violation of the box constraints at x^0. For the initial value of ε we use an heuristic that balances the weight of $f(x)$ and of the other terms in the merit function. The linear system for the direction was solved by a conjugate gradient method. Numerically, to prevent overflows, we also introduced a constant multiplying the last term φ.

In Tables 1 and 2 we report the results obtained by Algorithm PLA and by Lancelot (LANC) on the a selection of problems from the CUTE collection. In particular, Table

Table 1. Results on a selection of problem from CUTE collection (HS problems).

Problem name	n	m	Computing Code	ni	nf	ng
			LANC	1	1	2
HS21	2	1	PLA	2	2	2
			LANC	1	1	2
HS21MOD	7	1	PLA	2	2	2
			LANC	7	7	8
HS24	2	3	PLA	7	10	10
			LANC	6	6	7
HS35	3	1	PLA	2	3	3
			LANC	11	11	12
HS36	3	1	PLA	4	6	6
			LANC	15	15	16
HS37	3	2	PLA	6	7	7
			LANC	6	6	7
HS76	4	3	PLA	6	9	9
			LANC	17	17	18
HS86	5	10	PLA	20	26	26
			LANC	17	17	18
S268	5	5	PLA	7	3	3
			LANC	9	9	10
S277	4	4	PLA	1	1	1
			LANC	10	10	11
S278	6	6	PLA	1	1	1
			LANC	9	9	10
S279	8	8	PLA	1	1	1
			LANC	13	13	14
S280	10	10	PLA	1	1	1

1 refers to problems from the Hoch-Schittkowski (HS) collection. For each problem we report the dimensions n, m, the number of the overall iterations (ni), meant as the number of distinct points x^k generated by the algorithm, the number of function evaluations (nf) and the number of objective function gradient evaluations (ng).

As concerns the YM problem, the corresponding mathematical problem has the structure of Problem (1) with the additional feature of having a convex and separable objective function. Moreover, the starting point $x^0 = l$ is always feasible.

The test problems are randomly generated on the basis of 12 different data set corresponding to small ($m = 5,\ n = 15$) medium ($m = 10,\ n = 55$) and large ($m = 15,\ n = 120$)

Table 2. Results on a selection of problems from the CUTE collection.

Problem name	n	m	Computing Code	ni	nf	ng
			LANC	28	28	29
HIMMELBI	100	12	PLA	189	192	192
			LANC	8	8	9
PENTAGON	6	15	PLA	23	36	36
			LANC	17	17	18
POWELL20	10	10	PLA	4	5	5
			LANC	4	4	5
SIMPLLPA	2	2	PLA	5	5	5
			LANC	3	3	4
SIMPLLPB	2	3	PLA	2	2	2
			LANC	13	13	14
STANCMIN	3	2	PLA	119	121	121
			LANC	8	8	9
PRIMAL1	325	85	PLA	15	15	15
			LANC	7	7	9
PRIMAL2	649	96	PLA	25	36	36
			LANC	8	8	9
PRIMAL3	745	111	PLA	6	18	18
			LANC	910	910	906
PRIMALC1	230	9	PLA	12	12	12
			LANC	31	31	28
PRIMALC2	231	7	PLA	8	15	15
			LANC	32	32	33
PRIMALC5	287	8	PLA	15	20	20
			LANC	182	182	183
PRIMALC8 [1]	520	8	PLA	13	13	13
			LANC	4	4	5
ZECEVIC2	2	2	PLA	4	5	5

problems. From the data set we obtained randomly about 300 optimization problems that differs in at least a critical aspect and the simulation was repeated 100 times, so that the overall number of problems solved is about $3 \cdot 10^4$.

In Table 3 we report cumulative results for the YM problems, and a comparison with Lancelot and Minos in terms of average number of iterations and average CPU time required for the solution.

As we can see the algorithm PLA performs always better than Lancelot and Minos.

Table 3. Cumulative results on YM problems of MINOS, LANCELOT, PLA.

	Computing Code	average CPU time (msecs)	average # of iterations
Small YM Problems	PLA	32.79	12.59
	LANCELOT	60.41	31.77
	MINOS	55.42	33.07
Medium YM Problems	PLA	279.57	19.41
	LANCELOT	460.27	39.28
	MINOS	552.42	129.41
Large YM Problems	PLA	1150.10	24.95
	LANCELOT	1790.96	35.08
	MINOS	2034.66	230.05

Notes

1. The points found are different: Lancelot finds a better value

References

1. J. F . Bonnans and C. Pola, "A trust region interior point algorithm for linearly constrained optimization," SIAM J. on Optimization, vol. 7, pp. 717–731, 1997.
2. A. Ciancimino, G. Inzerillo, S. Lucidi and L. Palagi, "A mathematical programming approach for the solution of the railway yield management problem," Transportation Science (to appear).
3. F.H. Clarke, Optimization and Nonsmooth Analysis, John Wiley & Sons: New York, 1983.
4. A. Conn, N. Gould and P. Toint, Lancelot. A Fortan Package for Large-Scale Nonlinear Optimization (Release A), Springer-Verlag: Berlin, 1991.
5. A. R. Conn, N. Gould and Ph. L. Toint, "A primal-dual algorithm for minimizing a nonconvex function subject to bound and linear equality constraints," Technical Report 96/9, Dept. of Mathematics, FUNDP, Namur (B),1996.
6. A. R. Conn, N. Gould and Ph. L. Toint, "A globally convergent Lagrangian barrier algorithm for optimization with general inequality constraints and simple bounds," Technical Report 92/7(2nd revision), 1995.
7. R. S. Dembo and T. Steihaug, "Truncated Newton algorithms for large scale unconstrained optimization," Mathematical Programming, vol.26, pp.190-212, 1983.
8. G. Di Pillo, S. Lucidi and L. Palagi, "A superlinearly convergent algorithm for constrained nonlinear programming problems with bounded variables," Technical Report DIS 02.99, 1999.
9. G. Di Pillo, "Exact penalty methods," in Algorithms for Continuous Optimization: the State of the Art, pages 1-45. Kluwer Ac. Press, Boston, (1994).
10. G. Di Pillo and L. Grippo, "An augmented Lagrangian for inequality constraints in nonlinear programming problems," Journal of Optimization Theory and Applications, vol. 36, pp. 495–519, 1982.
11. G. Di Pillo and L. Grippo, "Exact penalty functions in constrained optimization," SIAM Journal on Control and Optimization, vol. 27, pp. 1333-1360, 1989.
12. G. Di Pillo, L. Grippo and S. Lucidi, "Globally convergent exact penalty algorithms for constrained optimization," Technical Report CNR-IASI no. 128, 1985.
13. G. Di Pillo and S. Lucidi, " On exact augmented Lagrangian functions in nonlinear programming," in Nonlinear optimization and applications: Proceedings of the International School of Mathematics, Erice, Italy, 1995.
14. G. Di Pillo and S. Lucidi, "An augmented Lagrangian function with improved exactness properties," Technical Report DIS 18.95, 1995.
15. G. Di Pillo, S. Lucidi and L. Palagi, "An exact penalty-Lagrangian approach for a class of constrained optimization problems with bounded variables," Optimization, vol. 28, pp. 129–148, 1993.

16. F. Facchinei and S. Lucidi, "A class of methods for optimization problems with simple bounds," Technical Report CNR-IASI no. 336, 1992.
17. F. Facchinei and S. Lucidi, "Quadratically and superlinearly convergent algorithms for the solution of inequality constrained minimization problem," J. of Optimization Theory and Application, vol. 85, pp. 265–289, 1995.
18. F. Facchinei, "Minimization of SC1 functions and the Maratos effect," Operations Research Letters, vol. 17, pp. 131–137, 1995.
19. A. Forsgren and W. Murray, "Newton methods for large-scale linear inequality-constrained minimization," SIAM J. on Optimization, vol. 7, pp. 162-176, 1997.
20. A. Forsgren and W. Murray, "Newton methods for large-scale linear equality-constrained minimization," SIAM J. Matrix Anal. and Appl., vol. 14, pp. 560-587, 1993.
21. P. E. Gill, W. Murray and M. A. Saunders, "SNOPT: An SQP Algorithm for large-scale constrained optimization," Technical Report NA 97-2, Dept. of Mathematics, University of California, San Diego, 1997.
22. L. Grippo, F. Lampariello and S. Lucidi, "A truncated Newton Method with nonmonotone line search for unconstrained optimization," J. of Optimization Theory and Application, vol. 60, 1989.
23. L. Grippo, F. Lampariello and S. Lucidi, "A class of nonmonotone stabilization methods in unconstrained optimization," Numer. Math., vol. 59, pp. 779-805, 1991.
24. J.B. Hiriart-Urruty, J.J. Strodiot and V.H. Nguyen, "Generalized Hessian matrix and second-order optimality conditions for problems with $C^{1,1}$ data," Applied Mathematics and Optimization, vol. 11, pp. 43–56, 1984.
25. D. Klatte and K. Tammer, "On second-order sufficient optimality conditions for $C^{1,1}$- optimization Problems," Optimization, vol. 19, pp. 169–179, 1988.
26. S. Lucidi, "New results on a continuously differentiable exact penalty function," J. of Optimization Theory and Application, vol. 2, pp. 558-574, 1992.
27. B. A. Murtagh and M. A. Saunders, "Large-scale linearly constrained optimization," Mathematical Programming, vol. 14, pp. 41-72, 1978.
28. L. Qi and J. Sun, "A nonsmooth version of Newton's method," Mathematical Programming, vol. 58, pp. 353–367, 1993.

Appendix A

A.1.

Proof of Proposition 2: For all $x \notin \mathcal{F}$ the matrix $M(x)$ is positive definite due to the presence of the term $r(x)E$. Assume now that $x \in \mathcal{F}$, so that $r(x) = 0$ and the matrix $M(x)$ can be put in the dyadic form:

$$\begin{bmatrix}\text{diag}(Ax-b) & A(L-X)(X-U)D(x)^{\frac{1}{2}}\end{bmatrix}\begin{bmatrix}\text{diag}(Ax-b)\\ D(x)^{\frac{1}{2}}(L-X)(X-U)A^T\end{bmatrix}.$$

Hence we must prove that the column of the $(n+m)\times m$ matrix

$$\begin{bmatrix}\text{diag}(Ax-b)\\ D(x)^{\frac{1}{2}}(L-X)(X-U)A^T\end{bmatrix}$$

are linearly independent. Let $\mathcal{I}_0(x)$, $\mathcal{L}_0(x)$, $\mathcal{U}_0(x)$ be the index sets defined in (16). The columns corresponding to non active linear constraints are linearly independent due to the presence of the diagonal submatrix $\text{diag}(Ax-b)$. Therefore let us consider the columns corresponding to active linear constraints. Hence the assertion follows, by proving that the submatrix

$$(L - X)(X - U)A_{\mathcal{I}_0(x)}^T$$

has full rank. This is equivalent to prove that the matrix $E_{\mathcal{N}(x)} A_{\mathcal{I}_0}(x)^T$ has full rank with $\mathcal{N}(x) = \{1, \ldots, n\} \setminus \mathcal{L}_0(x) \cup \mathcal{U}_0(x)$, that follows from the (LICQ) assumption. ■

A.2. Proof of Proposition 4.

Preliminarily we establish the following inequalities:

PROPOSITION 15

(i) *For every* $(x, \lambda) \in \mathcal{S} \times \mathrm{IR}^m$ *and for all* $\varepsilon > 0$ *we have:*

$$P(x, \lambda; \varepsilon) \leq f(x) + \lambda^T(Ax - b) + \rho(x, \lambda)^T(l - x) + \sigma(x, \lambda)^T(x - u) + \varphi(x, \lambda)$$
$$+\frac{1}{2\varepsilon}\left[\|Ax - b\|^2 + (l - x)^T(q(x)^{-1} \circ (l - x)) + (x - u)^T(p(x)^{-1} \circ (x - u))\right];$$

(ii) *for every* $(x, \lambda) \in \mathcal{F} \times \mathrm{IR}^m$ *and for all* $\varepsilon > 0$ *we have:*

$$P(x, \lambda; \varepsilon) \leq f(x) + \varphi(x, \lambda);$$

(iii) *for every* $(x, \lambda) \in \mathcal{S} \times \mathrm{IR}^m$ *and for all* $\varepsilon > 0$ *we have:*

$$P(x, \lambda; \varepsilon) \geq f(x) + \varphi(x, \lambda) - \frac{\varepsilon}{2}\left[\rho(x, \lambda)^T(q(x) \circ \rho) + \sigma(x, \lambda)^T(p(x) \circ \sigma) + \lambda^T \lambda\right].$$

Proof: Point (i) follows from the fact that

$$\phi(v, z) = 2v^T z + v^T v - \min\{0, v + z\}^T \min\{0, v + z\},$$
$$\phi_W(v, w \circ z) = 2v^T z + v^T(w^{-1} \circ v) - \min\{0, v + w \circ z\}^T(w^{-1} \circ \min\{0, v + w \circ z\}).$$

Point (ii) follows from the definition of ϕ, $\phi_{P(x)}$, $\phi_{Q(x)}$. Indeed, we have that, for every feasible point, it results

$$\phi(Ax - b, \varepsilon\lambda) \leq 0, \quad \phi_{P(x)}(x - u, \varepsilon p(x) \circ \sigma(x, \lambda)) \leq 0, \quad \phi_{Q(x)}(l - x, \varepsilon q(x) \circ \rho(x, \lambda)) \leq 0.$$

Point (iii) follows from the fact that a positive definite quadratic form $2a^T x + x^T K x$ takes the minimum value $-a^T K^{-1} a$. Hence the proof follows, by considering the quadratic forms ϕ, $\phi_{P(x)}$, $\phi_{Q(x)}$. ■

Actually, we prove a more general result than Proposition 4. Indeed, the following proposition holds:

PROPOSITION 16 *Let* (x^0, λ^0) *be a given point in* $\mathcal{S} \times \mathbb{R}^m$, *then there exists a value* $\bar{\varepsilon}$ *such that, for every* $\varepsilon \in (0, \bar{\varepsilon}]$, *the level set* $\Omega(x^0, \lambda^0; \varepsilon)$ *is compact. In particular, there exists positive constants* $c_1(x^0, \lambda^0)$ *and* $c_2(x_0, \lambda_0)$ *such that*

$$\Omega(x^0, \lambda^0; \varepsilon) \subseteq \left\{(x, \lambda) : x \in \mathcal{S}, \ \|\lambda\| \leq c_1(x^0, \lambda^0) + \frac{1}{\varepsilon^{1/2}} c_2(x^0, \lambda^0)\right\}.$$

In the case that $x_0 \in \mathcal{F}$ *it results* $c_2(x_0, \lambda_0) = 0$ *so that we get Proposition 4.*

Proof: We recall that, without loss of generality, we assumed $\varepsilon \leq \varepsilon_M$.

From the definition in (13), we note that $\Omega(x^0, \lambda^0; \varepsilon)$ is closed, since $P(x, \lambda; \varepsilon) \to \infty$ as $x \to \partial\mathcal{S}$ and $\lambda \in \mathbb{R}^m$. Hence we must prove that it is also bounded.

First of all, we note that for every $(x, \lambda) \in \mathcal{S} \times \mathbb{R}^m$ we have

$$\begin{aligned}\varphi(x, \lambda) \geq \mu^2_{\min}(M(x))\|\lambda\|^2 & -2\left\|M(x)A(l-x)^2 \circ d(x) \circ (x-u)^2 \circ \nabla f(x)\right\| \|\lambda\| \\ & - \left\|A(l-x)^2 \circ d(x) \circ (x-u)^2 \circ \nabla f(x)\right\|^2.\end{aligned}$$

Hence by point (iii) of Proposition 15, and the expression of the multiplier functions (6), we can write:

$$P(x, \lambda; \varepsilon) \geq \tilde{k}_0(x; \varepsilon) + \tilde{k}_1(x; \varepsilon)\|\lambda\| + \tilde{k}_2(x; \varepsilon)\|\lambda\|^2,$$

where

$$\begin{aligned}\tilde{k}_0(x;\varepsilon) &= f(x) - \frac{\varepsilon}{2}\|\nabla f(x)\|^2 \Big[\|Q(x)\|\|D(x)(X-U)^2\|^2 \\ &\qquad + \|P(x)\|\|D(x)(L-X)^2\|^2\Big] - \left\|A(l-x)^2 \circ d(x) \circ (x-u)^2 \circ \nabla f(x)\right\|^2, \\ \tilde{k}_1(x;\varepsilon) &= -2\left\|M(x)A(l-x)^2 \circ d(x) \circ (x-u)^2 \circ \nabla f(x)\right\| \\ &\quad -\varepsilon\left[\|Q(x)\|\|Ad(x)^2 \circ (x-u)^4 \circ \nabla f(x)\| + \|P(x)\|\|Ad(x)^2 \circ (l-x)^4 \circ \nabla f(x)\|\right], \\ \tilde{k}_2(x;\varepsilon) &= \mu^2_{\min}(M(x)) \\ &\quad -\frac{\varepsilon}{2}\left[1 + \|Q(x)\|\,\|AD(x)^2(X-U)^4A^T\| + \|P(x)\|\,\|AD(x)^2(L-X)^4A^T\|\right].\end{aligned}$$

Let us define $k_0(\varepsilon) = \min_{x \in \bar{\mathcal{S}}} \tilde{k}_0(x; \varepsilon)$, $k_1(\varepsilon) = \max_{x \in \bar{\mathcal{S}}} \tilde{k}_1(x; \varepsilon)$, and

$$\begin{aligned}k_2(\varepsilon) &= \min_{x \in \bar{\mathcal{S}}} \mu^2_{\min}(M(x)) \\ &\quad -\frac{\varepsilon}{2}\max_{x \in \bar{\mathcal{S}}}\left[\|Q(x)\|\,\left\|AD(x)(X-U)^2\right\|^2 + \|P(x)\|\,\left\|AD(x)(L-X)^2\right\|^2 + 1\right].\end{aligned}$$

If (LICQ) holds, the matrix $M(x)$ is positive definite on $\bar{\mathcal{S}}$ and we can choose a value $\bar{\varepsilon}$

$$\bar{\varepsilon} < \frac{2\min_{x \in \bar{\mathcal{S}}} \mu^2_{\min}(M(x))}{\max_{x \in \bar{\mathcal{S}}}\left[\|Q(x)\|\|AD(x)(X-U)^2\|^2 + \|P(x)\|\|AD(x)(L-X)^2) + 1\|^2\right]}$$

such that, for all $\varepsilon \in (0, \bar{\varepsilon}]$ we have $k_2(\varepsilon) > 0$ and

$$P(x, \lambda; \varepsilon) \geq k_0(\bar{\varepsilon}) + k_1(\bar{\varepsilon})\|\lambda\| + k_2(\bar{\varepsilon})\|\lambda\|^2.$$

Now, by the expression (10) of the merit function and by point (i) of Proposition 15, for every $(x, \lambda) \in \mathcal{S} \times \mathbb{R}^m$ and $\varepsilon > 0$ we have that

$$P(x, \lambda; \varepsilon) \leq h_1(x, \lambda) + \frac{1}{2\varepsilon} h_2(x),$$

where

$$h_1(x, \lambda) = f(x) + \lambda^T(Ax - b) + \rho(x, \lambda)^T(l - x) + \sigma(x, \lambda)^T(x - u) + \varphi(x, \lambda),$$

and

$$h_2(x) = \|Ax - b\|^2 + \|(l - x)\|^2_{Q(x)^{-1}} + \|(x - u)\|^2_{P(x)^{-1}}.$$

Hence we have, for every $(x, \lambda) \in \mathcal{S} \times \mathbb{R}^m$ and for all $\varepsilon \in (0, \bar{\varepsilon}]$,

$$\begin{aligned}\Omega(x^0, \lambda^0; \varepsilon) &\subseteq \{(x, \lambda) \in \mathcal{S} \times \mathbb{R}^m : \ P(x, \lambda; \varepsilon) \leq h_1(x_0, \lambda_0) + \tfrac{1}{2\varepsilon} h_2(x_0)\} \\ &\subseteq \Big\{(x, \lambda) \in \mathcal{S} \times \mathbb{R}^m : \ k_0(\bar{\varepsilon}) + k_1(\bar{\varepsilon})\|\lambda\| + k_2(\bar{\varepsilon})\|\lambda\|^2 \\ &\qquad \leq h_1(x_0, \lambda_0) + \frac{1}{2\varepsilon} h_2(x_0)\Big\},\end{aligned}$$

that yields

$$\Omega(x^0, \lambda^0; \varepsilon) \subseteq \left\{(x, \lambda) \in \mathcal{S} \times \mathbb{R}^m : \ \|\lambda\| \leq c_1(x_0, \lambda_0) + \frac{1}{2\varepsilon} c_2(x_0, \lambda_0)\right\}$$

with

$$c_1 = -\frac{k_1(\bar{\varepsilon})}{2k_2(\bar{\varepsilon})} > 0$$

and

$$c_2(x_0, \lambda_0; \bar{\varepsilon}) = \left[\frac{4h_2(x_0)}{k_2(\bar{\varepsilon})} + \varepsilon_M \frac{k_1(\bar{\varepsilon})^2 - 4k_2(\bar{\varepsilon})\left(k_0(\bar{\varepsilon}) - h_1(x_0, \lambda_0)\right)}{4k_2(\bar{\varepsilon})^2}\right]^{\frac{1}{2}}.$$

Hence we have proved that $\Omega(x^0, \lambda^0; \varepsilon)$ is bounded.

If x_0 is feasible, then from point (ii) of Proposition 15, we have that $P(x_0, \lambda_0; \varepsilon) \leq f(x_0) + \varphi(x_0, \lambda_0)$, and hence, for $\varepsilon \in (0, \bar{\varepsilon}]$, we have

$$\Omega(x^0, \lambda^0; \varepsilon) \subseteq \{(x, \lambda) \in \mathcal{S} \times \mathbb{R}^m : \ k_0(\bar{\varepsilon}) + k_1(\bar{\varepsilon})\|\lambda\| + k_2(\bar{\varepsilon})\|\lambda\|^2 \leq h_1(x_0, \lambda_0)\},$$

that yields:

$$\Omega(x^0, \lambda^0; \varepsilon) \subseteq \{(x, \lambda) \in \mathcal{S} \times \mathbb{R}^m : \ \|\lambda\| \leq c_1(x_0, \lambda_0)\}$$

with

$$c_1(x_0, \lambda_0) = -\frac{k_1(\bar{\varepsilon})}{2k_2(\bar{\varepsilon})} + \left[\frac{k_1(\bar{\varepsilon})^2 - 4k_2(\bar{\varepsilon})\left(k_0(\bar{\varepsilon}) - h_1(x_0, \lambda_0)\right)}{4k_2(\bar{\varepsilon})^2}\right]^{\frac{1}{2}}.$$

■

A.3. Proof of Proposition 6

In order to prove Proposition 6, we need some preliminary technical results. The first result easily follows from the property of the $\max\{\cdot,\cdot\}$ function.

LEMMA 1 *For every* $(x,\lambda) \in \mathcal{S} \times \mathrm{I\!R}^m$ *the following identities hold:*

$$\left[l - x - \max\{l-x, -\varepsilon q(x)\circ\rho\}\right] \circ \left[\rho + \frac{1}{\varepsilon} q(x)^{-1} \circ \max\{l-x, -\varepsilon q(x)\circ\rho\}\right] = 0, \quad (A.1)$$

$$\left[x - u - \max\{x-u, -\varepsilon p(x)\circ\sigma\}\right] \circ \left[\sigma + \frac{1}{\varepsilon} p(x)^{-1} \circ \max\{x-u, -\varepsilon p(x)\circ\sigma\}\right] = 0, \quad (A.2)$$

$$\left[Ax - b - \max\{Ax-b, -\varepsilon\lambda\}\right] \circ \left[\lambda + \frac{1}{\varepsilon} \max\{Ax-b, -\varepsilon\lambda\}\right] = 0. \quad (A.3)$$

Given $x \in \mathcal{S}$ let us define the set $\Theta(x) \subseteq \mathrm{I\!R}^m$ as follows:

$$\Theta(x) = \Bigg\{ \lambda \in \mathrm{I\!R}^m : \|\nabla_\lambda P(x,\lambda;\varepsilon)\| \le \theta \Big[\|\max\{Ax-b, -\varepsilon\lambda\}\| + \|\max\{l-x, -\varepsilon p(x)\circ\rho(x,\lambda)\}\| + \|\max\{x-u, -\varepsilon q(x)\circ\sigma(x,\lambda)\}\| \Big] \Bigg\}. \quad (A.4)$$

PROPOSITION 17 *For every* $x \in \mathcal{S}$, *there exists a value* $\hat{\varepsilon}$ *such that, for all* $\varepsilon \in (0,\hat{\varepsilon}]$, *it results* $\Theta(x) \subseteq C$ *where* C *is a compact subset of* $\mathrm{I\!R}^m$.

Proof: In the following we assume that $\varepsilon \le \varepsilon_M$ and for sake of simplicity we set $\theta = 1$. t Consider the expression $\nabla_\lambda P$ given by (12). We have:

$$\begin{aligned}\|\nabla_\lambda P(x,\lambda;\varepsilon)\| \ge\ & 2\mu^2_{\min}(M(x))\|\lambda\| - 2\|M(x)A(l-x)^2\circ d(x)\circ(x-u)^2\circ\nabla f(x)\| \\ & -\|\max\{Ax-b, -\varepsilon\lambda\}\| - \|AD(x)(X-U)^2\|\ \|\max\{l-x, -\varepsilon q(x)\circ\rho(x,\lambda)\}\| \\ & -\|AD(x)(L-X)^2\|\ \|\max\{x-u, -\varepsilon p(x)\circ\sigma(x,\lambda)\}\|.\end{aligned}$$

Moreover we have

$$\begin{aligned}\|\max\{Ax-b, -\varepsilon\lambda\}\| &\le \|Ax-b\| + \varepsilon\|\lambda\|, \\ \|\max\{l-x, -\varepsilon q(x)\circ\rho(x,\lambda)\}\| &\le \|l-x\| + \varepsilon_M\|q(x)\circ d(x)\circ(x-u)^2\circ\nabla f(x)\| \\ &\quad +\varepsilon\|Q(x)D(x)(X-U)^2A^T\|\ \|\lambda\|, \\ \|\max\{x-u, -\varepsilon p(x)\circ\sigma(x,\lambda)\}\| &\le \|x-u\| + \varepsilon_M\|p(x)\circ d(x)\circ(l-x)^2\circ\nabla f(x)\| \\ &\quad +\varepsilon\|P(x)D(x)(L-X)^2A^T\|\ \|\lambda\|.\end{aligned}$$

Hence, for $x \in \mathcal{S}$ the set $\Theta(x)$ is contained in the set defined by the following inequality:

$$\begin{aligned}\Big[2\mu^2_{\min}(M(x)) - \varepsilon\Big(1 + \|AD(x)(X-U)^2\|\|Q(x)D(x)(X-U)^2A^T\| \\ + \|AD(x)(L-X)^2\|\|P(x)D(x)(L-X)^2A^T\|\Big)\Big]\|\lambda\| - \|Ax-b\| \\ -\|AD(x)(X-U)^2\|\left(\|l-x\| + \varepsilon_M\|q(x)\circ d(x)\circ(x-u)^2\circ\nabla f(x)\|\right) \\ -\|AD(x)(L-X)^2\|\left(\|x-u\| + \varepsilon_M\|p(x)\circ d(x)\circ(l-x)^2\circ\nabla f(x)\|\right) \\ -2\|M(x)A(l-x)^2\circ d(x)\circ(x-u)^2\circ\nabla f(x)\| \le 0.\end{aligned} \tag{A.5}$$

Let us define the following constants

$$c_1 = 2\min_{x\in\bar{\mathcal{S}}} \mu^2_{\min}(M(x)) > 0,$$

$$\begin{aligned}c_2 = 1 + \max_{x\in\bar{\mathcal{S}}}\Big[\|AD(x)(X-U)^2\|\|Q(x)D(x)(X-U)^2A^T\| \\ + \|AD(x)(L-X)^2\|\|P(x)D(x)(L-X)^2A^T\|\Big] > 0,\end{aligned}$$

$$\begin{aligned}c_3 = \max_{x\in\bar{\mathcal{S}}}\Big[\|Ax-b\| + 2\|M(x)A(l-x)^2\circ d(x)\circ(x-u)^2\circ\nabla f(x)\| \\ +(1+\|AD(x)(X-U)^2\|)\left(\|l-x\| + \varepsilon_M\|q(x)\circ d(x)\circ(x-u)^2\circ\nabla f(x)\|\right) \\ +(1+\|AD(x)(L-X)^2\|)\left(\|x-u\| + \varepsilon_M\|p(x)\circ d(x)\circ(l-x)^2\circ\nabla f(x)\|\right)\Big].\end{aligned}$$

Now we have from (A.5)

$$(c_1 - \varepsilon c_2)\|\lambda\| \le c_3,$$

and hence for every $x \in \mathcal{S}$ and for $\varepsilon \le \hat{\varepsilon}$ with $\hat{\varepsilon} < c_1/c_2$ we have

$$\Theta(x) \subseteq \left\{\lambda \in \mathbb{R}^m : \|\lambda\| \le \frac{c_3}{c_1 - \hat{\varepsilon}c_2}\right\}.$$

■

PROPOSITION 18 *For every $\tilde{x} \in \mathcal{F}$, there exist values $\varepsilon(\tilde{x})$, $\delta_1(\tilde{x}) > 0$ and $\delta_2(\tilde{x}) > 0$ such that for all $\varepsilon \in (0, \varepsilon(\tilde{x})]$, for all $x \in \mathcal{S}$ satisfying $\|x - \tilde{x}\| \le \delta_1(\tilde{x})$ and for all $\lambda \in \Theta(x)$ the following inequality holds:*

$$\begin{aligned}\|\varepsilon\nabla_x P(x,\lambda;\varepsilon)\|^2 \ge \delta_2(\tilde{x})\Big(\|\max\{Ax-b, -\varepsilon\lambda\}\|^2 + \|\max\{l-x, -\varepsilon q(x)\circ\rho(x,\lambda)\}\|^2 \\ + \|\max\{x-u, -\varepsilon p(x)\circ\sigma(x,\lambda)\}\|^2\Big).\end{aligned}$$

Proof: In the following we assume that $\varepsilon \le \varepsilon_M$. For short notation, we denote

$$\begin{aligned}\pi(x,\lambda) &= M(x)\lambda + A(l-x)^2\circ d(x)\circ(x-u)^2\circ\nabla f(x) \\ &= (Ax-b)^2\circ\lambda + r(x)\lambda - A(x-u)^2\circ\sigma(x,\lambda)\end{aligned} \tag{A.6}$$

so that $\varphi(x,\lambda) = \|\pi(x,\lambda)\|^2$.

Moreover we introduce the following vector $v(x,\lambda;\varepsilon) \in \mathbb{R}^{m+2n}$:

$$v(x,\lambda;\varepsilon) = \begin{bmatrix} \max\{Ax-b, -\varepsilon\lambda\} \\ \max\{l-x, -\varepsilon q(x)\circ\rho(x,\lambda)\} \\ \max\{x-u, -\varepsilon p(x)\circ\sigma(x,\lambda)\} \end{bmatrix}.$$

First we observe that, since $M(x)$ is positive definite, by (12) we can write

$$\begin{aligned}\pi(x,\lambda) = \frac{1}{2}[M(x)]^{-1}\Big[&\nabla_\lambda P(x,\lambda;\varepsilon) - \max\{Ax-b, -\varepsilon\lambda\} \\ &-AD(x)(X-U)^2 \max\{l-x, -\varepsilon q(x)\circ\rho(x,\lambda)\} \\ &+AD(x)(L-X)^2 \max\{x-u, -\varepsilon p(x)\circ\sigma(x,\lambda)\}\Big].\end{aligned}$$

Hence for every $x \in S$ and $\lambda \in \Theta(x)$, there exists a constant c_4 such that

$$\varphi(x,\lambda) \le c_4\|v(x,\lambda;\varepsilon)\|^2.$$

On the other hand we can write:

$$\begin{aligned}\varepsilon\nabla_x P(x,\lambda;\varepsilon) = \ & A^T \max\{Ax-b, -\varepsilon\lambda\} \\ & -q(x)^{-1}\circ\max\{l-x, -\varepsilon q(x)\circ\rho(x,\lambda)\} \\ & -\frac{q(x)^{-2}}{2}\circ\max\{l-x, -\varepsilon q(x)\circ\rho(x,\lambda)\}^2 \\ & +p(x)^{-1}\circ\max\{x-u, -\varepsilon p(x)\circ\sigma(x,\lambda)\} \\ & +\frac{p(x)^{-2}}{2}\circ\max\{x-u, -\varepsilon p(x)\circ\sigma(x,\lambda)\}^2 \\ & +\varepsilon\Big[\nabla_x\rho(x,\lambda)\max\{l-x, -\varepsilon q(x)\circ\rho(x,\lambda)\} \\ & +\nabla_x\sigma(x,\lambda)\max\{x-u, -\varepsilon p(x)\circ\sigma(x,\lambda)\} \\ & +2\nabla_x\pi(x,\lambda)\pi(x,\lambda)\Big].\end{aligned}$$

Now, by making the Hadamard product between (A.1) and $(l - x)$, and between (A.2) and $(x - u)$, summing up and then subtracting the first of (5) we obtain the n equations

$$\begin{aligned} 0 = \; & (l-x)^2 \circ q(x)^{-1} \circ \max\{l - x, -\varepsilon q(x) \circ \rho(x,\lambda)\} \\ - \; & (l-x) \circ q(x)^{-1} \circ \max\{l - x, -\varepsilon q(x) \circ \rho(x,\lambda)\}^2 \\ + \; & (x-u)^2 \circ p(x)^{-1} \circ \max\{x - u, -\varepsilon p(x) \circ \sigma(x,\lambda)\} \\ - \; & (x-u) \circ p(x)^{-1} \circ \max\{x - u, -\varepsilon p(x) \circ \sigma(x,\lambda)\}^2 \\ - \; & \varepsilon \Big[(l-x) \circ \rho(x,\lambda) \circ \max\{l - x, -\varepsilon q(x) \circ \rho(x,\lambda)\} \\ + \; & (x-u) \circ \sigma(x,\lambda) \circ \max\{x - u, -\varepsilon p(x) \circ \sigma(x,\lambda)\} \Big]. \end{aligned}$$

By making the Hadamard product between (A.3) and $(Ax - b)$, by premultiplying (A.2) by $A(X - U)$, then subtracting and recalling (A.6), we obtain the m equations

$$\begin{aligned} 0 = \; & \varepsilon \pi(x,\lambda) - \varepsilon r(x) \lambda \\ & -(Ax - b) \circ [\varepsilon \lambda + \max\{Ax - b, -\varepsilon \lambda\} - (Ax - b)] \circ \max\{Ax - b, -\varepsilon \lambda\} \\ & + \varepsilon A \sigma(x,\lambda) \circ (x - u) \circ \max\{x - u, -\varepsilon p(x) \circ \sigma(x,\lambda)\} \\ & - A p(x)^{-1} \circ (x-u)^2 \max\{x - u, -\varepsilon p(x) \circ \sigma(x,\lambda)\} \\ & + A p(x)^{-1} \circ (x-u) \circ \max\{x - u, -\varepsilon p(x) \circ \sigma(x,\lambda)\}^2. \end{aligned}$$

Hence we obtain a system of $2n + m$ equations in the $2n + m$ variables $v(x, \lambda; \varepsilon)$ of the kind:

$$\begin{bmatrix} \varepsilon \nabla_x P(x,\lambda;\varepsilon) \\ 0 \\ 0 \end{bmatrix} = K_1(x,\lambda;\varepsilon) v(x,\lambda;\varepsilon) + \varepsilon K_2(x,\lambda) \pi(x,\lambda) - \varepsilon r(x) \begin{bmatrix} 0 \\ 0 \\ \lambda \end{bmatrix}.$$

For $\varepsilon = 0$ and $\tilde{x} \in \mathcal{F}$ we have that the matrix $K_1(\tilde{x}, \lambda; 0)$ is given by:

$$K_1(\tilde{x},\lambda;0) = \begin{bmatrix} A^T & -E & E \\ 0 & (L - \tilde{X})^2 & (\tilde{X} - U)^2 \\ \mathrm{diag}\{A\tilde{x} - b\}^2 & 0 & -A(\tilde{X} - U)^2 \end{bmatrix} \begin{bmatrix} E & 0 & 0 \\ 0 & Q(\tilde{x})^{-1} & 0 \\ 0 & 0 & P(\tilde{x})^{-1} \end{bmatrix},$$

where E and 0 are respectively the identity and the zero matrix of appropriate dimension. In order to prove that the matrix $K_1(\tilde{x}, \lambda; 0)$ is nonsingular, we show that

$$\begin{bmatrix} A^T & -E & E \\ 0 & (L - \tilde{X})^2 & (\tilde{X} - U)^2 \\ \mathrm{diag}\{A\tilde{x} - b\}^2 & 0 & -A(\tilde{X} - U)^2 \end{bmatrix} \begin{bmatrix} u \\ v \\ w \end{bmatrix} = 0 \quad \Longleftrightarrow \quad \begin{bmatrix} u \\ v \\ w \end{bmatrix} = 0.$$

By simple calculation, this is equivalent to show that $M(\tilde{x})w = 0 \iff w = 0$; this follows from the non singularity of the matrix $M(\tilde{x})$ that has been proved in Proposition 2.

By continuity, there exist a value $\varepsilon(\tilde{x})$, and a value $\delta_1(\tilde{x})$ such that $K_1(x, \lambda; \varepsilon)^2$ is positive definite for all $\varepsilon \leq \varepsilon(\tilde{x})$ and $\|x - \tilde{x}\| \leq \delta_1(\tilde{x})$, and we can write:

$$\|\varepsilon \nabla_x P(x, \lambda; \varepsilon)\|^2 \geq \left[\mu_{\min}(K_1(x, \lambda; \varepsilon)^2) - \varepsilon^2 \|K_2(x, \lambda)\|^2 c_4\right] \|v(x, \lambda; \varepsilon)\|^2 - \varepsilon^2 \|\lambda\|^2 r(x)^2$$

with $\mu_{\min}(K_1(x, \lambda; \varepsilon)^2) > 0$. By using the equivalence of the norms, there exists a constant $c_5 > 0$ such that

$$r(x)^2 = \|v(x, \lambda; 0)\|_p^{2p} \leq c_5 \|v(x, \lambda; 0)\|_2^{2p} \leq c_5 \|v(x, \lambda; 0)\|^{2(p-1)} \|v(x, \lambda; \varepsilon)\|^2.$$

Therefore we can write

$$\begin{aligned}
&\|\varepsilon \nabla_x P(x, \lambda; \varepsilon)\|^2 \\
&\geq \left[\mu_{\min}(K_1(x, \lambda; \varepsilon)^2) - \varepsilon^2 \left(c_4 \|K_2(x, \lambda)\|^2 + c_5 \|v(x, \lambda; 0)\|^{2(p-1)} \|\lambda\|^2\right)\right] \|v(x, \lambda; \varepsilon)\|^2 \\
&= \Big[\mu_{\min}(K_1(\tilde{x}, \lambda; 0)^2) + \left(\mu_{\min}(K_1(x, \lambda; \varepsilon)^2) - \mu_{\min}(K_1(\tilde{x}, \lambda; 0)^2)\right) \\
&\qquad - \varepsilon^2 \left(c_4 \|K_2(x, \lambda)\|^2 + c_5 \|v(x, \lambda; 0)\|^{2(p-1)} \|\lambda\|^2\right)\Big] \|v(x, \lambda; \varepsilon)\|^2
\end{aligned}$$

and finally, taking into account that λ stays in a compact set, (see Proposition 17) and that $\mu_{\min}(K_1(\tilde{x}, \lambda; 0)^2)$ does not depend on λ we get that there exists a constant $\delta_2(\tilde{x})$ such that

$$\|\varepsilon \nabla_x P(x, \lambda; \varepsilon\|^2 \geq \delta_2(\tilde{x}) \|v(x, \lambda; \varepsilon)\|^2,$$

for all $(x, \lambda; \varepsilon)$ such that $\varepsilon \in (0, \varepsilon(\tilde{x})]$, $x \in S$ and $\|x - \tilde{x}\| \leq \delta_1(\tilde{x})$, $\lambda \in \Theta(x)$; so that the proposition is proved. ■

PROPOSITION 19 *Let $\{\varepsilon^k\}$ be a sequence of positive numbers converging to zero, C be a compact subset of $\mathbb{R}^m$ and $\{x^k, \lambda^k\}$ be a sequence of points such that $(x^k, \lambda^k) \in S \times C$. Assume that:*

$$\lim_{k \to \infty} \varepsilon^k \nabla_x P(x^k, \lambda^k; \varepsilon^k) = 0. \tag{A.7}$$

Then either every limit point $(\tilde{x}, \tilde{\lambda})$ of the sequence $\{(x^k, \lambda^k)\}$ is such that $\tilde{x} \in \mathcal{F}$ or the feasible set $\mathcal{F}$ is empty.

Proof: Recalling the expression (11), we have that

$$\begin{aligned}
0 = \lim_{k \to \infty} \varepsilon^k \nabla_x P(x^k, \lambda^k; \varepsilon^k) &= A^T \max\{A\tilde{x} - b, 0\} \\
&\quad - \left[q(\tilde{x})^{-1} + \frac{q(\tilde{x})^{-2}}{2} \circ \max\{l - \tilde{x}, 0\}\right] \circ \max\{l - \tilde{x}, 0\} \\
&\quad + \left[p(\tilde{x})^{-1} + \tfrac{1}{2} p(\tilde{x})^{-2} \circ \max\{\tilde{x} - u, 0\}\right] \circ \max\{\tilde{x} - u, 0\}.
\end{aligned}$$

This implies that:

$$\begin{bmatrix} A^T & -E & E \end{bmatrix} \begin{bmatrix} \max\{A\tilde{x} - b, 0\} \\ \bar{v} \circ \max\{l - \tilde{x}, 0\} \\ \bar{w} \circ \max\{\tilde{x} - u, 0\} \end{bmatrix} = 0,$$

where $\bar{v}, \bar{w}$ are strictly positive vectors. Then, either $\max\{A\tilde{x} - b, 0\} = 0$, $\max\{l - \tilde{x}, 0\} = \max\{\tilde{x} - u, 0\} = 0$, that means that $\tilde{x} \in \mathcal{F}$, or at least one of these equalities does not hold, that means that $\tilde{x} \notin \mathcal{F}$. In this second case, let us denote by $\mathcal{I}_> = \{i : a_i^T \tilde{x} > b_i\}$, $\mathcal{L}_> = \{j : \tilde{x} < l_j\}$ and $\mathcal{U}_> = \{j : \tilde{x} > u_j\}$; then there exists a solution v of the system

$$\begin{bmatrix} A_{\mathcal{I}_>}^T & -E_{\mathcal{L}_>}^T & E_{\mathcal{U}_>}^T \end{bmatrix} v = 0 \quad v \geq 0 \quad v \neq 0.$$

By the Gordan theorem of the alternative we have that there is no solution $y \in \mathbb{R}^n$ to the system

$$\begin{bmatrix} A_{\mathcal{I}_>} \\ -E_{\mathcal{L}_>} \\ E_{\mathcal{U}_>} \end{bmatrix} y > 0, \quad \text{that is} \quad A_{\mathcal{I}_>} y > 0, \;\; y_{\mathcal{L}_>} < 0, \;\; y_{\mathcal{U}_>} > 0, \tag{A.8}$$

and this implies that $\mathcal{F}$ is empty. In fact, assume by contradiction that $\mathcal{F}$ is not empty and let $\hat{x} \in \mathcal{F}$. Hence, by definition, we have that

$$\begin{array}{lll} A_{\mathcal{I}_>} \tilde{x} > b_{\mathcal{I}_>} & \tilde{x}_{\mathcal{L}_>} < l_{\mathcal{L}_>} & \tilde{x}_{\mathcal{U}_>} > u_{\mathcal{U}_>} \\ A_{\mathcal{I}_>} \hat{x} \leq b_{\mathcal{I}_>} & \hat{x}_{\mathcal{L}_>} \geq l_{\mathcal{L}_>} & \tilde{x}_{\mathcal{U}_>} \leq u_{\mathcal{U}_>} \end{array}$$

and by subtracting we would get that the point $(\tilde{x} - \hat{x}) > 0$ satisfies (A.8) and this yield a contradiction. ■

COROLLARY 1 *Let $\{\varepsilon^k\}$ be a sequence of positive numbers converging to zero, $x^0 \in \mathcal{F}$ and $\{x^k, \lambda^k\}$ be a sequence of points such that $(x^k, \lambda^k) \in \Omega(x^0, \lambda^0)$. Then every limit point $(\tilde{x}, \tilde{\lambda})$ of the sequence $\{(x^k, \lambda^k)\}$ is such that $\tilde{x} \in \mathcal{F}$.*

Proof of Proposition 6: The proof is by contradiction. Suppose that the result is false; then there exists a value $\theta > 0$ and sequences $\{\varepsilon^k\}$ and $\{x^k, \lambda^k\}$ such that

$$\{\varepsilon^k\} \to 0, \tag{A.9}$$

$$\{x^k, \lambda^k\} \in \mathcal{S} \times \mathbb{R}^m, \tag{A.10}$$

$$\begin{aligned} \|\nabla_x P(x^k, \lambda^k; \varepsilon^k)\|^2 < \theta\Big(& \|\max\{Ax^k - b, -\varepsilon^k \lambda^k\}\|^2 \\ & + \|\max\{l - x^k, -\varepsilon^k q(x^k) \circ \rho(x^k, \lambda^k)\}\|^2 \\ & + \|\max\{x^k - u, -\varepsilon^k p(x^k) \circ \sigma(x^k, \lambda^k)\}\|^2 \Big), \end{aligned} \tag{A.11}$$

$$\begin{aligned} \|\nabla_\lambda P(x^k, \lambda^k; \varepsilon^k)\|^2 < \theta\Big(& \|\max\{Ax^k - b, -\varepsilon^k \lambda^k\}\|^2 \\ & + \|\max\{l - x^k, -\varepsilon^k q(x^k) \circ \rho(x^k, \lambda^k)\}\|^2 \\ & + \|\max\{x^k - u, -\varepsilon^k p(x^k) \circ \sigma(x^k, \lambda^k)\}\|^2 \Big). \end{aligned} \tag{A.12}$$

From (A.11) and Proposition 17 we have that there exists a value $\bar{k}$ such that, for $k \geq \bar{k}$, λ^k stays in a compact set. Hence, recalling that $\{x^k\} \in \mathcal{S}$, we obtain

$$\lim_{k\to\infty} \varepsilon^k \left[\|\max\{Ax^k - b, -\varepsilon^k\lambda^k\}\|^2 + \|\max\{l - x^k, -\varepsilon^k q(x^k)\circ\rho(x^k,\lambda^k)\}\|^2 + \|\max\{x^k - u, -\varepsilon^k p(x^k)\circ\sigma(x^k,\lambda^k)\}\|^2\right] = 0.$$

From (A.11) multiplying by ε^k we obtain also that:

$$\lim_{k\to\infty} \varepsilon^k \nabla_x P(x^k, \lambda^k; \varepsilon^k) = 0. \tag{A.13}$$

Since $\{x^k, \lambda^k\}$ is bounded, it admits a limit point $(\tilde{x}, \tilde{\lambda})$. Then from (A.13) and Proposition 19 we obtain that $\tilde{x} \in \mathcal{F}$. Now, by using Proposition 18 we get for sufficiently large k a contradiction with (A.11). ■

A.4.

Proof of Proposition 10: By Proposition 3, the pair $(\bar{x}, \bar{\lambda})$ is a stationary point for $P(x, \lambda; \varepsilon)$. Then by Proposition 9 and by the Caratheodory theorem we have that every matrix $\bar{H}(\bar{x}, \bar{\lambda}; \varepsilon)$ in $\tilde{\partial}^2 P(\bar{x}, \bar{\lambda}; \varepsilon)$ can be written in the form:

$$\bar{H}(\bar{x}, \bar{\lambda}; \varepsilon) = \sum_{i=1}^{t} \beta_i H(\bar{x}, \bar{\lambda}; \varepsilon, \mathcal{W}_i),$$

where $t \leq (n+m)^2+1$, $\beta_i \geq 0$, $\sum_{i=1}^{t} \beta_i = 1$. Then we need only to prove that $H(\bar{x}, \bar{\lambda}; \varepsilon, W_i)$ is positive definite for any $\mathcal{W}_i \in \mathcal{P}$.

Consider now the following quadratic form in $(v, w) \in \mathbb{R}^n \times \mathbb{R}^m$:

$$(v^T, w^T) H(\bar{x}, \bar{\lambda}; \varepsilon, W) \begin{pmatrix} v \\ w \end{pmatrix} = s(v, w; W) + \frac{1}{\varepsilon} q(v, w; W) + \varepsilon r(v, w; W), \tag{A.14}$$

where

$$\begin{aligned}
s(v, w; \mathcal{W}) &= v^T \nabla_x^2 f(\bar{x}) v + 2\|AD(\bar{x})(\bar{X} - U)^2 (L - \bar{X})^2 \nabla_x^2 f(\bar{x}) v + \tilde{M}(\bar{x}) w\|^2 \\
&\quad -2v_{\mathcal{L}}^T \left(\nabla_x \rho_{\mathcal{L}}^T v + \nabla_\lambda \rho_{\mathcal{L}}^T w\right) + 2v_{\mathcal{U}}^T \left(\nabla_x \sigma_{\mathcal{U}}^T v + \nabla_\lambda \sigma_{\mathcal{U}}^T w\right) + 2v^T A_{\mathcal{I}_a}^T w_{\mathcal{I}_a}, \\
q(v, w; \mathcal{W}) &= v_{\mathcal{L}}^T Q_{\mathcal{L},\mathcal{L}}^{-1} v_{\mathcal{L}} + v_{\mathcal{U}}^T P_{\mathcal{U},\mathcal{U}}^{-1} v_{\mathcal{U}} + \|A_{\mathcal{I}_a} v\|^2, \\
r(v, w; \mathcal{W}) &= -v^T \left(\nabla_x \rho_{\mathcal{N}} Q_{\mathcal{N},\mathcal{N}} \nabla_x \rho_{\mathcal{N}}^T + \nabla_x \sigma_{\mathcal{N}} P_{\mathcal{N},\mathcal{N}} \nabla_x \sigma_{\mathcal{N}}^T\right) v \\
&\quad -w^T \left(\nabla_\lambda \rho_{\mathcal{N}} Q_{\mathcal{N},\mathcal{N}} \nabla_\lambda \rho_{\mathcal{N}}^T + \nabla_\lambda \sigma_{\mathcal{N}} P_{\mathcal{N},\mathcal{N}} \nabla_\lambda \sigma_{\mathcal{N}}^T\right) w - w^T E_{\mathcal{I}_n}^T w_{\mathcal{I}_n} \\
&\quad -2v^T \left(\nabla_x \rho_{\mathcal{N}} Q_{\mathcal{N},\mathcal{N}} \nabla_\lambda \rho_{\mathcal{N}}^T + \nabla_x \sigma_{\mathcal{N}} P_{\mathcal{N},\mathcal{N}} \nabla_\lambda \sigma_{\mathcal{N}}^T\right) w.
\end{aligned}$$

Recalling known results on the sum of quadratic forms (see, for example, [10]) it is possible to assert that if for every $(v, w) \in \mathbb{R}^n \times \mathbb{R}^m$ we have that

(i) $q(v, w; W) \geq 0$,

(ii) $q(v, w; W) = 0$ and $s(v, w; W) \leq 0$ imply $v = 0$ and $w = 0$,

then there exists a value $\bar{\varepsilon}$ such that $s(v, w; W) + (1/\varepsilon)q(v, w; W) + \varepsilon r(v, w; W)$ is positive definite for all $\varepsilon \in (0, \bar{\varepsilon}]$.

We note that $q(v, w; W) \geq 0$ and hence (i) holds. As regards (ii), we observe that $q(v, w; W) = 0$ implies that

$$A_{I_+(\bar{x},\bar{\lambda})} v = 0, \quad v_{\mathcal{L}_+(\bar{x},\bar{\rho})} = 0, \quad v_{\mathcal{U}_+(\bar{x},\bar{\sigma})} = 0.$$

Therefore, since the point $(\bar{x}, \bar{\lambda}, \bar{\rho}, \bar{\sigma})$ satisfies the strong second order sufficient condition for Problem (1), $s(v, w; W) \leq 0$ and $q(v, w; W) = 0$ imply that

$$v = 0, \qquad \tilde{M}(\bar{x})w = 0.$$

Since, by the LICQ, the matrix $\tilde{M}(\bar{x})$ is nonsingular, this last equality implies that $w = 0$.

Hence, by (A.14), there exists a value $\bar{\varepsilon}$ such that every matrix in $\tilde{\partial}^2 P(\bar{x}, \bar{\lambda}; \varepsilon)$ is positive definite for every $\varepsilon \in (0, \bar{\varepsilon}]$. Then the pair $(\bar{x}, \bar{\lambda})$ satisfies also the second order sufficient condition to be an isolated minimum point for P, see [25]. ■

Computational Optimization and Applications 12, 189–205 (1999)

Arithmetic Continuation of Regular Roots of Formal Parametric Polynomial Systems

B. CURTIS EAVES
Department of EES and OR, Stanford University, Stanford, CA 94305-4022, USA

URIEL G. ROTHBLUM
Faculty of Industrial Engineering and Management, Technion—Israel Institute of Technology, Haifa 32000, Israel

Received February 21, 1998; Accepted June 7, 1998

Abstract. Given a regular system of polynomial equations with power series coefficients, an initial root is continued as a power series. With the ground domain as an arbitrary field, arithmetic alone is used for the root continuation over this field, and computation is quadratic in the number of computed coefficients. If the power series of the coefficients of the polynomial are geometrically bounded, then the coefficients of the power series of the root are also.

Keywords: polynomial systems, roots, arithmetic continuation, parametric roots, fields

Dedication: The authors dedicate this paper to Olvi Mangasarian for his friendship and his many contributions to the field of mathematical programming, both over many years.

1. Introduction

A parametric system $\mathbb{P}$ of n polynomials in n variables is considered; coefficients of the polynomials and coordinates of the roots are formal power series over an arbitrary field. Under an initial regularity condition, an initial root of the initial system $\mathbb{P}$ is continued uniquely to a parametric root of the parametric system. Succeeding coefficients of the parametric root are computed recursively using only arithmetic, in polynomial time. If the coefficients of the polynomial are geometrically bounded, then the components of the parametric root are also.

The coefficients of the parametric polynomials are formal power series in the parameter ω of the form

$$\sum_{i=0}^{+\infty} a(i)\omega^i$$

over an arbitrary field F; the $a(i)$'s are in F and the parameter ω is an indeterminate over F.

By the operator "initial" is meant replace ω with 0. Given a root r of the initial system our goal is to find a parametric root of the parametric system with r as the initial root. We say that r is a regular root of the initial system $\mathbb{P}$ if the derivative of the initial system with respect to its n variables evaluated at r is nonsingular. Assuming r is regular, we show that it can be continued to a parametric root $\sum_{i=0}^{+\infty} r(i)\omega^i$ of the parametric system. We show that succeeding coefficients $r(i)$ of the power series of the continued root can be computed using only arithmetic in F, with $Kni + n^2$ multiplications and additions, with K as the sum of the number of terms of the polynomial system and the maximum degree of the individual terms. In particular, we see that coefficients of the power series of the parametric root lie in the same field as those of the parametric system $\mathbb{P}$ and root r.

If the field F has an absolute value, we say that a formal power series is *geometrically bounded* if for some A and B its coefficients satisfy $|a(i)| \leq AB^i$ for each $i \geq 0$. When all power series in the parametric system are geometrically bounded, we show that the continued root has this property also.

For the special case where F is the real or complex numbers, geometric boundedness coincides with absolute convergence. The existence of our continuation then follows from the analytic form of the implicit function theorem; however, the arithmetic results do not follow. In [4], one finds a full discussion of the parametric root problem, its importance and its relationship to the implicit function theorem; see also [3] and references therein.

Preliminaries about domains and polynomials are summarized in Section 2 and explicit representations of expansions of valuations of polynomials over the parametric field at parametric elements are provided in Section 4. These expansions are used in Section 4 to develop a recursive formulae for parametric roots. Geometric boundedness of these roots is obtained in Section 5. Finally, in Section 6 we provide an example of a parametric polynomial system and its solution by our methods.

2. Domains and polynomials

The domains we use throughout are the nonnegative integers $Z_{\oplus}$, an arbitrary field F to which we refer as the *ground field*, the ring of formal power series $F\lfloor\omega\rfloor$ and the rings of polynomials $F[x]$ and $F\lfloor\omega\rfloor[x]$. Let n be a positive number and let $N \equiv \{1, \ldots, n\}$. Relative to F, let $\omega, x_1, \ldots, x_n$ be $n+1$ indeterminates and let $x \equiv (x_1, \ldots, x_n)$.

Let $F\lfloor\omega\rfloor$ be the formal commutative ring of power series in ω; elements in $F\lfloor\omega\rfloor$ have representation

$$\alpha = \sum_{j=0}^{+\infty} a(i)\omega^i \tag{2.1}$$

with coefficients $a(i)$ in the ground field F. The ω^0-coefficient of α, that is $a(0)$, is referred to as *initial* α and is sometimes denoted $\alpha_{\diamond}$. Addition in $F\lfloor\omega\rfloor$ is componentwise and multiplication is by standard convolution. We shall use the notation $\sum_i$ to abbreviate $\sum_{i=0}^{+\infty}$. An element $\alpha = (\alpha_1, \ldots, \alpha_n)$ in $(F\lfloor\omega\rfloor)^n$ is also written as $\alpha = \sum_i a(i)\omega^i$ with $a(i)$ in F^n for each i in $Z_{\oplus}$; in this case $\alpha_k = \sum_i a(i)_k\omega^i$.

Let $F[x]$ and $F\lfloor\omega\rfloor[x]$ be the commutative rings of polynomials in x over F and $F\lfloor\omega\rfloor$, respectively. As usual, we regard $F \subseteq F\lfloor\omega\rfloor \subseteq F\lfloor\omega\rfloor[x]$ and $F \subseteq F[x] \subseteq F\lfloor\omega\rfloor[x]$.

For n-vectors $\sigma = (\sigma_1, \ldots, \sigma_n)$ in $(F\lfloor\omega\rfloor[x])^n$ and $\mu = (\mu_1, \ldots, \mu_n)$ in $Z^n_\oplus$, let σ^μ denote the product (of the powers) $(\sigma_1)^{\mu_1}(\sigma_2)^{\mu_2} \cdots (\sigma_n)^{\mu_n}$, in particular, we have $x^\mu = (x_1)^{\mu_1}(x_2)^{\mu_2} \cdots (x_n)^{\mu_n}$. A polynomial $P = P(x)$ in $F\lfloor\omega\rfloor[x]$ then has a representation

$$P(x) = \sum_{\mu \in M} \alpha_\mu x^\mu \tag{2.2}$$

with M a finite subset of $Z^n_\oplus$ and α_μ in $F\lfloor\omega\rfloor$ for each μ in M. Such P is in $F[x]$ if and only if each α_μ is in F. As usual, given ρ in $(F\lfloor\omega\rfloor)^n$, we denote by $P(\rho)$ the value in $F\lfloor\omega\rfloor$ obtained by replacing x_k by ρ_k for each k in N. The polynomial $\sum_{\mu \in M} \alpha_{\mu\Diamond} x^\mu$ in $F[x]$ is referred to as *initial* P and is denoted $P_\Diamond$.

Let $P(x)$ in $F\lfloor\omega\rfloor[x]$ have a representation as in (2.2) and let ν be a vector in $Z^n_\oplus$. With the standard notation for Binomial coefficients, the (formal) *ν-derivative of $P(x)$ with respect to ν*, denoted $D_\nu P$, is the polynomial in $F\lfloor\omega\rfloor[x]$ defined by

$$D_\nu P(x) \equiv \sum_{\mu \in M, \mu \geq \nu} \alpha_\mu \left[\nu_1! \binom{\mu_1}{\nu_1} \cdots \nu_n! \binom{\mu_n}{\nu_n} \right] x^{\mu - \nu}. \tag{2.3}$$

The operator D_ν is linear on $F\lfloor\omega\rfloor[x]$. For ρ and γ in $F\lfloor\omega\rfloor$, we write $D_\nu P(\rho)\gamma$ for the product of $D_\nu P(\rho)$ and γ. For e^k as the kth unit vector in $Z^n_\oplus$, we have $D_k P \equiv D_{e^k} P$ as the usual *partial derivative of P with respect to x_k*. Let $DP \equiv (D_1 P, \ldots, D_n P)$ in $(F\lfloor\omega\rfloor)^n$ denote the usual *derivative of P*. For ρ and γ in $(F\lfloor\omega\rfloor)^n$, the derivative of P evaluated at ρ is denoted $DP(\rho)$, and the application of the linear operator $DP(\rho)$ to γ is written $DP(\rho)(\gamma)$. In particular, $D_\nu P(0) = \alpha_\nu [\prod_{k \in N} \nu_k!]$.

We refer to elements in $(F\lfloor\omega\rfloor[x])^n$ as *polynomial systems* and denote such elements by

$$\mathbb{P} = \mathbb{P}(x) = (P_1(x), \ldots, P_n(x)); \tag{2.4}$$

where each coordinate $P_k(x)$ is in $F\lfloor\omega\rfloor[x]$ for each k in N. Such, $\mathbb{P}(x)$ can also be written as $\sum_{\mu \in M} \alpha_\mu x^\mu$ with $M \subseteq Z^n_\oplus$ and α_μ in $(F\lfloor\omega\rfloor)^n$ for each μ in M. For $\mathbb{P}$ as in (2.4), let $\mathbb{P}_\Diamond(x) \equiv (P_{1\Diamond}(x), \ldots, P_{n\Diamond}(x))$. The *derivative* of $\mathbb{P}$, denoted $D\mathbb{P}$, can be represented by the matrix

$$(D\mathbb{P})(x) = \begin{pmatrix} D_1 P_1 & D_2 P_1 & \cdots & D_n P_1 \\ D_1 P_2 & D_2 P_2 & \cdots & D_n P_2 \\ & & \vdots & \\ D_1 P_n & D_2 P_n & \cdots & D_n P_n \end{pmatrix} \in (F\lfloor\omega\rfloor[x])^{n \times n}. \tag{2.5}$$

A *root* of $\mathbb{P}$ is a vector ρ in $(F\lfloor\omega\rfloor)^n$ with $\mathbb{P}(\rho) = 0$, that is, $P_k(\rho) = 0$ for k in N. Such a root is called *regular* if the matrix $D\mathbb{P}(\rho)$ is nonsingular.

We can now state the first part of our main result more formally. For a system of polynomials $\mathbb{P}$ and regular root r of $\mathbb{P}_\Diamond$, we compute the unique root $\rho = \sum_i r(i)\omega^i$ of $\mathbb{P}$ with $\rho_\Diamond = r$. The coefficients $r(1), r(2), \ldots$ are computed recursively with arithmetic in F and with polynomial bounds on the required effort. In particular, the root lies in

$(H\lfloor\omega\rfloor)^n$ where H is the smallest subfield of F which contains the coefficients of $\mathbb{P}$ and the coordinates of r.

3. Expansions

Let $P(x)$ be a polynomial in $F[x]\lfloor\omega\rfloor$ and let ρ be a vector in $(F\lfloor\omega\rfloor)^n$. We express the expansion of $P(\rho)$ in terms of the expansions of P and ρ. With this expansion we obtain necessary and sufficient arithmetic conditions for ρ to be a root of P. To this purpose we need expressions for sums of products of coefficients of elements ρ in $F\lfloor\omega\rfloor$ and $(F\lfloor\omega\rfloor)^n$, which we denote $\rho(h,i)$, $\rho(\mu,i)$, $\rho(h,i)$ and $\rho^-(\mu,i)$ for h and i in $Z_\oplus$ and μ in $Z^n_\oplus$; see (3.1), (3.2), (3.4) and (3.5).

First we expand powers of an element $\rho = \sum_i r(i)\omega^i$ in $F\lfloor\omega\rfloor$. For h and i in $Z_\oplus$ define

$$\rho(h,i) \equiv \sum_{\substack{i_1+\cdots+i_h=i\\ 0\le i_t\le i}} \prod_{t=1}^{h} r(i_t) \tag{3.1}$$

where $(i_1,\ldots,i_h)$ ranges over $Z^h_\oplus$. Observe that $\rho(0,0)=1$, $\rho(h,0)=(\rho_\diamond)^h$ for h in $Z_\oplus$ and $\rho(0,i)=0$ for $i=1,2,\ldots$. Also observe that $\rho(h,i)$ does not depend on any $r(j)$ with $j>i$. The next lemma provides an expansion for ρ^h in terms of the $\rho(h,i)$'s.

Lemma 3.1. *For ρ in $F\lfloor\omega\rfloor$ and h in $Z_\oplus$, we have $\rho^h = \sum_i \rho(h,i)\omega^i$.*

Next, for $\rho=(\rho_1,\ldots,\rho_n)$ in $(F\lfloor\omega\rfloor)^n$, μ in $Z^n_\oplus$ and i in $Z_\oplus$, let

$$\rho(\mu,i) \equiv \sum_{\substack{i_1+\cdots+i_n=i\\ 0\le i_k\le i}} \prod_{k\in N} \rho_k(\mu_k,i_k) \tag{3.2}$$

where $(i_1,\ldots,i_n)$ ranges over $Z^n_\oplus$. Observe that if $\rho_k=\sum_i r_k(i)\omega^i$ for k in N, we have that $\rho(\mu,0)=\prod_{k\in N}\rho_k(\mu_k,0)=\prod_{k\in N} r_k(0)^{\mu_k}=(\rho_\diamond)^\mu$. Also, $\rho(0,i)=0$ for $i=1,2,\ldots$. Notice that $\rho(\mu,j)$ does not depend on any $r_k(j)$ with $j>i$. The next lemma provide an expansion for ρ^μ in the $\rho(\mu,j)$'s.

Lemma 3.2. *For $\rho=(\rho_1,\ldots,\rho_n)$ in $(F\lfloor\omega\rfloor)^n$ and μ in $Z^n_\oplus$, we have $\rho^\mu=\sum_i\rho(\mu,i)\omega^i$.*

Proof: Using Lemma 3.1 we see that

$$\rho^\mu = \prod_{k\in N}(\rho_k)^{\mu_k} = \prod_{k\in N}\left[\sum_i \rho_k(\mu_k,i)\omega^i\right]$$

$$= \sum_i \sum_{\substack{i_1+\cdots+i_n=i\\ 0\le i_k\le i}} \left[\prod_{k\in N}\rho_k(\mu_k,i_k)\right]\omega^i = \sum_i \rho(\mu,i)\omega^i.$$

□

With $n = 1$ and $\mu = (h)$, Lemma 3.2 specializes Lemma 3.1.

The following theorem expands $P(\rho)$ in terms of the expansions of P and ρ.

Theorem 3.3. *Let $P(x) = \sum_{\mu \in M} \alpha_\mu x^\mu$ be in $F\lfloor\omega\rfloor[x]$ with $\alpha_\mu = \sum_i a_\mu(i)\omega^i$ in $F\lfloor\omega\rfloor$ for each μ in M, and let ρ in $(F\lfloor\omega\rfloor)^n$. Then*

$$P(\rho) = P_\Diamond(\rho_\Diamond) + \sum_{i=1}^{+\infty}\left[\sum_{j=0}^{i}\sum_{\mu\in M} a_\mu(j)\rho(\nu, i-j)\right]\omega^i. \tag{3.3}$$

Proof: Using Lemma 3.2, convolution and an exchange in the order of summation, we have

$$P(\rho) = \sum_{\mu\in M}\alpha_\mu\rho^\mu = \sum_{\mu\in M}\left[\sum_i a_\mu(i)\omega^i\right]\left[\sum_i \rho(\mu,i)\omega^i\right]$$

$$= \sum_{\mu\in M}\sum_i\left[\sum_{j=0}^{i} a_\mu(j)\rho(\mu,i-j)\right]\omega^i = \sum_i\left[\sum_{j=0}^{i}\sum_{\mu\in M} a_\mu(j)\rho(\mu,i-j)\right]\omega^i;$$

(3.3) now follows by isolating the ω^0-coefficient and observing that it equals $\sum_{\mu\in M} a_\mu(0)$ $\rho(\mu, 0) = \sum_{\mu\in M} a_\mu(0)(\rho_\Diamond)^\mu = P_\Diamond(\rho_\Diamond)$. □

We need expressions $\rho^-(h,i)$ and $\rho^-(\mu,i)$ which coincide with $\rho(h,i)$ and $\rho(\mu,i)$ except for the exclusion of the terms that depend on the $r(i)$'s.

For $\rho = \sum_i r(i)\omega^i$ in $F\lfloor\omega\rfloor$, h in $Z_\oplus$ and integer $i = 1, 2, \ldots$, define

$$\rho^-(h,i) \equiv \rho(h,i) - h(\rho_\Diamond)^{h-1}r(i) = \sum_{\substack{i_1+\cdots+i_h=i\\ 0\le i_t<i}}\prod_{t=1}^{h} r(i_t), \tag{3.4}$$

where $h(\rho_\Diamond)^{h-1}$ is defined as zero for $h = 0$ and $(i_1, \ldots, i_h)$ ranges over $Z_\oplus^h$. Observe that $\rho^-(0,i) = \rho^-(1,i) = 0$ for $i = 1, 2, \ldots$. It is important to see that the expression on the right-hand side of (3.4) does not depend on any $r(j)$ for $j > i$.

For $\rho = (\rho_1, \ldots, \rho_n) \in (F\lfloor\omega\rfloor)^n$ with $\rho_k = \sum_i r_k(i)\omega^i$ for $k \in N$, for μ in $Z_\oplus^n$ and for positive i in $Z_\oplus$, define

$$\rho^-(\mu,i) \equiv \rho(\mu,i) - D(x^\mu)(\rho_\Diamond)(r(i)). \tag{3.5}$$

The next lemma provides an alternative view of the $\rho^-(\mu,i)$'s.

Lemma 3.4. *Let $\rho = (\rho_1, \ldots, \rho_n)$ be in $(F\lfloor\omega\rfloor)^n$ with $\rho_k = \sum_i r_k(i)\omega^i$ for k in N, let μ be in $Z_\oplus^n$ and let i be a positive integer. Then*

$$\rho^-(\mu,i) = \sum_{\substack{i_1+\cdots+i_n=i\\ 0\le i_k<i}}\prod_{k\in N}\rho_k(\mu_k,i_k) + \sum_{k\in N}\prod_{m\in N\setminus k}(\rho_{\Diamond m})^{\mu_m}\rho_k^-(\mu_k,i). \tag{3.6}$$

where $(i_1, \ldots, i_n)$ ranges over $Z_\oplus^n$.

Proof: From Theorem 3.3, from hr^{h-1} taken as zero for $h = 0$ and r in F, and from $r_m(0) = \rho_{\Diamond m}$ for m in N, we see that

$$\sum_{\substack{i_1+\cdots+i_n=i \\ 0\le i_k<i}} \prod_{k\in N} \rho_k(\mu_k, i_k) + \sum_{k\in N} \prod_{m\in N\setminus k} (\rho_{\Diamond m})^{\mu_m} \rho_k^-(\mu_k, i)$$

$$= \sum_{\substack{i_1+\cdots+i_n=i \\ 0\le i_k<i}} \prod_{k\in N} \rho_k(\mu_k, i_k) + \sum_{k\in N} \prod_{m\in N\setminus k} r_m(0)^{\mu_m}[\rho_k(\mu_k, i) - \mu_k(\rho_{\Diamond k})^{\mu_k-1} r_k(i)]$$

$$= \sum_{\substack{i_1+\cdots+i_n=i \\ 0\le i_k\le i}} \prod_{k\in N} \rho_k(\mu_k, i_k) - \sum_{\substack{k\in N \\ \mu_k\ge 1}} \prod_{m\in N\setminus k} (\rho_{\Diamond m})^{\mu_m} \mu_k(\rho_{\Diamond k})^{\mu_k-1} r_k(i)$$

$$= \rho(\mu, i) - \sum_{k\in N} D_k(x^\mu)(\rho_\Diamond)(r_k(i)) = \rho(\mu, i) - D(x^\mu)(\rho_\Diamond)(r(i)).$$

□

The representation of the $\rho^-(\mu, i)$'s in Lemma 3.4 implies that $\rho^-(\mu, i) = 0$ for $i = 1, 2, \ldots$ and μ in $Z_\oplus^n$ with $\sum_{k\in N} \mu_k \le 1$, that is, for μ the zero vector or a unit vector. It is important to note that the right-hand side of (3.6) does not depend on any $r_k(j)$ for $j \ge i$, thus, the same holds for $\rho^-(\mu, i)$.

For a polynomial P in $F\lfloor\omega\rfloor[x]$ and for ρ in $(F\lfloor\omega\rfloor)^n$, Theorem 3.3 provides an expansion for the evaluation $P(\rho)$ of P at ρ in terms of expansions of P and ρ. Next, the $\rho^-(\mu, i)$'s are used to obtain a more revealing expansion for the evaluation.

Lemma 3.5. *Let $P(x) = \sum_{\mu\in M} \alpha_\mu x^\mu$ be in $F\lfloor\omega\rfloor[x]$ with $\alpha_\mu = \sum_i a_\mu(i)\omega^i$ in $F\lfloor\omega\rfloor$ for each μ in M, and let $\rho = \sum_i r(i)\omega^i$ be in $(F\lfloor\omega\rfloor)^n$ with $r(i)$ in F^n for each i in $Z_\oplus$. Then*

$$P(\rho) = P_\Diamond(\rho_\Diamond) + \sum_{i=1}^{+\infty} \left[\sum_{j=1}^{i} \sum_{\mu\in M} a_\mu(j)\rho(\mu, i-j) + \sum_{\mu\in M} a_\mu(0)\rho^-(\mu, i) + DP_\Diamond(\rho_\Diamond)(r(i)) \right] \omega^i. \tag{3.7}$$

Proof: With the expansion of $P(\rho)$ from Theorem 3.3, from (3.5) and from the linearity of D we see that

$$P(\rho) = P_\Diamond(\rho_\Diamond) + \sum_{i=1}^{+\infty} \left[\sum_{j=1}^{i} \sum_{\mu\in M} a_\mu(j)\rho(\mu, i-j) + \sum_{\mu\in M} \alpha_{\mu\Diamond}\rho(\mu, i) \right] \omega^i$$

$$= P_\Diamond(\rho_\Diamond) + \sum_{i=1}^{+\infty} \left\{ \sum_{j=1}^{i} \sum_{\mu\in M} a_\mu(j)\rho(\mu, i-j) + \sum_{\mu\in M} \alpha_{\mu\Diamond}[\rho^-(\mu, i) + D(x^\mu)(\rho_\Diamond)(r(i))] \right\} \omega^i$$

$$= P_\Diamond(\rho_\Diamond) + \sum_{i=1}^{+\infty}\left[\sum_{j=1}^{i}\sum_{\mu\in M} a_\mu(j)\rho(\mu, i-j) + \sum_{\mu\in M}\alpha_{\mu\Diamond}\rho^-(\mu, i) + D\left(\sum_{\mu\in M}\alpha_{\mu\Diamond}x^\mu\right)(\rho_\Diamond)(r(i))\right]\omega^i$$

$$= P_\Diamond(\rho_\Diamond) + \sum_{i=1}^{+\infty}\left[\sum_{j=1}^{i}\sum_{\mu\in M} a_\mu(j)\rho(\mu, i-j) + \sum_{\mu\in M}\alpha_{\mu\Diamond}\rho^-(\mu, i) + DP_\Diamond(\rho_\Diamond)(r(i))\right]\omega^i.$$

□

Lemma 3.5 provides necessary and sufficient conditions for a root of a polynomial in $F\lfloor\omega\rfloor[x]$.

Corollary 3.6. *Let $P(x) = \sum_{\mu\in M}\alpha_\mu x^\mu$ be in $F\lfloor\omega\rfloor[x]$ with $\alpha_\mu = \sum_i a_\mu(i)\omega^i$ in $F\lfloor\omega\rfloor$ for each μ in M, and let $\rho = \sum_i r(i)\omega^i$ be in $(F\lfloor\omega\rfloor)^n$ with $r(i)$ in F^n for each i in $Z_\oplus$. Then $P(\rho) = 0$ if and only if $P_\Diamond(\rho_\Diamond) = 0$ and*

$$DP_\Diamond(\rho_\Diamond)(r(i)) = -\sum_{\mu\in M}\sum_{j=1}^{i} a_\mu(j)\rho(\mu, i-j) - \sum_{\mu\in M}\alpha_{\mu\Diamond}\rho^-(\mu, i) \quad \text{for } i = 1, 2, \ldots. \tag{3.8}$$

It is important to see that the right-hand side of (3.8) is independent of $r(j)$ for $i = 1, 2, \ldots$ and $j \geq i$.

4. Continuing a regular root

Our expansions are used to present the main result on existence, uniqueness, characterization and computation of roots of polynomial systems that continue regular roots.

The technical heart of our results is contained in the following theorem.

Theorem 4.1. *Let $\mathbb{P}(x) = \sum_{\mu\in M}\alpha_\mu x^\mu$ in $(F\lfloor\omega\rfloor[x])^n$ be a polynomial system and let r be a regular root of $\mathbb{P}_\Diamond(x)$. Then there exists a unique root $\rho = \sum_i r(i)\omega^i$ of $\mathbb{P}$ with $\rho_\Diamond = r$. The coefficients $r(i)$ satisfy $r(0) = r$ and*

$$r(i) = -[DP_\Diamond(\rho_\Diamond)]^{-1}\left[\sum_{\mu\in M}\sum_{j=1}^{i} a_\mu(j)\rho(\mu, i-j) + \sum_{\mu\in M}\alpha_{\mu\Diamond}\rho^-(\mu, i)\right] \quad \text{for } i = 1, 2, \ldots. \tag{4.1}$$

Proof: For μ in M, let $\alpha_\mu = (\alpha_{1\mu}, \ldots, \alpha_{n\mu})$ with $\alpha_{k\mu}$ in $F\lfloor\omega\rfloor$ for each $k \in N$. The polynomial system $\mathbb{P}(x)$ can be written as $(P_1(x), \ldots, P_n(x))$ with $P_k(x) = \sum_{\mu \in M} \alpha_{k\mu} x^\mu$ for $k \in N$. By applying Corollary 3.6 to each of these polynomials we get that $\mathbb{P}(\rho) = 0$ if and only if $\mathbb{P}_\Diamond(\rho_\Diamond) = 0$ and (3.8) holds as a vector equation, the latter being equivalent to (4.1). This proves the necessity of (4.1) for $\mathbb{P}(\rho) = 0$. Further, the right-hand side of (4.1) does not depend on any $r(j)$ for $j > i$, indeed, it depends only on $\{a_\mu(j) : \mu \in M, j = 0, 1, \ldots, i\} \subseteq F^n$ and $\{r(j) : j = 0, 1, \ldots, i-1\} \subseteq F^n$. Thus, the $r(i)$'s can be constructed recursively from (4.1) beginning with $r(0) = r$; the resulting element $\rho = \sum_i r(i)$ in $(F\lfloor\omega\rfloor)^n$ satisfies the sufficient condition for $\mathbb{P}(\rho) = 0$. Thus, the proofs of existence, uniqueness and characterization of the desired root are complete. □

The right-hand side of (4.1) expresses $r(i)$ in terms of $\{a_\mu(j) : \mu \in M, j = 0, 1, \ldots, i\} \subseteq F^n$ and $\{r(j) : j = 0, 1, \ldots, i-1\} \subseteq F^n$, the latter through $\{\rho(\mu, j) : j = 0, \ldots, i-1\}$ and $\{\rho^-(\mu, j) : j = 0, 1, \ldots, i\}$. Thus, with the $a_\mu(j)$'s given, we have a recursion for computing the $r(i)$'s. A complexity bound for the recursive computation of the $r(i)$'s through (4.1) is next determined.

For a finite subset M of $Z^n_\oplus$, let $|M|$ denote the cardinality of M, let $\|M\|$ be the integer $\max\{\mu_k : \mu \in M \text{ and } k \in N\}$ and let $\bar{M}$ be the subset $\{1, \ldots, \|M\|\}$ of $Z_\oplus$.

Let $\mathbb{P}$ and r be as in Theorem 4.1. In iteration $i-1$ we compute $r_k(i-1)$ for $k \in N$ from (4.1) using data $a_{k\mu}(j)$ for $k \in N, \mu \in M$ and $j = 1, \ldots, i-1$ and from the previously computed quantities

(a) $\rho_k(h, j)$ for $k \in N, h \in \bar{M}$ and $j = 0, \ldots, i-2$,
(b) $\rho_k^-(h, j)$ for $k \in N, h \in \bar{M}$ and $j = 0, \ldots, i-1$,
(c) $\rho(\mu, j)$ for $\mu \in M$ and $j = 0, \ldots, i-2$, and
(d) $\rho^-(\mu, j)$ for $\mu \in M$ and $j = 0, \ldots, i-1$.

Now suppose all these quantities have been calculated in iteration $i-1$ and consider iteration i in which $r(i)$ is to be calculated. The required new quantities are then:

(a′) $\rho_k(h, i-1)$ for $k \in N$ and $h \in \bar{M}$,
(b′) $\rho_k^-(h, i)$ for $k \in N$ and $h \in \bar{M}$,
(c′) $\rho(\mu, i-1)$ for $\mu \in M$, and
(d′) $\rho^-(\mu, i)$ for $\mu \in M$.

Given these new quantities, the $r_k(i)$'s are then available from (4.1). For each k in N, computing (a') and (b') is executed recursively for $h = 1, \ldots, \|M\|$ using the following formulae:

$$\rho_k(h, i-1) = \rho_k^-(h, i-1) + hr(i-1)_k, \tag{4.2}$$

and

$$\rho_k^-(h, i) = \sum_{j=1}^{i} r_k(j)\rho_k(h-1, i-j) + r_k(0)\rho_k^-(h-1, i), \tag{4.3}$$

Table 1. Complexity bounds for $r(i)$.

Quantities	Parameter values	Formula	Multiplications	Additions
$\rho_k(h,i)$	$k \in N, h \in \bar{M}$	(4.2)	$n\|M\|$	$n\|M\|$
$\rho_k^-(h,i)$	$k \in N, h \in \bar{M}$	(4.3)	$in\|M\|$	$(i-1)n\|M\|$
$\rho(\mu,i)$	$\mu \in M$	(4.4)	$\lvert M\rvert$	$\lvert M\rvert$
$\rho^-(\mu,i,t)$	$\mu \in M, t \in N$	(4.6)	$in\lvert M\rvert$	$(i-1)n\lvert M\rvert$
$r(i)$	$k \in N$	(4.1)	$(i-1)n\lvert M\rvert + n^2$	$i\lvert nM\rvert + n(n-1)$

formula (4.2) follows from (3.4) while the verification of (4.3) is left to the reader. Next, let μ be in M. From (3.5) we have that

$$\rho(\mu, i-1) = \rho^-(\mu, i-1) + D(x^\mu)(\rho_\Diamond)(r(i-1)); \tag{4.4}$$

with

$$[D(x^\mu)(\rho_\Diamond)]_k = D_k(x^\mu)(\rho_\Diamond) = \left[\prod_{m\in N\setminus k} (\rho_{\Diamond m})^{\mu_m}\right] \mu_k(\rho_{\Diamond k})^{-1} \tag{4.5}$$

For $t \in N$, let $\rho^-(\mu, i, t) \equiv [\rho|_t]^-[(\mu|_t, i+1)]$, where $|_t$ restricts n-vectors to their first t coordinates. Then $\rho^-(\mu, i, 1) = \rho^-(\mu_1, i)$ and using Lemma 3.4 for $t \in N\setminus 1$,

$$\rho^-(\mu, i, t) = \sum_{j=1}^{i} \rho_t(\mu_t, j)\rho(\mu, i-j, t-1) + \rho_t(\mu_t, 0)\rho^-(\mu, i, t-1) \tag{4.6}$$

Computation of $DP_\Diamond(\rho_\Diamond)$ requires at most $n^2|M|(\|M\|+1)$ multiplication and additions, and its inversion requires at most n^3 additional multiplications and additions. Using (4.5) and available information from the computation of $DP_\Diamond(\rho_\Diamond)$, the computation of $D(x^\mu)(\rho_\Diamond)$ requires at most $2n$ additional multiplications. A bound on the number of arithmetic operations required for $r(i)$ is recorded in Table 1.

In summary, $r(i)$ requires at most $in(|M| + \|M\|) + n^2$ multiplications and additions and n divisions. (The term $n(|M| + \|M\|)$ can be replaced by $\sum_{k\in N}(|M_k| + H_k)$, where $P_k(x) = \sum_{\mu\in M_k} \alpha_{k\mu}x^\mu$, $M \equiv \bigcup_{k\in N} M_k$ and $H_k \equiv \max\{\mu_k : \mu \in M\}$.)

5. Geometric boundedness

The formal power series of the ring $F\lfloor\omega\rfloor$ are very general; here we concern ourselves with the subring of geometrically bounded power series. If the system is geometrically bounded we show that the continued root is also. Our technique in establishing geometric boundedness is to demonstrate that partial sums of power series are uniformly bounded; see [2] for the use of this approach for parametric polynomials with one variable. The arguments we use have the virtue of being algebraic and self contained.

An *absolute value* on the field F is a function $|\cdot| : F \to E$, where E is an ordered subfield of F, with the following properties

(i) $|a| = 0$ if and only if $a = 0$,
(ii) $|a| = a$ for each $a \geq 0$ in E,
(iii) $|a+b| \leq |a| + |b|$, and
(iv) $|ab| = |a||b|$,

for all a and b in F. For example, the rationals, the reals and the complex numbers have an absolute value. Throughout this section we assume that the ground field F has an absolute value $|\cdot|$. An element $\alpha = \sum_i a(i)\omega^i$ in $F\lfloor\omega\rfloor$ is defined to be *geometrically bounded* if there exist A and B in E with

$$|a(i)| \leq AB^i \quad \text{for each } i \text{ in } Z_\oplus. \tag{5.1}$$

Elements in $F\lfloor\omega\rfloor[x]$ and $(F\lfloor\omega\rfloor[x])^n$ are called *geometrically bounded* if all their coefficients in $F\lfloor\omega\rfloor$ are geometrically bounded. In this section we show that if the system $\mathbb{P}$ is geometrically bounded, then the unique root which continues a regular value of $\mathbb{P}_\Diamond$ is also geometrically bounded.

The next three lemmas record standard facts about geometric boundedness; we establish only the third.

Lemma 5.1. *The set of geometrically bounded elements in $F\lfloor\omega\rfloor$ is a subring of $F\lfloor\omega\rfloor$.*

Lemma 5.2. *Let P and Q be a geometrically bounded polynomials in $F\lfloor\omega\rfloor[x]$. Then:*
(a) *$P + Q$ is a geometrically bounded,*
(b) *PQ is a geometrically bounded,*
(c) *for ν in $z^n_\oplus$, the derivative $D_\nu P$ is geometrically bounded, and*
(d) *if ρ in $(F\lfloor\omega\rfloor)^n$ is geometrically bounded, then $P(\rho)$ is geometrically bounded.*

For a polynomial P in $F\lfloor\omega\rfloor[x]$ as in (2.2) and ρ in $(F\lfloor\omega\rfloor)^n$, $P(x+\rho)$ is the polynomial $\sum_{\mu\in M} \alpha_\mu(\rho+x)^\mu$.

Lemma 5.3. *Let P in $F\lfloor\omega\rfloor[x]$ and ρ in $(F\lfloor\omega\rfloor)^n$ be geometrically bounded. Then $P(\rho+x)$ in $F\lfloor\omega\rfloor[x]$ is geometrically bounded.*

Proof: Let $P(x)$ be as in (2.2). Using the Taylor expansion of Lemma A.2 (see Appendix), we have

$$P(\rho + x) = \sum_{\substack{\nu\in Z^n_\oplus \\ 1\leq\nu_k\leq\|M\|}} \frac{1}{\nu_1!\cdots\nu_n!} D_\mu P(\rho)x^\mu.$$

From Lemma 5.2, each $D_\nu P(\rho)$ is geometrically bounded, hence, is Q also. □

Proposition 5.4. *Suppose $\mathbb{P}(x)$ in $(F\lfloor\omega\rfloor[x])^n$ is geometrically bounded, $\mathbb{P}_\Diamond(0) = 0$ and $D\mathbb{P}_\Diamond(0)$ as the identity I in $(F\lfloor\omega\rfloor[x])^{n\times n}$. Then $\mathbb{P}$ has a unique root ρ with $\rho_\Diamond = 0$ and ρ is geometrically bounded.*

Proof: Let $\mathbb{P}(x) = \sum_{\mu\in M} \alpha_\mu x^\mu$ with $M \subseteq Z^n_\oplus$ and $\alpha_\mu = (\alpha_{1\mu}, \ldots, \alpha_{n\mu})$ in $(F\lfloor\omega\rfloor)^n$ for each μ in M. As $\mathbb{P}(x)$ is geometrically bounded there exist positive elements A and B in F with $a_{k\mu}(i) \le AB^i$ for k in N, $\mu \in M$ and $i = 0, 1, \ldots$.

For $\mu \in Z^n_\oplus$, let $\|\mu\| \equiv \sum_{k\in N} \mu_k$ and define

$$V \equiv (1 + 2|M|A)^{-1}, \tag{5.2}$$

and

$$W \equiv B\left[2A\left(\sum_{\mu\in M} V^{\|\mu\|-1}\right) + 1\right]. \tag{5.3}$$

From Theorem 4.1, the system $\mathbb{P}$ has a unique root ρ in $(F\lfloor\omega\rfloor)^n$ with $\rho_\Diamond = 0$. Let $\rho = (\rho_1, \ldots, \rho_n)$ with $\rho_k = \sum_i r_k(i)\omega^i$ for $k \in N$. We prove the geometric boundedness of ρ by showing for $p = 0, 1, \ldots$

$$\sum_{i=0}^{p} |r_k(i)W^{-i}| \le V \quad \text{for each } k \in N. \tag{5.4}$$

We establish (5.4) by induction on p. The case for $p = 0$ follows as $r_k(0) = \rho_{k\Diamond} = 0$ for each $k \in N$. Next, assume that (5.4) holds for p and we move to establish the inequality for $p + 1$.

We proceed by showing in (5.5)–(5.8) that for h in $Z_\oplus$ and μ in $Z^n_\oplus$, $\sum_{i=0}^p |\rho_k(h, i)W^{-i}|$, $\sum_{i=0}^p |\rho(\mu, i)W^{-i}|$, $\sum_{i=0}^p |\rho_k^-(h, i)W^{-i}|$ and $\sum_{i=0}^p |\rho^-(\mu, i)|$ are bounded by V^h, $V^{\|\mu\|}$, V^h and $V^{\|\mu\|}$, respectively. The induction assumption implies that for each $k \in N$ and $h = 0, 1, \ldots$

$$\begin{aligned}\sum_{i=0}^{p} |\rho_k(h, i)W^{-i}| &\le \sum_{i=0}^{p} \sum_{\substack{i_1+\cdots+i_h=i\\ 0\le i_t\le i}} \prod_{t=1}^{h} |r_k(i_t)W^{-i_t}| \\ &\le \sum_{\substack{(i_1,\ldots,i_h)\\ 0\le i_t\le p}} \prod_{t=1}^{h} |r_k(i_t)W^{-i_t}| = \prod_{t=1}^{h}\left[\sum_{i=0}^{p} |r_k(i)W^{-i}|\right] \le V^h,\end{aligned} \tag{5.5}$$

consequently for μ in $Z^n_\oplus$,

$$\sum_{i=0}^{p} |\rho(\mu, i)W^{-i}| \le \sum_{i=0}^{p} \sum_{\substack{i_1+\cdots+i_n=i\\ 0\le i_k\le i}} \prod_{k\in N} |\rho_k(\mu_k, i_k)W^{-i_k}|$$

$$\leq \sum_{\substack{(i_1,\ldots,i_n)\\ 0\leq i_k\leq p}} \prod_{k\in N} |\rho_k(\mu_k, i_k)W^{-i_k}| = \prod_{k\in N}\left[\sum_{i=0}^{p} |\rho_k(\mu_k, i)W^{-i}|\right]$$

$$\leq \prod_{k\in N} V^{\mu_k} = V^{\|\mu\|}. \tag{5.6}$$

Similarly, for $k \in N$ and $h = 1, 2, \ldots$

$$\sum_{i=1}^{p+1} |\rho_k^-(h, i)W^{-i}| \leq \sum_{i=1}^{p+1} \sum_{\substack{i_1+\cdots+i_h=i\\ 0\leq i_t<i}} \prod_{t=1}^{h} |r_k(i_t)W^{-i_t}| \leq \sum_{\substack{(i_1+\cdots+i_h)\\ 0\leq i_t\leq p}} \prod_{t=1}^{h} |r_k(i_t)W^{-i_t}|$$

$$= \prod_{t=1}^{h}\left[\sum_{i=0}^{p} |r_k(i)W^{-i}|\right] \leq V^h. \tag{5.7}$$

Next, for $\mu \in Z_{\oplus}^n$ we bound $\sum_{i=0}^{p+1} |\rho_k^-(\mu, i)W^{-i}|$ by considering two cases. First, if μ is not a scale of a unit vector, the assertion $\rho_\Diamond = 0$ implies that $\prod_{m\in N|\{k\}}(\rho_{\Diamond r})^{\mu_m} = 0$ for each $k \in N$. Hence,

$$\sum_{i=1}^{p+1} |\rho^-(\mu, i)W^{-i}| \leq \sum_{i=1}^{p+1} \sum_{\substack{i_1+\cdots+i_n=i\\ 0\leq i_k<i}} \prod_{k\in N} |\rho_k(\mu_k, i_k)W^{-i_k}|$$

$$+ \sum_{k\in N} \prod_{\substack{m\in N\\ m\neq k}} (\rho_{\Diamond m})^{\mu_m} |\rho_k^-(\mu_k, i)W^{-i}|$$

$$\leq \sum_{0\leq i_k\leq p} \prod_{k\in N} |\rho_k(\mu_k, i_k)W^{-i_k}| = \prod_{k\in N}\left[\sum_{i=0}^{p} |\rho_k(\mu_k, i)W^{-i}|\right]$$

$$\leq \prod_{k\in N} V^{\mu_k} = V^{\|\mu\|}. \tag{5.8a}$$

Second, if μ is a scale of the rth unit vector, then $\prod_{m\in N\setminus\{r\}}(\rho_{\Diamond m})^{\mu_m} = 1$, $\prod_{m\in N\setminus k}(\rho_{\Diamond m})^{\mu_m} = 0$ for each $k \in N\setminus r$ and there exist no $(i_1, \ldots, i_n)$ with $\sum_{k\in N} i_k = i$ and $i_k < i$ for each k; consequently

$$\sum_{i=1}^{p+1} |\rho^-(\mu, i)W^{-i}| \leq \sum_{i=1}^{p+1} \sum_{\substack{i_1+\cdots+i_n=i\\ 0\leq i_k<i}} \prod_{k\in N} |\rho_k(\mu_k, i_k)W^{-i_k}|$$

$$+ \sum_{k\in N} \prod_{\substack{m\in N\\ m\neq k}} (\rho_{\Diamond m})^{\mu_m} |\rho_k^-(\mu_k, i)W^{-i}|$$

$$= \sum_{i=1}^{p+1} |\rho_r^-(\mu_r, i)W^{-i}| \leq V^{\mu_r} = V^{\|\mu\|}. \tag{5.8b}$$

From the selection $V = (1 + 2|M|A)^{-1}$ we have that $V \le 1$, $VMA \le \frac{1}{2}$ and

$$\sum_{\substack{\mu\in M\\ \|\mu\|\ge 2}} AV^{\|\mu\|} \le V^2\left[\sum_{\substack{\mu\in M\\ \|\mu\|\ge 2}} A\right] \le V[V|M|A] \le \frac{V}{2}. \tag{5.9}$$

Also, from the definition of W we have that

$$\sum_{\mu\in M}\left[\sum_{j=1}^{p+1} AB^j W^{-j}\right]V^{\|\mu\|} \le A\left[\sum_{\mu\in M} V^{\|\mu\|}\right]\left(\frac{B}{W}\right)\left(\frac{1}{1-\frac{B}{W}}\right)$$

$$= A\left[\sum_{\mu\in M} V^{\|\mu\|}\right]\left(\frac{1}{\frac{W}{B}-1}\right) = \frac{V}{2}. \tag{5.10}$$

Also, from Corollary 3.6 and $DP_\diamond(\rho_\diamond) = DP_\diamond(0) = I$, for $k \in N$ and $i = 1, 2, \ldots$ we have

$$r_k(i) = -\sum_{j=1}^{i}\sum_{\mu\in M} a_{k\mu}(j)\rho(\mu, i-j) - \sum_{\mu\in M} \alpha_{k\mu\diamond}\rho^-(\mu, i). \tag{5.11}$$

Now, from $\rho_\diamond = 0$, an interchange in the order of summation, $|a_{k\mu}(i)| \le AB^i$ (in particular, $|\alpha_{k\mu\diamond}| = |a_{k\mu}(0)| \le A$), $\rho^-(\mu, i) = 0$ for $i \ge 1$ and $\mu \in Z^n_\oplus$ with $\sum_{k\in N}\mu_k \le 1$, (5.6), (5.8)–(5.10) we conclude that

$$\sum_{i=0}^{p+1}|r_k(i)W^{-i}| = \sum_{i=1}^{p+1}|r_k(i)W^{-i}| \le \sum_{i=1}^{p+1}\left[\sum_{j=1}^{i}\sum_{\mu\in M}|a_{k\mu}(j)\rho(\mu, i-j)W^{-i}|\right]$$

$$+\sum_{i=1}^{p+1}\left[\sum_{\mu\in M}|\alpha_{k\mu\diamond}\rho^-(\mu, i)|W^{-i}\right] \le \sum_{\mu\in M}\left[\sum_{j=1}^{p+1}|a_{k\mu}(j)W^{-j}|\right]$$

$$\times\left[\sum_{i=j}^{p+1}|\rho(\mu, i-j)W^{-(i-j)}|\right] + \sum_{\substack{\mu\in M\\ \|\mu\|\ge 2}} A\left[\sum_{i=1}^{p+1}|\rho^-(\mu, i)|W^{-i}\right]$$

$$\le \sum_{\mu\in M}\left[\sum_{j=1}^{p+1}|a_{k\mu}(j)W^{-j}|\right]\left[\sum_{u=0}^{p}|\rho(\mu, u)W^{-u}|\right]$$

$$+\sum_{\substack{\mu\in M\\ \|\mu\|\geq 2}} A\left[\sum_{i=1}^{p+1}|\rho^-(\mu,i)|W^{-i}\right] \leq \sum_{\mu\in M}\left[\sum_{j=1}^{p+1}AB^jW^{-j}\right]V^{\|\mu\|}$$

$$+\sum_{\substack{\mu\in M\\ \|\mu\|\geq 2}} AV^{\|\mu\|} \leq \frac{V}{2}+\frac{V}{2} = V,$$

establishing the induction hypothesis with $p+1$ replacing p. □

Our main result on geometrically boundedness follows, namely, that the continued root of a geometrically bounded polynomial system is geometrically bounded.

Theorem 5.5. *Let $\mathbb{P}(x)$ in $(F\lfloor\omega\rfloor[x])^n$ be geometrically bounded, and let r in F^n be a regular root of $\mathbb{P}_\Diamond$. Then the unique root ρ of $\mathbb{P}$ with $\rho_\Diamond = r$ is geometrically bounded.*

Proof: As r is a regular root of $\mathbb{P}_\Diamond$, $\mathbb{P}_\Diamond(r) = 0$ and $D\mathbb{P}_\Diamond(r)$ is invertible. Premultiplying $\mathbb{P}$ by $[D\mathbb{P}_\Diamond(r)]^{-1}$ we obtain a polynomial system $\mathbb{P}^*$ with the same set of roots as $\mathbb{P}$ and with $\mathbb{P}^*_\Diamond = [D\mathbb{P}_\Diamond(r)]^{-1}\mathbb{P}_\Diamond$ having the same set of roots as $\mathbb{P}_\Diamond$. Lemma 5.2 assures that $\mathbb{P}^*$ is geometrically bounded. Next, from $\mathbb{P}^*_\Diamond = ((D\mathbb{P}_\Diamond(r))^{-1}\mathbb{P})_\Diamond = (D\mathbb{P}_\Diamond(r))^{-1}\mathbb{P}_\Diamond$ and from the linearity of the derivative operators and evaluations, we have

$$D\mathbb{P}^*_\Diamond(r) = D((D\mathbb{P}_\Diamond(r))^{-1}\mathbb{P}_\Diamond)(r) = (D\mathbb{P}_\Diamond(r))^{-1}D\mathbb{P}_\Diamond(r) = I \in F^{n\times n}.$$

Lemma 5.3 assures that $\mathbb{P}^\#(x) \equiv \mathbb{P}^*(x+r)$ is geometrically bounded. Further, $\mathbb{P}^\#_\Diamond(x) = \mathbb{P}^*_\Diamond(x+r)$ and, under the transformation $x \to x+r$, the roots of $\mathbb{P}^\#$ and $\mathbb{P}^\#_\Diamond$ are in one-to-one correspondence with those of $\mathbb{P}^*$ and $\mathbb{P}^*_\Diamond$, respectively. In particular, 0 is a root of $\mathbb{P}^*_\Diamond$ and Corollary A.3 gives $(D\mathbb{P}^\#_\Diamond)(0) = (D\mathbb{P}^*_\Diamond)(r) = I$.

As $\mathbb{P}^\#$ satisfies the asumptions of Proposition 5.3 we have that it has a unique root $\rho^\#$ with $\rho^\#_\Diamond = 0$ and this root is geometrically bounded. It follows immediately that $\rho = r + \rho^\#$ is geometrically bounded. Further, ρ is a root of $\mathbb{P}^*$ and therefore of $\mathbb{P}$, and $\rho_\Diamond = r + \rho^\#_\Diamond = r$. By our earlier results, $\mathbb{P}$ has exactly one root satisfying $\rho_\Diamond = r$; we conclude that this root is ρ and that it is geometrically bounded. □

With the *order topology* on F, it is easy to see that absolute convergence implies geometric boundedness, but not the converse; see [1]. For the complex and real fields, it is well known that geometric boundedness and absolute convergence coincide.

6. Examples

We develop two short examples which illustrate our results; the first is one-dimensional and the second is multi-dimensional.

Consider the scalar polynomial $\mathbb{P}(x) = x^2 - \alpha$ where $\alpha = \sum_i a(i)\omega^i$ is a positive element in $F\lfloor\omega\rfloor$, thus the task is to determine the square root of α. The element $r \in F$ is a root of $P_\Diamond(x)$ if and only if $r^2 = a(0)$. As $DP_\Diamond(r) = 2r$, such a root r is regular if and only if

$r \neq 0$, or equivalently, $a(0) \neq 0$. From our results we have that if $a(0) \neq 0$ and $r^2 = a(0)$ one can obtain the "square root" ρ of α satisfying $\rho_{\diamond} = r$ with a recursive formula that uses only arithmetic and with quadratic complexity bounds; see Section 4. We note that a power series expansion of a square root of α need not be possible in the irregular cases, that is, when $a(0) \neq 0$; in such cases fractional power series in ω are required, e.g., when $\alpha = \omega$ we have that $\rho = \pm\omega^{1/2}$ (see [2] or [3]).

Next consider n n-dimensional spheres in F^n and r in their intersection. Suppose that the centers and radii of these spheres are perturbed by continuations. The task is then to determine a continuation of r that lies in the intersection of the perturbed spheres.

With the interpretation of $\| \ \|$ as the l_2 norm, we write $\|\delta\| = \gamma$ for $\sum_{s=1}^{n}(\delta_s)^2 = \gamma^2$ over δ in $(F\lfloor\omega\rfloor)^n$ and γ in $F\lfloor\omega\rfloor$. Now, assume that the perturbed sphere-centers $\beta^1, \ldots, \beta^n$ in $(F\lfloor\omega\rfloor)^n$ and the perturbed sphere-radii $\gamma^1, \ldots, \gamma^n$ in $F\lfloor\omega\rfloor$ are given, and r satisfies $\|(\beta^j)_{\diamond} - r\| = (\gamma^j)_{\diamond}$ for $j = 1, \ldots, n$. The task is then to find a vector $\rho \in (F\lfloor\omega\rfloor)^n$ satisfying $\rho_{\diamond} = r$ and $\|\beta^j - \rho\| = \gamma^j$ for $j = 1, \ldots, n$, that is, to solve the n-dimensional polynomial system $\mathbb{P}(x) = 0$ given by

$$\sum_{s=1}^{n}[x_s - (\beta^j)_s]^2 - (\gamma^j)^2 = 0 \quad \text{for } j = 1, \ldots, n. \tag{6.1}$$

As $P_{\diamond}(x) = 2x - 2\beta^j$, we have that $DP\,(r)$ is the $n \times n$ matrix whose jth column is $r - (\beta^j)_{\diamond}$; of course, this matrix is nonsingular if and only the vectors $r - (\beta^1)_{\diamond}, \ldots, r - (\beta^n)_{\diamond}$ are linearly independent. In such cases, our results yield a solution $\sum_i \rho_i \in (F\lfloor\omega\rfloor)^n$ of (6.1) to prescribed accuracy, with arithmetic only and with effort that is proportional to the square of the number of coefficients in the prescribed accuracy. Our results do not apply when the regularity condition is violated; this situation is much more complicated.

Acknowledgments

Research was supported in part by NSF Grant DMS 95-02892 and ONR Grant N00014-92-J1142.

Appendix

An exact multivariable Taylor expansion for polynomials over commutative rings is established. Throughout this appendix, let G be a commutative ring.

Lemma A.1 (Multidimensional Binomial Expansion). *For μ in $Z_{\oplus}^n$ and σ and τ in G^n, we have*

$$(\sigma + \tau)^{\mu} = \sum_{\substack{\nu \in Z_{\oplus}^n \\ \nu < \mu}} \left[\prod_{k \in N} \binom{\mu_k}{\nu_k}\right] \sigma^{\mu - \nu} \tau^{\nu}. \tag{A.1}$$

Proof: As G is a commutative ring, we have from the standard Binomial expansion

$$
\begin{aligned}
(\sigma+\tau)^{\mu} &= \prod_{k=1}^{n}[(\sigma+\tau)_k]^{\mu_k} = \prod_{k\in N}(\sigma_k+\tau_k)^{\mu_k} \\
&= \prod_{k\in N}\left[\sum_{\nu_k=0}^{\mu_k}\binom{\mu_k}{\nu_k}(\sigma_k)^{\mu_k-\nu_k}(\tau_k)^{\mu_k}\right] \\
&= \sum_{\substack{\nu\in Z_{\oplus}^{n}\\ \nu\le\mu}}\left[\prod_{k\in N}\binom{\mu_k}{\nu_k}(\sigma_k)^{\mu_k-\nu_k}(\tau_k)^{\mu_k}\right] \\
&= \sum_{\substack{\nu\in Z_{\oplus}^{n}\\ \nu\le\mu}}\left[\prod_{k\in N}\binom{\mu_k}{\nu_k}\right]\left[\prod_{k\in N}(\sigma_k)^{\mu_k-\nu_k}\right]\left[\prod_{k\in N}(\tau_k)^{\mu_k}\right] \\
&= \sum_{\substack{\nu\in Z_{\oplus}^{n}\\ \nu\le\mu}}\left[\prod_{k\in N}\binom{\mu_k}{\nu_k}\right]\sigma^{\mu-\nu}\tau^{\nu}.
\end{aligned}
$$

□

For a polynomial $P(x)$ in $F\lfloor\omega\rfloor[x]$ with representation as in (2.2) and for vectors σ and τ in $(F\lfloor\omega\rfloor[x])^n$, $P(\sigma+\tau)$ is the polynomial $\sum_{\mu\in M}\alpha_\mu(\sigma+\tau)^\mu$ (in $F\lfloor\omega\rfloor[x]$). Recall that $\|M\| = \max_{\mu\in M, k\in N}\mu_k$.

Lemma A.2 (Taylor Expansion for Polynomials). *Let $P(x)$ be a polynomial in $G[x]$ as in (2.2) and let σ and τ be in $(G[x])^n$. Then*

$$
P(\sigma+\tau) = \sum_{\substack{\nu\in Z_{\oplus}^{n}\\ 1\le\nu_k\le\|M\|}} \frac{1}{\nu_1!\cdots\nu_n!}D_\nu P(\sigma)\tau^{\nu}. \tag{A.2}
$$

Proof: Lemma A.1 shows that for each μ in $Z_{\oplus}^{n}$

$$
(\sigma+\tau)^{\mu} = \sum_{\substack{\nu\in Z_{\oplus}^{n}\\ \nu\le\mu}}\left[\prod_{k=1}^{n}\binom{\mu_k}{\nu_k}\right]\sigma^{\mu-\nu}\tau^{\nu} = \sum_{\substack{\nu\in Z_{\oplus}^{n}\\ \nu\le\mu}}\left(\frac{1}{\nu_1!\cdots\nu_n!}\right)D_\nu(x^\mu)(\sigma)\tau^{\nu}. \tag{A.3}
$$

From (A.3), from $D_\nu(x^\mu)=0$ for ν and μ in $Z_{\oplus}^{n}$ with $\nu\not\le\mu$, from $\mu_k\le\|M\|$ for each k in N, and from the linearity of taking derivatives and evaluations, we conclude that

$$P(\sigma+\tau)=\sum_{\mu\in M}\alpha_\mu(\sigma+\tau)^\mu=\sum_{\mu\in M}\alpha_\mu\left[\sum_{\substack{\nu\in Z^n_\oplus\\ 1\le\nu_k\le\|M\|}}\frac{1}{\nu_1!\cdots\nu_n!}D_\nu(x^\mu)(\sigma)\tau^\nu\right]$$

$$=\sum_{\substack{\nu\in Z^n_\oplus\\ 1\le\nu_k\le\|M\|}}\frac{1}{\nu_1!\cdots\nu_n!}D_\nu\left(\sum_{\mu\in M}\alpha_\mu x^\mu\right)(\sigma)\tau^\nu$$

$$=\sum_{\substack{\nu\in Z^n_\oplus\\ 1\le\nu_k\le\|M\|}}\frac{1}{\nu_1!\cdots\nu_n!}D_\nu P(\sigma)\tau^\nu.$$

□

Corollary A.3. *Let $P(x)$ be a polynomial in $G[x]$, let ρ be a vector in G^n and let $Q(x)=P(\rho+x)$. Then for each ν in $Z^n_\oplus$, $D_\nu Q(0)=(D_\nu P)(\rho)$.*

Proof: From Lemma A.2,

$$Q(x)=\sum_{\substack{\nu\in Z^n_\oplus\\ 1\le\nu_k\le\|M\|}}\frac{1}{\nu_1!\cdots\nu_n!}D_\mu P(\rho)x^\mu.$$

Thus, the coefficient of x^ν in $Q(x)$ is $D_\nu P(\rho)/[\prod_{k\in N}\nu_k!]$; as $D_\nu Q(0)$ is the product of this coefficient and $[\prod_{k\in N}\nu_k!]$, we conclude that $D_\nu Q(0)=D_\nu P(\rho)$. □

References

1. B.C. Eaves and U.G. Rothblum, "A theory on extending algorithms for parametric problems," Mathematics of Operations Research, vol. 14, pp. 502–533, 1989.
2. B.C. Eaves and U.G. Rothblum, "Polynomial time root computation of parametric polynomials with Newton's polygon process," 1996, unpublished manuscript.
3. H.T. Kung and J.F. Traub, "All algebraic functions can be computed fast," Journal of the Association for Computing Machinery, vol. 25, pp. 245–260, 1978.
4. M.M. Vainberg and V.A. Trenogin, Theory of Branching of Solutions of Non-linear Equations, Noordhoff International Publishing: Leyden, The Netherlands, 1974.

Computational Optimization and Applications 12, 207–227 (1999)

Interfaces to PATH 3.0: Design, Implementation and Usage*

MICHAEL C. FERRIS ferris@cs.wisc.edu
TODD S. MUNSON tmunson@cs.wisc.edu
Computer Sciences Department, University of Wisconsin-Madison, 1210 West Dayton Street, Madison, Wisconsin 53706

Received November 2, 1997; Accepted June 1, 1998

Abstract. Several new interfaces have recently been developed requiring PATH to solve a mixed complementarity problem. To overcome the necessity of maintaining a different version of PATH for each interface, the code was reorganized using object-oriented design techniques. At the same time, robustness issues were considered and enhancements made to the algorithm. In this paper, we document the external interfaces to the PATH code and describe some of the new utilities using PATH. We then discuss the enhancements made and compare the results obtained from PATH 2.9 to the new version.

Keywords: complementarity problems, variational inequalities, algorithms

Dedication: Olvi Mangasarian has always been a great friend, mentor, and "unparalleled" leader for the optimization group in Madison. He continues to educate, inspire and make fundamental contributions to our work and the mathematical programming field in general. It is a great pleasure to dedicate this paper to him, for without his insight and support, constant help and encouragement, and clear vision, none of this would have been possible.

1. Introduction

The PATH solver [12] for mixed complementarity problems (MCPs) was introduced in 1995 and has since become the standard against which new MCP solvers are compared. However, the main user group for PATH continues to be economists using the MPSGE preprocessor [36]. While developing the new PATH implementation, we had two goals: to make the solver accessible to a broad audience and to improve the effectiveness of the code on large, complex problems. Therefore, this paper is split into two main parts, each discussing one of these issues.

We completely redesigned the PATH implementation using object-oriented design techniques in order to easily maintain the code, rapidly build interfaces to the solver, and quickly test new ideas regarding the algorithm. The relevant components of the PATH library, necessary to add new interfaces and port the code to different architectures, are the subject of Section 2. A discussion of some of the interfaces built using this functionality

*This material is based on research supported by National Science Foundation Grant CCR-9619765 and GAMS Corporation.

and how they influenced design decisions can be found in Section 3. Some of the interfaces now supported include links to AMPL [23], GAMS [6], MATLAB, and NEOS [9], as well as a callable subroutine library version. To further broaden the user base, implementations for some of the interfaces presented are available online at

`http://www.cs.wisc.edu/cpnet/path.html`

and the PATH library can be freely obtained by contacting one of the authors.

Two significant changes were previously made to the code described in [12, 13] for improved reliability. One was the introduction of a crashing technique [14] to quickly identify an active set from the user-supplied starting point. The other was the addition of a proximal perturbation scheme [1, 2] to overcome problems with a singular basis matrix. The new implementation has further improved numerical properties and restarts when a stationary point of the merit function is found. These changes have led to a more effective version of the code. Section 4 describes the new extensions made to the algorithm for improved robustness. Finally, Section 5 reports on a comparison made between PATH 3.0 and PATH 2.9 on a suite of test problems.

Some notation and definitions are in order before proceeding. The mixed complementarity problem (MCP) is defined using lower bounds, $l \in \{\Re \cup \{-\infty\}\}^n$, and upper bounds, $u \in \{\Re \cup \{\infty\}\}^n$, such that $l < u$. Let $B := \{z \in \Re^n \mid l \le z \le u\}$, and $F : B \to \Re^n$. The vector $z \in B$ is a solution to MCP(F, B), if and only if exactly one of the following holds for each $i \in \{1, \ldots, n\}$:

$$\begin{aligned} l_i \le z_i \le u_i &\text{ and } F_i(z) = 0 \\ z_i = l_i &\text{ and } F_i(z) > 0 \\ z_i = u_i &\text{ and } F_i(z) < 0. \end{aligned}$$

The core PATH algorithm [12] uses a nonsmooth Newton method [34] to find a zero of the normal map [33] associated with the MCP. The normal map for the MCP is given by

$$F(\pi(x)) + x - \pi(x)$$

where $\pi(x)$ represents the projection of x onto B in the Euclidean norm. It is well known that if x is a zero of the normal map, then $\pi(x)$ solves the MCP. A non-monotone pathsearch [13, 20] using the residual of the normal map

$$\|F(\pi(x)) + x - \pi(x)\|$$

as a merit function is used to improve robustness. A proof of convergence and rate of convergence results can be found in [32]. The enhancements described in this paper do not change the basic properties of the algorithm; they only modify the implementation. Hence, the theory already developed [12, 32] still applies.

2. Component interfaces

Object-oriented design techniques were used to completely restructure the PATH code. The basic premise is to encapsulate data and functions together into objects. The implementation for each of the objects is hidden from the user. Virtual classes are used for system-dependent code and the basis package. The exact implementation for each of these objects is chosen at link time without making any modifications to PATH. C was chosen as the implementation medium so that we can easily port the code to any platform. In this section, we discuss the external interfaces required in order to link the code with new utilities.

The overriding concern faced when developing the component interfaces was flexibility. The goal was to easily support any environment, architecture, or programming language using the same interfaces. In order to achieve this goal, the component interfaces are split into two parts: the system-dependent components, which include memory allocation, timing mechanisms, error reporting, and output, and the interface-specific functions, which provide information about the problem and mechanisms to perform function and Jacobian evaluations. The interface writer also provides a driver program that performs initializations and calls the main PATH routine. We describe the system-dependent objects required, mechanisms needed to provide problem data, and the driver in the following subsections.

2.1. System-dependent objects

The system-dependent objects provide an abstract view of the machine. Whenever the PATH code is ported to a new architecture, only implementations for these virtual base classes need to be coded. The four objects comprising the system-dependent interface, memory, time, error, and output, are now described and the required functionality given.

2.1.1. Memory. The memory object consists of all necessary functions to allocate and relinquish memory. We distinguish between two different types of memory allocation: general memory allocation and the allocation necessary for the basis factors. The general memory allocation routine is frequently called upon to allocate relatively small pieces of memory. The factor allocation requires a single large section of memory to be obtained. Mechanisms optimized for these differing types of memory request patterns can be coded. Table 1 describes the functionality required of the memory object.

We guarantee that within the core PATH algorithm only one set of factors will be allocated at a time. However, the sequence Memory_AllocateFactors(), Memory_FreeFactors() may be repeated within the code.

A standard implementation for this object would use the routines malloc() and free(). More sophisticated implementations are possible. For example, the GAMS implementation of these routines places the factors on the GAMS heap, a portion of memory previously allocated as workspace for the MPSGE preprocessor.

2.1.2. Time. The time object measures the amount of time spent in particular sections of the PATH code. The object requires the definition of a structure, struct_Time, containing the private data members for the class. An implementation using the clock() standard function

Table 1. Memory object functions.

Function	Description
void *Memory_Allocate (long int)	Allocate the specified number of bytes from memory and return a pointer to the allocated memory
void *Memory_AllocateFactors (long int)	Allocate the specified number of bytes of memory and return a pointer to the allocated memory. The amount of memory requested in Memory_AllocateFactors() is typically much greater than that requested in Memory_Allocate()
void Memory_Free (void *)	Free the indicated memory allocated with Memory_Allocate()
void Memory_FreeFactors (void *)	Free the indicated memory allocated with Memory_AllocateFactors()

Table 2. Time object functions.

Function	Description
Time *Time_Create ()	Allocate and return a time structure
void Time_Destroy (Time *)	Free the indicated time structure
void Time_Start (Time *)	Place the correct value for the current time in the indicated structure
double Time_Query (Time *)	Return the number of seconds elapsed since the indicated structure has been started

might define the structure as follows:

```
struct _Time { clock_t time; };
```

where clock_t is declared in the standard header file time.h. The necessary functions for the time object are found in Table 2.

System-dependent routines such as rusage() on a UNIX platform can be used to implement the time object. On the DOS/Windows platform the clock() function can be used. We refer the reader to the code available online for these implementations.

2.1.3. Error. The functions contained in this package provide the user with information concerning warnings and errors. A warning tells the user about difficulties or nonstandard events encountered. At the end of the warning, control should be returned to the PATH algorithm. An error on the other hand is fatal. Execution should stop at the end of the error routine. The requisite functions are in Table 3.

An interesting contrast in the error function implementation can be found in the online codes by comparing the standard implementation using exit() with the MATLAB interface error routine, which uses mexErrMsgTxt(). When exit() is used, the operating system frees all previously allocated memory and the program terminates. In the case of MATLAB, we need to relinquish all allocated memory before returning control back to MATLAB, otherwise, the MATLAB session leaks memory. The mexErrMsgTxt() routine deallocates

Table 3. Error object functions.

Function	Description
Warning (char *, . . .)	A warning has been generated. Do something with the information and return to the code
Error (char *, . . .)	An error has been generated. Do something with the information and exit from the program

the necessary memory for us before returning control back to MATLAB. The use of exit() is entirely inappropriate within the MATLAB session.

2.1.4. Output. Output is a key object because support for output from both C and FORTRAN is required. We have an enumerated type, Output_Mode, that indicates the necessary type of output. This enumerated type has three elements:

Output_Log. The log is used to demonstrate that PATH is making progress.
Output_Status. The status file is used for debugging purposes and records the essential information for this task.
Output_Copy. Output specified for copy is placed into both the log file and the status file.

Our use of these modes is motivated by GAMS standards. A brief description of the required functionality can be found in Table 4.

A typical C implementation uses the standard functions vprintf(), vfprintf(), and vsprintf(). FORTRAN output is considerably more difficult and depends upon the system and compiler. Details on FORTRAN output can be found in the online stand-alone FORTRAN code. However, the PATH code remains unchanged even if FORTRAN output is used.

2.2. *Problem-specific interface*

To accommodate all of the available interfaces, we have abstracted problem-specific information from the PATH algorithm. Five functions are required in the problem-specific interface implementation and are given in Table 5. The function_evaluation() and jacobian_evaluation() functions should return the number of domain violations encountered in the evaluation. A domain violation occurs, for example, when we attempt to take the log of a negative number or we encounter division by zero.

Table 4. Output object functions.

Function	Description
Output_Printf (Output_Mode, char *, . . .)	Output the indicated message of type Output_Mode to the correct place
Output_VPrintf (Output_Mode, char *, va_list)	Output the indicated message to the correct file

Table 5. Problem-specific interface functions.

Function	Description
void problem_size (int *n, int *nnz)	Give the size of the problem and number of nonzeros in the Jacobian
void bounds (int n, double *z, double *lb, double *ub)	Give the lower and upper bounds and a starting point for the problem
int function_evaluation (int n, double *z, double *f)	Evaluate the function for a point $z \in B$. Return the number of domain violations
int jacobian_evaluation (int n, double *z, int wantf, double *f, int *nnz, int *col, int *len, int *row, double *data)	Evaluates the function and Jacobian for a point $z \in B$. Return the number of domain violations. The Jacobian is stored in a compressed column format. Initially nnz is the allocated size for the row and data arrays. The actual number of nonzeros in the Jacobian should be supplied before returning
void variable_name (int var, char *buf, int bufSize)	Provide a name for the specified variable
void constraint_name (int con, char *buf, int bufSize)	Provide a name for the specified constraint

For the sake of compatibility, indices are in the FORTRAN format. i.e., an index goes from one to the number of indices. The sparse matrix is stored in compressed column format. Furthermore, the structure of the sparse matrix need only be determined and allocated once. We guarantee that the PATH code will not alter the structure. Finally, we also note that contrary to previous versions of the code, the addition of a diagonal element to the sparse Jacobian is no longer required.

2.3. *Driver*

Once the interfaces have been written, a main driver routine must be specified. Pseudocode for such a routine follows:

```
1. Initialize system-dependent parts.
2. Call Options_Default().
3. Create an initial Path structure
   using Path_Create().
4. Invoke Path_Solve().
5. Do something with the results.
6. Finish system-dependent parts.
```

If special setups need to be performed for the system-dependent parts of the code, they are done at the beginning. Options_Default() ensures that all defaults have been set and must be issued before solving the problem. An options file may be read using the command Options_Read(). Users who repeatedly invoke the algorithm can directly allocate required workspaces and reuse that workspace from one PATH solve to the next. This is a key point for applications solving a sequence of MCPs, as is done in [15]. The argument lists for the aforementioned routines can be found in the online version of the PATH header files.

These component interfaces have been successfully used to construct links to the GAMS and AMPL modeling languages. We have also developed MATLAB and NEOS hookups using the same library. Details on these interfaces and how to invoke PATH directly from C and FORTRAN codes are now given.

3. Supported interfaces

A number of new interfaces are now supported by the PATH implementation. We briefly discuss some of them and describe how their unique characteristics influenced the component interface design.

3.1. Matlab

The MATLAB interface (for MATLAB versions 5.0 and above) to PATH consists of a single mex file. Function and Jacobian evaluations are provided by the user as MATLAB m-functions. Hence, the PATH interface routines function_evaluation() and jacobian_evaluation() are implemented via the mexCallMATLAB() function. The problem size and bounds are passed directly to the mex function. The memory and error routines are implemented using the mxCalloc() and mexErrMsgTxt() routines provided by the MATLAB API. The abstract view of the machine provided by the new PATH library permits this and gives distinct advantages over previous implementations when a user interrupts the PATH code from within MATLAB. A MATLAB user invokes PATH with the following syntax:

```
z = pathsol(z, l, u, cpfunjac)
```

The name pathsol is used to avoid conflicts with the MATLAB defined variable path. Here, l and u are the bounds on the variables, and cpfunjac is the name of an m-file for evaluating the function F and its Jacobian J. The corresponding file cpfunjac.m contains the definition of

```
function [F, J, domerr] = cpfunjac(z, jacflag)
```

that computes the function F and if jacflag = 1 the sparse Jacobian J at the point z. domerr returns the number of domain violations encountered during the evaluation.

If the fourth input argument to pathsol is omitted, the code takes a function cpfunjac as its default. If l, u are also omitted, z is assumed non-negative. If PATH fails to solve the problem, then a MATLAB error is generated. This is part of the pathsol.m file, so a user could modify the termination behavior if desired.

3.2. Callable subroutines

To encourage other users to invoke PATH from directly within their applications we have developed two interfaces allowing the code to be called from either a C or FORTRAN program. The aim of our design was to be as simple as possible in order to eliminate most of the errors that may occur while coding a particular application to use PATH. Therefore, the standalone interface is even simpler than that described in Section 2.2.

The C interface consists of three routines:

```
void pathMain(int n, int nnz, int *status,
              double *z, double *F,
              double *l, double *u);
int funcEval(int n, double *z, double *F)
int jacEval(int n, int nnz, double *z, int *col,
            int *len, int *row, double *data);
```

The user has the responsibility of writing the latter two routines and linking them with the library of PATH routines. The first evaluates the nonlinear function F at $z \in B$. Contrary to many codes for MCP, we guarantee that the function F will only be evaluated within the box, B. The second fills in the relevant information about the Jacobian of F at z in compressed column format. The data regarding bounds and size are passed to the pathMain() routine, which returns the solution z and $F(z)$ along with a termination criterion status. The values for status are available in the header file 'Path.h'; a value of 1 indicates successful completion.

The FORTRAN interface is almost identical:

```
subroutine pathmain(n, nnz, status, z, F, l, u)
integer function funceval(n, z, F)
integer function jaceval(n, nnz, z,
                         col, len, row, data)
```

The PATH libraries for certain machine architectures are now freely available by contacting one of the authors.

3.3. NEOS

NEOS [9] enables users to submit optimization problems across the Internet to the NEOS server. The server contacts a client, transmits the problem information to the client, which then attempts to solve the problem. Results are sent back to the original submitter. The PATH solver is hooked up to NEOS via a suite of interface routines [21].

When a user submits an MCP problem to NEOS, the user specifies FORTRAN functions initpt(), xbounds(), and fcn(), along with an integer to represent n. The interface uses ADIFOR [5] to compute the Jacobian for the FORTRAN function representing the problem. The sparse structure and number of nonzeros in the Jacobian are also generated. The PATH problem-specific interface routines of Section 2.2 are coded to use the sparse Jacobian and the data supplied by the other routines passed to NEOS. All these routines are linked with the PATH library to produce an executable that is run on one of the machines in the Condor pool [30] available at Wisconsin.

Current work involves generating a new C interface using ADOL-C [27] and allowing the PATH solver to be called across the Internet as a subroutine in a manner similar to that outlined in Section 3.2.

3.4. GAMS

The major users of PATH continue to be economists, many of whom use the code [13] for solving MCP models generated with the MPSGE preprocessor [36, 37] of the GAMS modeling language [6]. CPLIB [16], a suite of routines giving a solver access to problem-specific information, including function and Jacobian evaluations, facilitates the linkage of PATH and other solvers [1, 4, 7, 35] to GAMS. The GAMS/MCP interface to PATH [16] remains essentially unchanged. However, several new uses of the code from within GAMS were made possible because of the added flexibility provided by the object-oriented design, the reuse of memory, and access to the workspace. Two examples are the GAMS/CNS interface, and the bundle method for solving MPEC, which uses PATH as a subroutine to solve MCP subproblems. The latter usage is described in [15].

Recently, GAMS added the constrained nonlinear system, CNS, model type to their language. The constrained nonlinear system is defined by a set of bounds, B and a function $F : B \to \Re^n$. A solution to CNS(F, B) is such that $x \in B$ and $F(x) = 0$. The most popular approach in GAMS to solving this problem has been to set up a dummy objective function and solve

$$\min 0 \quad \text{subject to } F(x) = 0, \quad x \in B. \tag{1}$$

The new CNS model type allows solvers such as CONOPT [17] and MINOS [31] to set up alternative internal models to solve (1) such as

$$\min \| F(x) \|_2^2 \quad \text{subject to } x \in B. \tag{2}$$

Our work internally reformulates the CNS model as an MCP. One choice would be to solve the Karush-Kuhn-Tucker conditions of (2) as an MCP. This is essentially a Gauss-Newton approach (involving a function $\nabla F(x)^T F(x)$) to the problem and appears to be computationally less effective than the following Newton approach. The particular MCP that our implementation uses, is defined by the following functions and bounds:

$$G(x, y) = \begin{bmatrix} -y \\ F(x) \end{bmatrix}, \quad L = \begin{bmatrix} l \\ -\infty \end{bmatrix}, \quad U = \begin{bmatrix} u \\ \infty \end{bmatrix}. \tag{3}$$

Let $\bar{B} := \{z \in \Re^{2n} | L \le z \le U\}$. If $\bar{x} \in \Re^n$ solves CNS(F, B), then $(\bar{x}, 0)$ solves MCP(G, $\bar{B}$). Furthermore, if $(\bar{x}, \bar{y})$ solves MCP(G, $\bar{B}$), then $\bar{x}$ solves CNS(F, B). These implications follow directly from the definitions of the MCP and CNS. The implementation of the CNS solver reuses the memory, time, error, and output packages from the GAMS/MCP interface. However, the problem-specific interface routines were modified to construct (3) and the Jacobian of G from the GAMS provided routines for CNS(F, B).

3.5. AMPL

New syntax for expressing complementarity relationships has been added to the AMPL modeling language [19]. In order to test this syntax, the PATH solver has been hooked up to AMPL [23] using the AMPL library routines described in [24]. These routines allow easy implementation of the PATH problem-specific routines outlined in Section 2.2. The AMPL interface routines exploit the fact that a function evaluation may have already been carried out (thus providing partial derivative information) when a call to the Jacobian evaluator is made. We have updated the PATH code to allow for efficiencies to be exploited when this occurs (see the parameters to jacobian_evaluation in Section 2.2).

4. Enhancements

During the reconstruction of the PATH implementation, robustness issues were once again addressed. Utilizing experience gained from solving practical problems, three areas were targeted for improvement. These areas, the linear model, pathsearch, and restarts, are the subject of subsequent subsections.

When the PATH implementation was rewritten, the names of several options were changed. We give a list of the user-available options along with their defaults and meaning in Tables 6 and 7.

Table 6. PATH options.

Option	Default	Explanation
convergence_tolerance	1e-6	Stopping criterion
crash_iteration_limit	50	Maximum iterations allowed in crash
crash_method	pnewton	pnewton or none
crash_minimum_dimension	1	Minimum problem dimension to perform crash
crash_nbchange_limit	1	Number of changes to the basis allowed
cumulative_iteration_limit	10000	Maximum minor iterations allowed
lemke_start	automatic	Frequency of lemke starts (automatic, first, always)
major_iteration_limit	500	Maximum major iterations allowed
minor_iteration_limit	1000	Maximum minor iterations allowed in each major iteration
nms	yes	Allow line/path searching, watchdoging, and non-monotone descent
nms_initial_reference_factor	20	Controls size of initial reference value
nms_memory_size	10	Number of reference values kept
nms_mstep_frequency	10	Frequency at which m steps are performed
nms_searchtype	path	Search type to use (path or line)

Table 7. PATH options (cont).

Option	Default	Explanation
output	yes	Turn output on or off. If output is turned off, selected parts can be turned back on
output_crash_iterations	yes	Output information on crash iterations
output_crash_iterations_frequency	1	Frequency at which crash iteration log is printed
output_errors	yes	Output error messages
output_linear_model	no	Output linear model each major iteration
output_major_iterations	yes	Output information on major iterations
output_major_iterations_frequency	1	Frequency at which major iteration log is printed
output_minor_iterations	yes	Output information on minor iterations
output_minor_iterations_frequency	500	Frequency at which minor iteration log is printed
output_options	no	Output all options and their values
output_warnings	no	Output warning messages
proximal_perturbation	0	Initial perturbation
restart_limit	3	Maximum number of restarts (0–3)
time_limit	3600	Maximum number of seconds algorithm is allowed to run

In particular, PATH can emulate Lemke's method [8, 29] for LCP with the following options:

```
crash_method none;
major_iteration_limit 1;
lemke_start first;
nms no;
```

If instead, PATH is to imitate the Josephy Newton method [28] for NCP with an Armijo style linesearch on the normal map residual, then the options to use are:

```
crash_method none;
lemke_start always;
nms_initial_reference_factor 1;
nms_memory size 1;
nms_mstep_frequency 1;
nms_searchtype line;
```

Note that `nms_memory_size 1` and `nms_initial_reference_factor 1` turn off the nonmonotone linesearch, while `nms_mstep_frequency 1` turns off watchdoging. `nms_searchtype line` forces PATH to search the line segment between the initial point and the solution to the linear model, as opposed to the default pathsearch (see Section 4.3).

4.1. *Basis package*

The core PATH code has been recoded in an object-oriented fashion as well. One key feature of the new design is the notion of a basis package. PATH requires a basis package to provide certain functionality expressed via the following functions:

Basis_Factor(). Factor the given basis matrix.
Basis_Solve(). Use the computed factors to solve a linear system of equations.
Basis_Replace(). Replace the indicated column of the basis matrix.
Basis_SingularVariable(). Tell whether the indicated variable is linearly dependent.

Currently, two interchangeable basis packages have been implemented. The choice of which package to use is done at link time without making any modifications to the code PATH code. One package uses a FORTRAN version of LUSOL [25] (based on a Markovitz factorization and Bartels-Golub updates). These routines are a key component of the MINOS [31] nonlinear programming package. The second is a dense matrix implementation, written in C that uses the Fletcher-Matthews [22] updating procedure. A freely distributed version of the PATH library contains the dense factorization. A library without any factorization software is also available. In this case, the user can obtain source code for MINOS from Stanford University and incorporate the appropriate object code into the library.

A future basis object for the MATLAB implementation will use mexCallMATLAB() to invoke the MATLAB LU routine, along with an update procedure provided via the Schur-Complement described in [18].

4.2. *The linear model*

Let $M \in \Re^{n\times n}$, $q \in \Re^n$, and $B = [l, u]$ be given. $(\bar{z}, \bar{w}, \bar{v})$ solves the linear mixed complementarity problem defined by M, q, and B if and only if it satisfies the following constrained system of equations:

$$Mz - w + v + q = 0 \tag{4}$$
$$w^T(z - l) = 0 \tag{5}$$
$$v^T(u - z) = 0 \tag{6}$$
$$z \in B, \quad w \in \Re^n_+, \quad v \in \Re^n_+, \tag{7}$$

where $x + \infty = \infty$ for all $x \in \Re$ and $0 \cdot \infty = 0$ by convention. A triple, $(\hat{z}, \hat{w}, \hat{v})$, satisfying Eqs. (5)–(7) is called a complementary triple.

The objective of the linear model solver is to construct a path from a given complementary triple $(\hat{z}, \hat{w}, \hat{v})$ to a solution $(\bar{z}, \bar{w}, \bar{v})$. The algorithm used to solve the linear problem is identical to that given in [10]; however, artificial variables are incorporated into the model. The augmented model is then:

$$Mz - w + v + a + (1 - t)r + q = 0 \tag{8}$$
$$w^T(z - l) = 0 \tag{9}$$
$$v^T(u - z) = 0 \tag{10}$$
$$z \in B, \quad w \in \Re^n_+, \quad v \in \Re^n_+, \quad a \equiv 0, \quad t \in [0, 1] \tag{11}$$

where r is the residual, t is the path parameter, and a is a vector of artificial variables. The residual is scaled in the code to improve numerical stability.

The addition of artificial variables enables us to construct an initial invertible basis consistent with the given starting point even under rank deficiency. The procedure consists of two parts: constructing an initial guess as to the basis and then recovering from rank deficiency to obtain an invertible basis. The crash technique gives a good approximation to the active set. The first phase of the algorithm uses this information to construct a basis by partitioning the variables into three sets:

1. $W = \{i \in \{1, \ldots, n\} \mid \hat{z}_i = l_i \text{ and } \hat{w}_i > 0\}$
2. $V = \{i \in \{1, \ldots, n\} \mid \hat{z}_i = u_i \text{ and } \hat{v}_i > 0\}$
3. $Z = \{1, \ldots, n\} \backslash W \cup V$

Since $(\hat{z}, \hat{w}, \hat{v})$ is a complementary triple, $Z \cap W \cap V = \emptyset$ and $Z \cup W \cup V = \{1, \ldots, n\}$.

Using the above guess, we can recover an invertible basis consistent with the starting point using the following procedure:

1. Let $A = \{\emptyset\}$.
2. While $M_{Z.Z}^{-1}$ does not exist:
   (a) Choose $k \in Z$ such that $M_{Z.k}$ is a linearly dependent column of $M_{Z.Z}$.
   (b) If $\hat{z}_k = l_k$, then $W = W \cup \{k\}$ and $Z = Z \backslash \{k\}$.
   (c) If $\hat{z}_k = u_k$, then $V = V \cup \{k\}$ and $Z = Z \backslash \{k\}$.
   (d) Otherwise, $A = A \cup \{k\}$ and $Z = Z \backslash \{k\}$.

The technique relies upon the factorization to tell the linearly dependent columns of $M_{Z.Z}$. LUSOL [25] and our dense factorization provide this information (see Section 4.1). Some of the variables may be nonbasic, but not at their bounds. For such variables, the corresponding artificial will be basic. The above procedure is guaranteed to terminate because Z is a finite set and we remove one element from it each iteration.

The invertible basis constructed is then:

$$H = [M_{\cdot,Z} \quad -I_{\cdot,W} \quad I_{\cdot,V} \quad I_{\cdot,A}]$$

To demonstrate that an inverse exists, we rearrange the rows and columns to obtain the following matrix, which is easily shown to be invertible because $M_{Z.Z}^{-1}$ is known to exist:

$$\hat{H} = \begin{bmatrix} -I_{W.W} & & & M_{W.Z} \\ & I_{V.V} & & M_{V.Z} \\ & & I_{A.A} & M_{A.Z} \\ & & & M_{Z.Z} \end{bmatrix}$$

We use a modified version of EXPAND [26] to perform the ratio test. Variables are prioritized as follows:

1. t leaving at its upper bound.
2. Any artificial variable.
3. Any z, w, or v variable.

If a choice as to the leaving variable can be made while maintaining numerical stability and sparsity, we choose the variable with the highest priority (lowest number above).

When an artificial variable leaves the basis and a z-type variable enters, we have the choice of either increasing or decreasing that entering variable because it is nonbasic but not at a bound. The determination is made such that t increases and stability is preserved.

If the code is forced to use a ray start at each iteration (`lemke_start always`), then the code carries out Lemke's method, which is known [8] not to cycle. However, by default, we use a regular start to guarantee that the generated path emanates from the current iterate. Under appropriate conditions, this guarantees a decrease in the nonlinear residual. However, it is then possible for the pivot sequence in the linear model to cycle. To prevent this undesirable outcome, PATH 2.9 uses a minor iteration limit. In PATH 3.0, we attempt to detect the formation of a cycle with the heuristic that if a variable enters the basis more than a given number of times, we are cycling. If this occurs, we terminate the linear solver at the largest value of t and return to the nonlinear pathsearch code.

Another heuristic is added when the linear code terminates on a ray. The returned point in this case is not the base of the ray. We move a slight distance up the ray and return this new point. If we fail to solve the linear subproblem five times in a row, a Lemke ray start will be performed in an attempt to solve the linear subproblem. Computational experience has shown this to be an effective heuristic and generally results in solving the linear model. Using a Lemke ray start is not the default mode, since typically many more pivots are required.

The resulting linear solver as modified above is robust and has the desired property that we start from $(\hat{z}, \hat{w}, \hat{v})$ and construct a path to a solution. This path is used by the nonlinear code for the implementation of pathsearching.

4.3. *Pathsearching*

An interpolating pathsearch that preserves stability replaces the pathsearch found in PATH 2.9. Instead of only using breakpoints, as outlined in [10], we allow points between the breakpoints to be used.

The EXPAND rules enforce a minimum stepsize. Therefore, some steps, and the first in particular, can be quite small. The breakpoint method has the undesirable property that a very small step may be taken in the backtrack. Therefore, we implemented a pathsearch that performs interpolation. We determine the section of the path where we want to find a point. We stably construct the two endpoints and interpolate between them to find the desired point.

Stability is achieved by using checkpoints, pivots where the basis matrix was refactorized and the basic variable values recomputed. At such a checkpoint, we know the basic variables, can reconstruct the basis matrix, factorize that matrix, and compute exactly the same basic variable values found during the original solve. Then we can perform the pivot sequence in order from the checkpoint. Once we have determined the two desired endpoints, we go to the appropriate checkpoint and regenerate the path followed to the desired endpoint, obtaining the exact values computed by the linear solve for those points. Careful bookkeeping minimizes the number of factorizations performed.

We note that difficulties can arise because t is not guaranteed to be monotonically increasing. In fact, it can decrease. To overcome this difficulty, we select a subsequence of the path over which t is monotonically increasing. We use this as the path over which we perform our nonlinear pathsearch.

In contrast, PATH 2.9 carried out backtracking by performing the sequence of pivots in reverse. This technique can generate additional numerical rounding error, leading on occasion to a failure in the backtracking procedure. The interpolating pathsearch removes these errors, is more stable, and furthermore checks a more varied sequence of iterates along the path.

4.4. *Restarts*

The PATH code attempts to fully utilize the resources provided by the user to solve the problem. PATH 3.0 is much more aggressive in its determination that a stationary point of the residual of the normal map has been encountered. When we determine that no progress is being made, we restart the problem from the user-supplied initial point with a different set of parameters. The three sets of parameter choices used during restarts are given in Table 8.

These restarts give us the flexibility to change the algorithm in the hopes that the modified algorithm leads to a solution. The ordering and nature of the restarts were determined by empirical evidence based on tests performed on real-world problems.

This technique can be contrasted with that used PATH 2.9, in which significant efforts were made in an attempt to move away from a stationary point by modifying algorithm parameters within a major iteration and only restarting if these modifications failed to make

Table 8. Restart definitions.

Restart number	Parameter values
1	crash_method none
	nms_initial_reference_factor 2
	proximal_perturbation 1e-2*initial_residual
2	crash_method none
	proximal_perturbation 0
3	crash_method pnewton
	crash_nbchange_limit 10
	nms_initial_reference_factor 2
	nms_searchtype line

Table 9. Step type codes.

Code	Meaning
B	A backtracking search was performed from the current iterate to the Newton point in order to obtain sufficient decrease in the merit function
D	The step was accepted because the distance between the current iterate and the Newton point was small
I	Initial information concerning the problem is displayed
M	The step was accepted because the merit function value is smaller than the nonmonotone reference value
O	A step that satisfies both the distance and merit function tests
R	A restart was carried out
W	A watchdog step was performed in which we returned to the last point encountered with a better merit function value than the nonmonotone reference value (M, O, or B step), regenerated the Newton point, and performed a backtracking search

progress. In some cases, the heuristics used were successful. More often they simply used resources that could have been better spent by restarting. We believe that finding a stationary point and restarting, as done in PATH 3.0, is a significantly better strategy than the ad-hoc heuristics of PATH 2.9.

At each iteration of the algorithm, several different step types can be taken. In order to help the PATH 3.0 user, we have added a code letter indicating the step type to the log file. Table 9 explains the meaning of each code letter.

5. Computational results

In this section, we compare the results obtained from PATH 2.9 and PATH 3.0 using their standard default options on the problems comprising MCPLIB [11]. We do not compare PATH 3.0 to other codes since this was carried out in [3]. However, the results of [3] seem to indicate that PATH was the most robust and efficient code at that time. The problems are available online (as GAMS files) as documented in [11], along with some recent additions. The computations were carried out on a Sun UltraSparc 300 MHz processor with 256 MB RAM. We report the numbers of successes (Suc), failures (Fail), function evaluations (F), and total time (Time) in seconds used by each algorithm on a given problem in the accompanying tables. The times reported are rounded to the nearest 1/10 s, resulting in several problems having solution times of 0.0 s. The size of the first instance of each problem is also given in these tables; note that several of the problems are modified within the GAMS code to include more variables or equations on subsequent runs. The termination criterion for both codes is identical, requiring the norm of the residual in the normal map to be less than 10^{-6}. We do not report results for GAMSLIB [6], since both algorithms solve all of the problems in the suite on defaults. The results displayed in Tables 10 and 11 indicate that PATH 3.0 is significantly more robust than PATH 2.9. We believe the main reasons for improved robustness are increased numerical accuracy (see comments in Section 4.3) and more aggressive restart procedures (see remarks in Section 4.4).

Table 10. Comparison of PATH 3.0 to PATH 2.9.

		PATH 3.0			PATH 2.9		
Problem	Size	Suc (Fail)	F	Time	Suc (Fail)	F	Time
asean9a	10199	1 (0)	4	2.9	1 (0)	4	2.4
badfree	5	1 (0)	4	0.0	0 (1)	4080	0.9
bert_oc	5000	4 (0)	16	1.9	4 (0)	16	1.5
bertsekas	15	6 (0)	79	0.0	6 (0)	94	0.1
billups	1	0 (3)	557	0.1	0 (3)	13584	1.8
bishop	1645	1 (0)	43	51.3	1 (0)	40	15.5
box	44	355 (6)	7875	10.4	351 (10)	31298	43.6
bratu	5625	1 (0)	48	18.8	1 (0)	48	18.1
choi	13	1 (0)	5	0.5	1 (0)	5	0.4
colvdual	20	4 (0)	91	0.1	4 (0)	57	0.2
colvnlp	15	6 (0)	44	0.0	6 (0)	43	0.0
cycle	1	1 (0)	5	0.0	1 (0)	60	0.0
degen	2	1 (0)	7	0.0	1 (0)	2	0.0
denmark	2093	40 (0)	1556	532.3	36 (4)	1504	836.1
duopoly	63	1 (0)	340	0.9	0 (1)	575	2.4
ehl_k40	41	2 (1)	2346	25.1	2 (1)	6194	60.0
ehl_k60	61	3 (0)	1009	19.9	3 (0)	108	7.5
ehl_k80	81	3 (0)	449	19.9	3 (0)	99	11.9
ehl_kost	101	3 (0)	36	4.4	3 (0)	36	4.6
electric	158	1 (0)	578	1.3	1 (0)	297	1.2
eppa	1269	8 (0)	45	11.9	8 (0)	45	13.4
eta2100	296	1 (0)	16	0.4	1 (0)	30	0.6
explcp	16	1 (0)	6	0.0	1 (0)	8	0.0
forcebsm	184	1 (0)	11	0.1	1 (0)	11	0.1
forcedsa	186	1 (0)	11	0.1	1 (0)	11	0.1
freebert	15	7 (0)	63	0.0	7 (0)	73	0.1
gafni	5	3 (0)	15	0.0	3 (0)	13	0.0
games	16	23 (2)	1970	5.6	22 (3)	9205	31.9
hanskoop	14	10 (0)	128	0.1	10 (0)	146	0.1
hanson	487	2 (0)	231	1.6	2 (0)	118	4.7
hydroc06	29	1 (0)	7	0.0	1 (0)	8	0.0
hydroc20	99	1 (0)	10	0.1	1 (0)	11	0.1
jel	6	2 (0)	11	0.0	2 (0)	11	0.0
jmu	2253	1 (0)	38	15.9	1 (0)	40	14.0
josephy	4	8 (0)	96	0.0	8 (0)	87	0.1
kojshin	4	8 (0)	132	0.0	8 (0)	54	0.1
lincont	419	1 (0)	15	1.3	1 (0)	16	1.3

Table 11. Comparison of PATH 3.0 to PATH 2.9.

		PATH 3.0			PATH 2.9		
Problem	Size	Suc (Fail)	F	Time	Suc (Fail)	F	Time
mathinum	3	6 (0)	59	0.0	6 (0)	60	0.1
mathisum	4	7 (0)	512	0.1	7 (0)	73	0.1
methan08	31	1 (0)	5	0.0	1 (0)	5	0.0
multi-v	48	3 (0)	48	0.2	2 (1)	1581	3.2
nash	10	4 (0)	24	0.0	4 (0)	24	0.1
ne-hard	3	1 (0)	37	0.0	0 (1)	4	0.0
obstacle	2500	8 (0)	104	8.7	8 (0)	104	7.5
olg	249	1 (0)	1396	10.5	0 (1)	2574	16.3
opt_cont	288	1 (0)	6	0.1	1 (0)	6	0.1
opt_cont127	4096	1 (0)	6	1.7	1 (0)	6	1.2
opt_cont255	8192	1 (0)	6	3.7	1 (0)	6	2.6
opt_cont31	1024	1 (0)	6	0.3	1 (0)	6	0.2
opt_cont511	16384	1 (0)	6	8.3	1 (0)	6	5.2
pgvon105	105	3 (1)	181	1.8	3 (1)	323	2.8
pgvon106	106	3 (3)	4447	22.6	1 (5)	5763	38.7
pies	42	1 (0)	15	0.0	1 (0)	13	0.0
powell	16	6 (0)	57	0.1	5 (1)	6561	4.7
powell_mcp	8	6 (0)	94	0.1	6 (0)	51	0.1
qp	4	1 (0)	3	0.0	1 (0)	2	0.0
romer	214	0 (1)	325	1.9	0 (1)	142	4.1
scarfanum	13	4 (0)	30	0.0	4 (0)	33	0.1
scarfasum	14	4 (0)	21	0.0	4 (0)	23	0.0
scarfbnum	39	2 (0)	47	0.1	2 (0)	49	0.1
scarfbsum	40	2 (0)	128	0.7	1 (1)	216	2.3
shubik	30	48 (0)	6241	4.2	40 (8)	13064	11.9
simple-ex	17	0 (1)	732	0.6	0 (1)	370	0.7
simple-red	13	1 (0)	11	0.0	1 (0)	11	0.0
sppe	27	3 (0)	27	0.0	3 (0)	29	0.0
tinloi	146	63 (1)	1638	12.2	56 (8)	9632	77.2
tobin	42	4 (0)	41	0.1	4 (0)	44	0.1
trade12	600	2 (0)	26	3.0	2 (0)	30	3.9
trafelas	2376	2 (0)	141	16.9	2 (0)	146	15.2
uruguay	2281	7 (0)	27	796.0	7 (0)	27	1477.3
Total		712 (19)	34344	1568.6	679 (52)	109015	2694.1

The CPU times in the tables show that in general, the increase in robustness is not due to more resources being expended, but by using the given resources more effectively. In general, PATH 3.0 takes about the same time as PATH 2.9, with some substantial improvements in CPU time on several problems. Because of the significant differences between the two codes, PATH 3.0 is sometimes slower than PATH 2.9. In some of these cases, PATH 2.9 exits prematurely because of numerical problems, i.e., the basis matrix becomes singular while backtracking. At other times, the extra basis factorizations performed by PATH 3.0 to determine an initial feasible basis and to backtrack stably account for the discrepancy. We also note that the aggressive restarts in PATH 3.0 can cause increased time because we restart the problem from the user-supplied starting point while PATH 2.9 may succeed quickly with one of its ad-hoc heuristics.

We firmly believe that computational times are the best indication of the efficiency of the code. However, we have also included the number of function evaluations in the tables. For most of the problems the costs associated with linear algebra dominate the costs of function evaluations. Of particular note in this class is the hanson problem, which due to our backtracking procedure takes more function evaluations, but spends less time on linear algebra leading to a faster computation. Our more careful backtracking procedure, while increasing robustness of the code, can lead to more function evaluations being performed. In particular, on the ehl_k* problems, in which function evaluations dominate linear algebra costs, we perform significantly more function evaluations than PATH 2.9. However, in summary we point out that PATH 3.0 is more robust, takes less time, and performs fewer function evaluations when we average over the entire test set.

6. Conclusion

We have described a redesign of the PATH solver that facilitates easier interfacing to a variety of computing environments and shown how such interfacing has been effected. The redesign has also improved robustness of the code, without a corresponding increase in computational resources used. This paper has carefully described the major enhancements to the implementation and provided some numerical justification of improved robustness.

References

1. S.C. Billups, "Algorithms for complementarity problems and generalized equations," Ph.D. Thesis, University of Wisconsin-Madison, Madison, Wisconsin, 1995.
2. S.C. Billups, "Improving the robustness of complementarity solvers using proximal perturbations," UCD/CCM Report No. 96, University of Colorado at Denver, Denver, Colorado, 1996.
3. S.C. Billups, S.P. Dirkse, and M.C. Ferris, "A comparison of large scale mixed complementarity problem solvers," Computational Optimization and Applications, vol. 7, pp. 3–25, 1997.
4. S.C. Billups and M.C. Ferris, "QPCOMP: A quadratic program based solver for mixed complementarity problems," Mathematical Programming, vol. 76, pp. 533–562, 1997.
5. C. Bischof, A. Carle, P. Khademi, A. Mauer, and P. Hovland, "ADIFOR 2.0 user's guide," Mathematics and Computer Science Division Report ANL/MCS-TM-192, Argonne National Laboratory, Argonne, IL, 1995.
6. A. Brooke, D. Kendrick, and A. Meeraus, GAMS: A User's Guide, The Scientific Press: South San Francisco, CA, 1988.

7. Chunhui Chen and O.L. Mangasarian, "A class of smoothing functions for nonlinear and mixed complementarity problems," Computational Optimization and Applications, vol. 5, pp. 97–138, 1996.
8. R.W. Cottle and G.B. Dantzig, "Complementary pivot theory of mathematical programming," Linear Algebra and Its Applications, vol. 1, pp. 103–125, 1968.
9. J. Czyzyk, M.P. Mesnier, and J.J. Moré, "The network-enabled optimization server," Preprint MCS-P615-0996, Argonne National Laboratory, Argonne, Illinois, 1996.
10. S.P. Dirkse, "Robust solution of mixed complementarity problems," Ph.D. Thesis, Computer Sciences Department, University of Wisconsin, Madison, Wisconsin, 1994. Available from ftp://ftp.cs.wisc.edu/math-prog/tech-reports/.
11. S.P. Dirkse and M.C. Ferris, "MCPLIB: A collection of nonlinear mixed complementarity problems," Optimization Methods and Software, vol. 5, pp. 319–345, 1995.
12. S.P. Dirkse and M.C. Ferris, "The PATH solver: A non-monotone stabilization scheme for mixed complementarity problems," Optimization Methods and Software, vol. 5, pp. 123–156, 1995.
13. S.P. Dirkse and M.C. Ferris, "A pathsearch damped Newton method for computing general equilibria," Annals of Operations Research, 1996.
14. S.P. Dirkse and M.C. Ferris, "Crash techniques for large-scale complementarity problems," in Complementarity and Variational Problems: State of the Art, M.C. Ferris and J.S. Pang (Eds.), SIAM Publications: Philadelphia, PA, 1997.
15. S.P. Dirkse and M.C. Ferris, "Modeling and solution environments for MPEC: GAMS & MATLAB," in Reformulation—Nonsmooth, Piecewise Smooth, Semismooth and Smoothing Methods, M. Fukushima and L. Qi (Eds.), Kluwer Academic Publishers, 1998, pp. 127–148, forthcoming.
16. S.P. Dirkse, M.C. Ferris, P.V. Preckel, and T. Rutherford, "The GAMS callable program library for variational and complementarity solvers," Mathematical Programming Technical Report 94-07, Computer Sciences Department, University of Wisconsin, Madison, Wisconsin, 1994.
17. A. Drud, "CONOPT: A GRG code for large sparse dynamic nonlinear optimization problems," Mathematical Programming, vol. 31, pp. 153–191, 1985.
18. S.K. Eldersveld and M.A. Saunders, "A block-LU update for large-scale linear programming," SIAM Journal on Matrix Analysis and Applications, vol. 13, pp. 191–201, 1992.
19. M.C. Ferris, R. Fourer, and D.M. Gay, "Expressing complementarity problems and communicating them to solvers," Mathematical Programming Technical Report 98-02, Computer Sciences Department, University of Wisconsin, Madison, Wisconsin, 1998.
20. M.C. Ferris and S. Lucidi, "Nonmonotone stabilization methods for nonlinear equations," Journal of Optimization Theory and Applications, vol. 81, pp. 53–71, 1994.
21. M.C. Ferris, M.P. Mesnier, and J. Moré, "NEOS and condor: Solving nonlinear optimization problems over the Internet," Mathematical Programming Technical Report 96-08, Computer Sciences Department, University of Wisconsin, Madison, Wisconsin, 1996 (revised 1998). Also available as ANL/MCS-P708-0398, Mathematics and Computer Science Division, Argonne National Laboratory.
22. R. Fletcher and S.P.J. Matthews, "Stable modifications of explicit LU factors for simplex updates," Mathematical Programming, vol. 30, pp. 267–284, 1984.
23. R. Fourer, D.M. Gay, and B.W. Kernighan, AMPL: A Modeling Language for Mathematical Programming, Duxbury Press, 1993.
24. D.M. Gay, "Hooking your solver to AMPL," Technical Report, Bell Laboratories, Murray Hill, NJ, 1997. (revised 1994, 1997).
25. P.E. Gill, W. Murray, M.A. Saunders, and M.H. Wright, "Maintaining LU factors of a general sparse matrix," Linear Algebra and Its Applications, vol. 88/89, pp. 239–270, 1987.
26. P.E. Gill, W. Murray, M.A. Saunders, and M.H. Wright, "A practical anti-cycling procedure for linearly constrained optimization," Mathematical Programming, vol. 45, pp. 437–474, 1989.
27. A. Griewank, D. Juedes, and J. Utke, "ADOL-C: A package for the automatic differentiation of algorithms written in C/C++," ACM Transactions on Mathematical Software, 1996.
28. N.H. Josephy, "Newton's method for generalized equations," Technical Summary Report 1965, Mathematics Research Center, University of Wisconsin, Madison, Wisconsin, 1979.
29. C.E. Lemke and J.T. Howson, "Equilibrium points of bimatrix games," SIAM Journal on Applied Mathematics, vol. 12, pp. 413–423, 1964.

30. M.J. Litzkow, M. Livny, and M.W. Mutka, "Condor: A hunter of idle workstations," in Proceedings of the 8th International Conference on Distributed Computing Systems, June 1988, pp. 104–111.
31. B.A. Murtagh and M.A. Saunders, "MINOS 5.0 user's guide," Technical Report SOL 83.20, Stanford University, Stanford, CA, 1983.
32. D. Ralph, "Global convergence of damped Newton's method for nonsmooth equations, via the path search," Mathematics of Operations Research, vol. 19, pp. 352–389, 1994.
33. S.M. Robinson, "Normal maps induced by linear transformations," Mathematics of Operations Research, vol. 17, pp. 691–714, 1992.
34. S.M. Robinson, "Newton's method for a class of nonsmooth functions," Set Valued Analysis, vol. 2, pp. 291–305, 1994.
35. T.F. Rutherford, "MILES: A mixed inequality and nonlinear equation solver," Working Paper, Department of Economics, University of Colorado, Boulder, 1993.
36. T.F. Rutherford, "Extensions of GAMS for complementarity problems arising in applied economic analysis," Journal of Economic Dynamics and Control, vol. 19, pp. 1299–1324, 1995.
37. T.F. Rutherford, "Applied general equilibrium modeling with MPSGE as a GAMS subsystem: An overview of the modeling framework and syntax," Computational Economics, 1998, forthcoming.

Computational Optimization and Applications, 12, 229–251 (1999)

Existence and Limiting Behavior of Trajectories Associated with $\mathbf{P}_0$-equations

M. SEETHARAMA GOWDA gowda@math.umbc.edu
M. A. TAWHID tawhid@math.umbc.edu
Department of Mathematics and Statistics, University of Maryland, Baltimore County, Baltimore, MD 21250, USA

Received November 23, 1997; Accepted June 1, 1998

Abstract. Given a continuous $\mathbf{P}_0$-function $F : R^n \to R^n$, we describe a method of constructing trajectories associated with the $\mathbf{P}_0$-equation $F(x) = 0$. Various well known equation-based reformulations of the nonlinear complementarity problem and the box variational inequality problem corresponding to a continuous $\mathbf{P}_0$-function lead to $\mathbf{P}_0$-equations. In particular, reformulations via (a) the Fischer function for the NCP, (b) the min function for the NCP, (c) the fixed point map for a BVI, and (d) the normal map for a BVI give raise to $\mathbf{P}_0$-equations when the underlying function is $\mathbf{P}_0$. To generate the trajectories, we perturb the given $\mathbf{P}_0$-function F to a $\mathbf{P}$-function $F(x, \varepsilon)$; unique solutions of $F(x, \varepsilon) = 0$ as ε varies over an interval in $(0, \infty)$ then define the trajectory. We prove general results on the existence and limiting behavior of such trajectories. As special cases we study the interior point trajectory, trajectories based on the fixed point map of a BVI, trajectories based on the normal map of a BVI, and a trajectory based on the aggregate function of a vertical nonlinear complementarity problem.

Keywords: nonlinear complementarity problem, box variational inequality problem, coercive, $\mathbf{P}_0$-function, min function, fixed point map, normal map, trajectory, interior point trajectory

Dedication: This paper is dedicated to Professor Olvi L. Mangasarian on the occasion of his 65th birthday. Prof. Mangasarian's contributions to optimization, particularly to the area of complementarity problems, is well-known. His articles, talks, and ideas have inspired a generation of researchers, including the first author of this paper. He has a way of putting people at ease and getting the best out of them. The first author still vividly remembers how Prof. Mangasarian, on one spring morning in 1980, standing in front of Union South in the University of Wisconsin-Madison campus, persuaded him to change his field from Several Complex Variables to Optimization.

Olvi, thanks to your suggestion, I am now 'at home' with Optimization. You are a great teacher, a fine mentor, and a true friend. Best wishes for a long, healthy, and fruitful life. – M.S.G

1. Introduction

Given a continuous function $F : R^n \to R^n$, we say that F is a $\mathbf{P}_0$ ($\mathbf{P}$)-function if for any $x \neq y$ in R^n,

$$\max_{\{i: x_i \neq y_i\}} (x - y)_i [F(x) - F(y)]_i \geq 0 \ (> 0). \tag{1}$$

The objective of this paper is to study the existence and the limiting behavior of certain continuous curves—to be called trajectories—associated with (the solution set of) the equation

$$F(x) = 0$$

where F is assumed to be a $\mathbf{P}_0$-function. As we see below, interesting and important equation-based reformulations of the nonlinear complementarity problem and the box variational inequality problem corresponding to a $\mathbf{P}_0$-function lead to $\mathbf{P}_0$-equations. To generate a trajectory of a $\mathbf{P}_0$-equation $F(x) = 0$, we perturb $F(x)$ to a $\mathbf{P}$-function $F(x, \varepsilon)$ where ε is a positive perturbation parameter. Since $\mathbf{P}$-functions are one-to-one, the equation $F(x, \varepsilon) = 0$ will have at most one solution. As ε varies, the solutions of $F(x, \varepsilon) = 0$ then describe, depending on the nature of $F(x, \varepsilon)$, a short trajectory $\{x(\varepsilon) : 0 < \varepsilon < \bar{\varepsilon}\}$ or a long trajectory $\{x(\varepsilon) : 0 < \varepsilon < \infty\}$. By means of such trajectories, one could then devise computational schemes to solve the equation $F(x) = 0$ either by tracing the trajectory as $\varepsilon \to 0$ or by constructing a sequence of points whose limit solves the given equation.

The motivation for our study comes from some recent results in complementarity and box variational inequality problems. Given a function $f : R^n \to R^n$ and a rectangular box K in R^n, the *box variational inequality problem* $\mathrm{BVI}(f, K)$ (also known as the mixed complementarity problem) of finding a vector $x^* \in K$ such that

$$\langle f(x^*), x - x^* \rangle \geq 0 \quad (\forall x \in K). \tag{2}$$

When K is the nonnegative orthant R^n_+ in R^n, this problem reduces to the *nonlinear complementarity problem* $\mathrm{NCP}(f)$ of finding an $x^* \in R^n$ such that

$$x^* \geq 0, \quad f(x^*) \geq 0 \quad \text{and} \quad \langle f(x^*), x^* \rangle = 0. \tag{3}$$

The importance of these problems in optimization, engineering, economics and other areas has been well documented in the literature, see e.g., [15], [16], [18], [19], [27], [44], and the references therein.

It is well known that nonlinear complementarity problems and more generally variational inequality problems can be reformulated as equations in many different ways. For illustration purposes, we consider few such reformulations.

(1) For the NCP, consider the so-called Fischer function [20] given by

$$F(x) := x + f(x) - \sqrt{x^2 + f(x)^2} \tag{4}$$

where all algebraic operations are performed componentwise. Then $F(x^*) = 0$ if and only if x^* solves $\mathrm{NCP}(f)$. It was shown recently by Ravindran and Gowda [49] that this F is a continuous $\mathbf{P}_0$-function when f is a continuous $\mathbf{P}_0$-function. Moreover, it can be easily shown (see Section 5) that the perturbation $F(x, \varepsilon)$ of $F(x)$ given by

$$F(x, \varepsilon) := x + f(x) - \sqrt{x^2 + f(x)^2 + 2\varepsilon e} \tag{5}$$

where e is the vector of ones in R^n and $\varepsilon \geq 0$, is a $\mathbf{P}$-function for $\varepsilon > 0$. The unique solution $x(\varepsilon)$ (if exists) of $F(x, \varepsilon) = 0$ will satisfy the conditions, see [31],

$$x(\varepsilon) > 0, \quad f(x(\varepsilon)) > 0, \quad \text{and} \quad x(\varepsilon)_i f_i(x(\varepsilon)) = \varepsilon \ (\forall i). \tag{6}$$

The trajectory induced by this $F(x, \varepsilon)$ is precisely the *interior point trajectory* or the *central path* well known in the interior point literature. Depending on the nature of f, the trajectory will be short (e.g., when $\mathrm{NCP}(f)$ has a nonempty bounded solution set) or long (e.g., when

f is monotone and NCP(f) has a strictly feasible solution). We shall say more about the interior point trajectory and its limiting behavior in Section 5.

(**2**) Once again consider NCP(f) corresponding to a continuous $\mathbf{P}_0$-function f. It is well known that NCP(f) is equivalent to finding a zero of

$$F(x) := \min\{x, f(x)\} \tag{7}$$

where the 'min' is taken componentwise. It was shown in [49] that F is a $\mathbf{P}_0$-function and its perturbation

$$F(x, \varepsilon) := \min\{x, f(x) + \varepsilon x\} \tag{8}$$

for each $\varepsilon > 0$ is a $\mathbf{P}$-function. This perturbation leads to a long trajectory called the trajectory of regularized solutions of NCP(f). When f is monotone, the entire trajectory converges to the least norm solution of the NCP(f) [50]. In the general $\mathbf{P}_0$ case, such a result is not possible, yet, when the solution set of NCP(f) is nonempty and bounded, it is shown in [49] that the distance between $x(\varepsilon)$ and the solution set of NCP(f) goes to zero as $\varepsilon \rightarrow 0$ (In [17], such a result was proved earlier for a continuously differentiable $\mathbf{P}_0$-function based on the Fischer function (4)). In the linear case, it is shown in [52] that the entire trajectory converges to a point which is a weak Pareto minimal point of the solution set of NCP(f).

(3) Consider BVI(f, K) where f is continuous $\mathbf{P}_0$-function. It is well known that solutions of BVI(f, K) are precisely the zeros of the fixed point map

$$F(x) := x - \Pi_K(x - f(x)) \tag{9}$$

where $\Pi_K(u)$ is the (orthogonal) projection of u onto K. It was noted in [49] that $F(x)$ is a $\mathbf{P}_0$-function. If we consider the perturbation

$$F(x, \varepsilon) := x - \Pi_K(x - f(x) - \varepsilon x) \tag{10}$$

then $F(x, \varepsilon)$ is a $\mathbf{P}$-function for $\varepsilon > 0$ (see Section 6) and we get the trajectory of regularized solutions of BVI(f, K). If we take a $p_\lambda(x)$ from the Gabriel-Moré family of smooth approximations to Π_K [21], we get the perturbation

$$F_\lambda(x) := x - p_\lambda(x - f(x)). \tag{11}$$

Under certain assumptions, even this perturbation turns out to be a $\mathbf{P}$-function for $\lambda > 0$, see Section 6. Many recent algorithmic studies pertain to this perturbation.

(4) Once again consider BVI(f, K) with a continuous $\mathbf{P}_0$-function f. Instead of the fixed point map, we now consider the normal map defined by

$$F(x) := f(\Pi_K(x)) + x - \Pi_K(x). \tag{12}$$

It turns out, see Section 7, that F is a $\mathbf{P}_0$-function. Under certain assumptions, for this F, the perturbations $F(x, \varepsilon)$ and $F_\lambda(x)$ turn out to be $\mathbf{P}$-functions, where for $\varepsilon > 0$,

$$F(x, \varepsilon) := f(\Pi_K(x)) + \varepsilon\Pi_K(x) + x - \Pi_K(x) \tag{13}$$

and for p_λ defined earlier,

$$F_\lambda(x) := f(p_\lambda(x)) + x - p_\lambda(x). \tag{14}$$

See [48] and [45] for numerical studies based on F_λ.

Having shown that $\mathbf{P}_0$-equations arise naturally in complementarity and (box) variational inequality problems, we turn our attention to the existence of trajectories. Well known ways of generating trajectories include minimization methods [38], [41], [32], maximal monotone methods [37], [23], minimization methods combined with the mountain pass theorem [8], etc. Extending the earlier work of Kojima, Mizuno, and Noma [35], [36], in [33], Kojima, Megiddo, and Noma use homeomorphism techniques to prove the existence of a trajectory in the context of an NCP(f) corresponding to a continuous $\mathbf{P}_0$-function satisfying strict feasibility and properness conditions. This paper has been highly influential in that in many subsequent papers, a similar methodology was adopted [39], [28], [10]. As in the Kojima-Megiddo-Noma approach [33], we also rely on homeomorphism ideas. But, instead of working in some nonnegative orthant and using the domain of invariance theorem, we work in R^n and use the so-called Banach-Mazur theorem (that a locally one-to-one, coercive continuous function on R^n is a homeomorphism of R^n), degree theory, and a recently discovered upper semicontinuity property of weakly univalent functions. We explicitly note that differentiability is not needed in our approach. In addition to discussing existence of trajectories, we also consider the limiting behavior of some of the trajectories.

The organization of the paper is as follows. In Section 2, we present basic definitions and results needed in the paper. In Section 3, we state and prove the main results. In subsequent sections, we apply these main results to study the interior point trajectory, trajectories corresponding to the fixed point map of a BVI, trajectories corresponding to the normal map of a BVI, trajectory arising from an aggregate function of a vertical nonlinear complementarity problem, etc.

2. Preliminaries

2.1. *The Banach-Mazur Theorem*

Consider a continuous function $g : R^n \to R^n$. Then it is easily seen that the following statements are equivalent:

- g is *coercive*, i.e., for any sequence $\{x^k\}$ in R^n with $||x^k|| \to \infty$, we have $||g(x^k)|| \to \infty$
  (or equivalently, for any sequence $\{x^k\}$ in R^n with $||x^k|| \to \infty$, there is a subsequence $\{x^{k_l}\}$ such that $||g(x^{k_l})|| \to \infty$ as $l \to \infty$).
- g is *proper*, i.e., the inverse image $g^{-1}(C)$ of any compact set C in R^n is compact.
- The level sets of the mapping $x \mapsto ||g(x)||$ are bounded, i.e., for any constant $\beta \geq 0$, the set $\{x : ||g(x)|| \leq \beta\}$ is bounded.

We now state a classical result of Banach and Mazur ([1] or Thm. 5.1.4, [2]) in the form that is pertinent to our discussion.

THEOREM 1 *(Banach-Mazur) If $g : R^n \to R^n$ is continuous, locally one-to-one, and coercive, then g is a (onto) homeomorphism of R^n.*

2.2. *An upper semicontinuity property*

We say that $g : R^n \to R^n$ is *univalent* if it is continuous and one-to-one, and *weakly univalent* if there exist one-to-one continuous functions $g_k : R^n \to R^n$ such that $g_k \to g$ uniformly on every bounded subset of R^n. g is said to be *monotone* if

$$\langle g(x) - g(y), x - y\rangle \geq 0 \quad \text{for all } x, y \in R^n.$$

We note that

- every continuous $\mathbf{P}$-function is univalent,
- every continuous $\mathbf{P}_0$-function is weakly univalent,
- every monotone function is a $\mathbf{P}_0$-function.

(The second statement follows from the observation that *if g is a $\mathbf{P}_0$-function, then $g(x)+\varepsilon x$ is a $\mathbf{P}$-function for every $\varepsilon > 0$.*)

We will need the following upper semicontinuity result from [49].

THEOREM 2 *Let $g : R^n \to R^n$ be weakly univalent and suppose that for a $q^* \in R^n$,*

$$g^{-1}(q^*) \text{ is nonempty and compact.}$$

Then for any given $\varepsilon > 0$, there exists a $\delta > 0$ such that for every weakly univalent function h and for every vector q with

$$\sup_{\overline{\Omega}} ||h(x) - g(x)|| < \delta, \quad ||q - q^*|| < \delta \tag{15}$$

we have

$$\emptyset \neq h^{-1}(q) \subseteq g^{-1}(q^*) + \varepsilon\mathcal{B} \tag{16}$$

where $\mathcal{B}$ denotes the open unit ball in R^n and $\Omega := g^{-1}(q^*) + \varepsilon\mathcal{B}$. Moreover, $h^{-1}(q)$ and $g^{-1}(q)$ are nonempty, connected, and uniformly bounded for q in a neighborhood of q^*.

2.3. *A result from algebraic geometry*

A subset of R^n is called *semi-algebraic* if it is a finite union of sets of the form

$$E = \{x \in R^n : p_j(x) \diamond_j 0, \ j = 1, \ldots, N\}$$

where for each $j = 1, 2 \ldots, N$, $p_j : R^n \to R$ is a polynomial and $\diamond_j \in \{=, \leq, <\}$.

THEOREM 3 *(Thm. 2.2.1, [3]) Every semi-algebraic set in R^n has a finite number of (connected) components.*

2.4. *Trajectories*

For a continuous function $F : R^n \to R^n$, let $F(x,\varepsilon)$ be a continuous perturbation, i.e., $F(x,\varepsilon) : R^n \times R_+ \to R^n$ is continuous and $F(x,0) \equiv F(x)$. A *long trajectory* associated with the solution set of $F(x) = 0$ or (by abuse of language) with the function F is a continuous curve $\{x(\varepsilon) : 0 < \varepsilon < \infty\}$ where each $x(\varepsilon)$ is a solution of $F(x,\varepsilon) = 0$ for every $\varepsilon > 0$. A *short trajectory* is a continuous curve $\{x(\varepsilon) : 0 < \varepsilon < \bar{\varepsilon}\}$ where each $x(\varepsilon)$ is a solution of $F(x,\varepsilon) = 0$ for $0 < \varepsilon < \bar{\varepsilon}$.

2.5. *The subsequence lemma*

The following result is quite useful. Its proof has been used in various works, see e.g., [53], [17], [49].

LEMMA 1 *Let $f : R^n \to R^n$ be a continuous $\mathbf{P}_0$-function and $\{u^k\}$ be a sequence in R^n such that $||u^k|| \to \infty$. Then there exists a subsequence, which we write without loss of generality as $\{u^k\}$, and an index i such that, either*

(a) $u_i^k \to \infty$ and $\{f_i(u^k)\}$ is bounded below, or

(b) $u_i^k \to -\infty$ and $\{f_i(u^k)\}$ is bounded above.

Proof. Since $||u^k|| \to \infty$, passing through a subsequence, if necessary, we may suppose that there exists an index set J such that for each $i \in J$, $|u_i^k| \to \infty$ as $k \to \infty$ and for $i \notin J$, $\{u_i^k\}$ is bounded. Define a bounded sequence $\{v^k\}$ as follows:

$$v_i^k := \begin{cases} 0 & \text{if } i \in J \\ u_i^k & \text{if } i \notin J. \end{cases}$$

Since $u^k \neq v^k$ for all large k, we can use the $\mathbf{P}_0$-property of f to get an index $i \in J$ so that without loss of generality,

$$u_i^k[f_i(u^k) - f_i(v^k)] = (u_i^k - v_i^k)[f_i(u^k) - f_i(v^k)] \geq 0$$

for all k. For simplicity, we may take $i = 1$ and note that $|u_1^k| \to \infty$. We assume without loss of generality that u_1^k converges either to ∞ or to $-\infty$. Suppose that u_1^k goes to ∞. Then the above inequality shows (assuming $u_1^k > 0$ for all k) that $f_1(u^k)$ is bounded below by $\inf_k f_1(v^k)$. When u_1^k goes to $-\infty$, the above inequality shows (assuming $u_1^k < 0$ for all k) that $f_1(u^k)$ is bounded above by $\sup_k f_1(v^k)$. This completes the proof. □

3. The main results

THEOREM 4 *(Existence of trajectories)*
Let $F(x) : R^n \to R^n$ be a continuous $\mathbf{P}_0$-function and $F(x,\varepsilon)$ be a continuous perturbation of F, that is, $F(x,\varepsilon) : R^n \times R_+ \to R^n$ is continuous and

$$F(x,0) \equiv F(x).$$

(1) Suppose that $F(x,\varepsilon)$ is a **P**-*function in x for each $\varepsilon > 0$ and either*

(i) $F(x,\varepsilon)$ is coercive in x for each $\varepsilon > 0$, or

(ii) there exists an $\varepsilon^* > 0$ such that the equation $F(x,\varepsilon^*) = 0$ has a solution and for every ε_1 and ε_2 with $0 < \varepsilon_1 < \varepsilon^* < \varepsilon_2$, the set

$$\{x : F(x,\varepsilon) = 0,\ \varepsilon_1 < \varepsilon < \varepsilon_2\}$$

is bounded.

Then $F(x,\varepsilon) = 0$ will have a unique solution $x(\varepsilon)$ for every $\varepsilon > 0$ and the mapping $\varepsilon \mapsto x(\varepsilon)$ is continuous on $(0,\infty)$.

(2) If $F(x,\varepsilon)$ is a **P**-*function in x for each ε positive and small, and if the solution set S of $F(x) = 0$ is nonempty and bounded, then*

(a) there exists an $\widehat{\varepsilon} > 0$ such that for any $0 < \varepsilon < \widehat{\varepsilon}$, $F(x,\varepsilon) = 0$ will have a unique solution $x(\varepsilon)$, the mapping $\varepsilon \mapsto x(\varepsilon)$ is continuous on $(0,\widehat{\varepsilon})$, and

(b) $dist\,(x(\varepsilon), S) \to 0$ as $\varepsilon \to 0$.

Remarks The theorem above gives in (1) conditions for the existence of a long trajectory and in (2) the existence of a short trajectory. Also, every limit $x^* = \lim_{\varepsilon_k \to 0} x(\varepsilon_k)$ (whenever defined) is a solution of $F(x) = 0$.

Proof. (1) Suppose that condition (i) holds. Fix any $\varepsilon > 0$. Since **P**-functions are one-to-one, by Theorem 1, the equation $F(x,\varepsilon) = 0$ will have a unique solution, say, $x(\varepsilon)$. By Theorem 2 applied to $g(x) := F(x,\widehat{\varepsilon})$ and $h(x) := F(x,\varepsilon)$, we see that $\varepsilon \mapsto x(\varepsilon)$ is continuous at any $\widehat{\varepsilon} > 0$.

Now suppose that condition (ii) holds. Since $F(x,\varepsilon^*)$ is a **P**-function, it is one-to-one. By Theorem 2, for all ε close to ε^*, the equation $F(x,\varepsilon) = 0$ will have a solution. Now consider

$$\alpha = \inf\{\overline{\varepsilon} : \forall \varepsilon \in [\overline{\varepsilon},\varepsilon^*],\ F(x,\varepsilon) = 0 \text{ has a solution}\}$$

and

$$\beta = \sup\{\overline{\varepsilon} : \forall \varepsilon \in [\varepsilon^*,\overline{\varepsilon}],\ F(x,\varepsilon) = 0 \text{ has a solution}\}.$$

We have $0 \le \alpha < \beta \le \infty$. We claim that $\alpha = 0$ and $\beta = \infty$ which means that $F(x,\varepsilon) = 0$ has a solution for all $\varepsilon > 0$. Suppose if possible that $\alpha > 0$. Then by condition (ii) and the continuity of $F(x,\varepsilon)$ we have $F(u,\alpha) = 0$ for some u. By Theorem 2, for all ε less than α and close to α, the equation $F(x,\varepsilon) = 0$ will have a solution. We reach a contradiction to the definition of α. Thus $\alpha = 0$. By a similar argument we can show that $\beta = \infty$.

(2) Suppose the equation $F(x) = 0$ has a nonempty and bounded solution set. Applying Theorem 2 to $g(x) := F(x)$ and $h(x) := F(x,\varepsilon)$ where ε is positive and small, say, $0 < \varepsilon < \widehat{\varepsilon}$, we see that $F(x,\varepsilon) = 0$ has a (unique) solution $x(\varepsilon)$ for each ε. The continuity of $x(\varepsilon)$ on $(0,\widehat{\varepsilon})$ follows once again, from Theorem 2. Another application of Theorem 2 proves part (b). □

Remark It is possible to formulate a result similar to the above for weakly univalent functions: one has to replace the continuous $\mathbf{P}_0$-function F by a weakly univalent function G and replace the continuous **P**-function $F(x,\varepsilon)$ by a univalent function $G(x,\varepsilon)$.

THEOREM 5 *(The limiting behavior of a trajectory)*
Consider a continuous function $\varepsilon \mapsto x(\varepsilon) : (0, \widehat{\varepsilon}) \to R^n$. *Suppose that*

(i) there exists a sequence $\{\varepsilon_k\} \downarrow 0$ *such that* $x(\varepsilon_k) \to x^* \in R^n$, *and*

(ii) the set $\{(x(\varepsilon), \varepsilon) : 0 < \varepsilon < \widehat{\varepsilon}\}$ *is semi-algebraic.*

Then $x(\varepsilon) \to x^*$ *as* $\varepsilon \to 0$.

The proof is similar to the one given in [52] for the convergence of regularized trajectory of polynomial NCPs. We repeat the proof for completeness.
Proof. We claim that $x(\varepsilon) \to x^*$ as $\varepsilon \to 0$. For any $\delta > 0$, let

$$B_\delta := \{x : -\delta \leq x_i - x_i^* \leq \delta,\ i = 1, 2, \ldots, n\}$$

be a box around x^*. To prove the convergence $x(\varepsilon) \to x^*$, we show that for any such δ, there exists $\bar{\varepsilon} > 0$ such that for $0 < \varepsilon < \bar{\varepsilon}$, $x(\varepsilon) \in B_\delta$. Assume the contrary that for some $\delta > 0$ there exists a sequence $\{\bar{\varepsilon}_l\} \downarrow 0$ such that $x(\bar{\varepsilon}_l) \notin B_\delta$. By renaming the sequences, if necessary, we may assume that

(a) $x(\varepsilon_k) \in B_\delta$, $x(\bar{\varepsilon}_k) \notin B_\delta$ for all k and

(b) $\widehat{\varepsilon} > \varepsilon_1 > \bar{\varepsilon}_1 > \varepsilon_2 > \bar{\varepsilon}_2 > \varepsilon_3 > \cdots$.

Now consider the set

$$E := \{(x(\varepsilon), \varepsilon) :\ 0 < \varepsilon < \widehat{\varepsilon}\}. \tag{17}$$

Then the sets E and $E \cap B_\delta \times (0, \widehat{\varepsilon})$ are semi-algebraic.

We show that the semi-algebraic set $E \cap B_\delta \times (0, \widehat{\varepsilon})$ has infinitely many components, thus reaching a contradiction to Theorem 3. For each $j = 1, 2, \ldots$, let C_j be the (connected) component of the set $E \cap B_\delta \times (0, \widehat{\varepsilon})$ containing $(x(\varepsilon_j), \varepsilon_j)$. Since connected sets in R are intervals, C_j must be of the form $\{(x(\varepsilon), \varepsilon) : \varepsilon \in I_j\}$ where I_j is an interval in $(0, \widehat{\varepsilon})$ containing ε_j. Since $x(\bar{\varepsilon}_l) \notin B_\delta$ for all l, we see that no $\bar{\varepsilon}_l$ can be in any I_j. By the interlacing property (b) above, the intervals I_j $(j = 1, 2 \ldots)$ and hence the components C_j $(j = 1, 2, \ldots)$ are pairwise disjoint. Thus we see that there are infinitely many components in $E \cap B_\delta \times (0, \widehat{\varepsilon})$ contradicting Theorem 3. This proves that $x(\varepsilon) \to x^*$. □

Remark Suppose that the assumptions of Theorem 4 (item 2) are in place and that the set $\{(x(\varepsilon), \varepsilon) : 0 < \varepsilon < \bar{\varepsilon}\}$ is semi-algebraic. Then the trajectory converges to a solution of the equation $F(x) = 0$ as $\varepsilon \downarrow 0$. This result is particularly useful while studying linear or polynomial complementarity problems.

4. Trajectory via a simple perturbation

THEOREM 6 *Let* F *be a continuous* $\mathbf{P}_0$*-function, and let*

$$F(x, \varepsilon) := F(x) + \varepsilon x. \tag{18}$$

Then for each $\varepsilon > 0$, $F(x, \varepsilon)$ is a continuous $\mathbf{P}$-function which is also coercive.

Proof. The continuity and the $\mathbf{P}$-property of $F(x, \varepsilon)$ are obvious. We verify the coercivity of $F(x, \varepsilon)$ in x. Fix $\varepsilon > 0$. Let $\{x^k\}$ be a sequence such that $||x^k|| \to \infty$. In view of Lemma 1, we may assume without loss of generality that for some index i, say $i = 1$, either $x_1^k \to \infty$ and $\{F_1(x^k)\}$ is bounded below, say, by α or that $x_1^k \to -\infty$ and $\{F_1(x^k)\}$ is bounded above, say, by β. In the former case, $F_1(x^k, \varepsilon) \geq \alpha + \varepsilon x_1^k$ and so $F_1(x^k, \varepsilon) \to \infty$. In the latter case, $F_1(x^k, \varepsilon) \leq \beta + \varepsilon x_1^k$ and so $F_1(x^k, \varepsilon) \to -\infty$. Hence $||F(x^k, \varepsilon)|| \to \infty$. This proves the coercivity of $F(x, \varepsilon)$. □

Remark By applying Theorem 4 we see that $F(x)$ has a long trajectory induced by the perturbation $F(x, \varepsilon) = F(x) + \varepsilon x$.

5. Interior point trajectory

In this section, we consider the interior point trajectory for the nonlinear complementarity problem NCP(f) defined by (3). We recall the Fischer function $F(x)$ and its perturbation $F(x, \varepsilon)$ from the Introduction:

$$F(x) := x + f(x) - \sqrt{x^2 + f(x)^2} \tag{19}$$

and

$$F(x, \varepsilon) := x + f(x) - \sqrt{x^2 + f(x)^2 + 2\varepsilon e}. \tag{20}$$

We also recall that $F(u) = 0$ if and only if u solves NCP(f) and $F(u, \varepsilon) = 0$ if and only if

$$u > 0, \quad f(u) > 0, \quad \text{and} \quad u_i f_i(u) = \varepsilon \ (\forall i). \tag{21}$$

In the result below, we prove that $F(x, \varepsilon)$ is coercive under the assumption of f being an $\mathbf{R}_0$-function. This $\mathbf{R}_0$ concept is introduced in [10] and is a generalization of an $\mathbf{R}_0$-matrix [15]. The proof of coercivity follows along standard lines.

THEOREM 7 *Let $f : R^n \to R^n$ be a continuous $\mathbf{P}_0$-function. Define $F(x)$ and $F(x, \varepsilon)$ by (19) and (20). Then the following statements hold:*

(a) $F(x, \varepsilon)$ is continuous on $R^n \times [0, \infty)$,

(b) $F(x)$ is a $\mathbf{P}_0$-function,

(c) $F(x, \varepsilon)$ is a $\mathbf{P}$-function in x for each $\varepsilon > 0$,

(d) $F(x, \varepsilon)$ is coercive in x for each $\varepsilon > 0$ under any one of the following:

- *f is an $\mathbf{R}_0$-function, i.e., for any sequence $\{x^k\}$ with $||x^k|| \to \infty$ and*

$$\liminf_{k\to\infty} \frac{\min_i x_i^k}{||x^k||} \geq 0, \quad \text{and} \quad \liminf_{k\to\infty} \frac{\min_i f_i(x^k)}{||x^k||} \geq 0$$

there exists an index j such that $x_j^k \to \infty$ and $f_j(x^k) \to \infty$.

- f *is a uniformly* **P**-*function, i.e., there exists a constant* $\Delta > 0$ *such that for all* $x, y \in R^n$,

$$\max_i (x_i - y_i)[f_i(x) - f_i(y)] \geq \Delta ||x - y||^2.$$

Proof. Item (a) is obvious. Item (b) follows from Remark 2 in [49]; it also follows from (c) by letting $\varepsilon \downarrow 0$. Item (c) is seen as follows. Fix $\varepsilon > 0$ and let $x \neq y$. Then there exists an index i such that $x_i \neq y_i$ and $(x_i - y_i)(f_i(x) - f_i(y)) \geq 0$. We show that for the same index i, $(x_i - y_i)(F_i(x, \varepsilon) - F_i(y, \varepsilon)) > 0$. Without loss of generality, we assume that $x_i > y_i$ and show $F_i(x, \varepsilon) - F_i(y, \varepsilon) > 0$. Assume the contrary, $F_i(x, \varepsilon) \leq F_i(y, \varepsilon)$ and let, for simplicity, $\alpha = x_i$, $\beta = y_i$, $\gamma = f_i(x)$, $\delta = f_i(y)$. We have $\alpha > \beta$ and $\gamma \geq \delta$. Now $F_i(x, \varepsilon) \leq F_i(y, \varepsilon)$ leads to $(\alpha - \beta) + (\gamma - \delta) \leq \sqrt{\alpha^2 + \gamma^2 + 2\varepsilon} - \sqrt{\beta^2 + \delta^2 + 2\varepsilon}$. Squaring, simplifying, and using the inequality $(\alpha - \beta)(\gamma - \delta) \geq 0$, we get

$$(\alpha\delta - \beta\gamma)^2 + 2\varepsilon[(\alpha - \beta)^2 + (\gamma - \delta)^2] \leq 0$$

which is false since $\varepsilon > 0$ and $\alpha > \beta$. Hence we have Item (c).
(d) Suppose that f is an $\mathbf{R}_0$-function and fix $\varepsilon > 0$. Let $\{x^k\}$ be a sequence such that $||x^k|| \to \infty$. We have to show that for some index j, $|F_j(x, \varepsilon)| \to \infty$ along some subsequence of $\{x^k\}$. If there is a subsequence $\{x^{k_l}\}$ and an index i such that either $x_i^{k_l} \to -\infty$ or $f_i(x^{k_l}) \to -\infty$, then the inequalities

$$F(x, \varepsilon) \leq F(x) \leq 2 \min\{x, f(x)\}$$

show that $F_i(x^{k_l}, \varepsilon) \to -\infty$. So we assume that for each index i, the sequences $\{x_i^k\}$ and $\{f_i(x^k)\}$ are bounded below. Then, by the $\mathbf{R}_0$-condition imposed on f, we get an index j such that $x_j^k \to \infty$ and $f_j(x^k) \to \infty$. In this case, it is easily seen that $F_j(x^k, \varepsilon) \to \infty$. This proves the coercivity of $F(x, \varepsilon)$ under the $\mathbf{R}_0$-condition on f. A similar proof works when f is a uniform **P**-function; it also follows from Proposition 3.11 in [10] and the previous argument. □

Remarks In the above theorem, we studied the properties of the Fischer function and its perturbation. A similar analysis can be carried out for the following functions:

$$F(x) := x + f(x) - \sqrt{(x - f(x))^2} \; (= 2 \min\{x, f(x)\})$$

and

$$F(x, \varepsilon) := x + f(x) - \sqrt{(x - f(x))^2 + 4\varepsilon\, e}.$$

Note that $F(u) = 0$ if and only if u solves NCP(f) and $F(u, \varepsilon) = 0$ if and only if (21) holds.

We are now ready to prove the existence of interior point trajectories.

THEOREM 8 *(Existence of short interior point trajectory) Suppose that* $f : R^n \to R^n$ *be a continuous* $\mathbf{P}_0$-*function. If NCP(f) has a nonempty and bounded solution set, then it will have a short interior point trajectory, that is, there exists an* $\overline{\varepsilon} > 0$ *such that for each* $0 < \varepsilon < \overline{\varepsilon}$ *there is a unique* $x(\varepsilon)$ *satisfying (21) and the mapping* $\varepsilon \mapsto x(\varepsilon)$ *is continuous on* $(0, \overline{\varepsilon})$.

Proof. This follows from Theorem 4 with $F(x)$ and $F(x, \varepsilon)$ given by (19) and (20). □

Remarks The above theorem extends a result of Chen, Chen, and Kanzow [8] on the existence of short interior point trajectory proved for a continuously differentiable $\mathbf{P}_0$-function via the Mountain Pass Theorem. We further note that when f is a polynomial function, Theorem 5 is applicable for the interior point trajectory.

THEOREM 9 *(Existence of long interior point trajectory) Suppose that $f : R^n \to R^n$ be a continuous $\mathbf{P}_0$-function. Then NCP(f) will have a long interior point trajectory when any one of the following conditions hold.*

(1) For any sequence $\{x^k\}$ with $||x^k|| \to \infty$ and

$$\liminf_{k\to\infty} \frac{\min_i x_i^k}{||x^k||} \geq 0, \quad \text{and} \quad \liminf_{k\to\infty} \frac{\min_i f_i(x^k)}{||x^k||} \geq 0$$

there exists an index j and a subsequence $\{x^{k_l}\}$ such that $x_j^{k_l} f_j(x^{k_l}) \to \infty$.

(2) For any sequence $\{x^k\}$ with

$$||x^k|| \to \infty, \quad x^k \geq c, \quad f(x^k) \geq c \quad \text{for some vector } c$$

there is an index j and a subsequence $\{x^{k_l}\}$ such that $[x_j^{k_l}]_+ [f_j(x^{k_l})]_+ \to \infty$.

(3) f is monotone with a strictly feasible point.

(4) f is an $\mathbf{R}_0$-function.

(5) f is a uniform $\mathbf{P}$-function.

Proof. Suppose Condition (1) holds. We verify Condition $1(ii)$ in Theorem 4. First we establish the existence of a ε^* such that the equation $F(x, \varepsilon^*) = 0$ has a solution. Consider

$$G(x) := \min\{x, f(x)\}$$

and

$$G(x, \varepsilon) := \min\{x, f(x) + \varepsilon x\}.$$

It was proved in (Corollary 6, [49]) that $G(x, \varepsilon) = 0$ has a solution for all $\varepsilon > 0$. Let x^k be the solution of $G(x, \frac{1}{k}) = 0$ for any k. From $x^k \geq 0$, $f(x^k) + \varepsilon x^k \geq 0$, $x_i^k[f_i(x^k)] = -\varepsilon[x_i^k]^2 \leq 0$ for all i and Condition (1) we see that the sequence $\{x^k\}$ is bounded. Any subsequential limit of this sequence will solve NCP(f). It is easy to verify that Condition (1) implies the boundedness of the solution set of NCP(f). Now we use the previous theorem to get an ε^* such that the equation $F(x, \varepsilon^*) = 0$ has a solution. For any $0 < \varepsilon_1 < \varepsilon^* < \varepsilon_2$, the boundedness of the set

$$\{x : F(x, \varepsilon) = 0, \ \varepsilon_1 < \varepsilon < \varepsilon_2\}$$

follows from Condition (1) and (21). Clearly Condition (2) implies Condition (1). Since Condition (5) implies Condition (4) [10] and Conditions (3) and (4) both imply Condition (2) [8], we see that NCP(f) has a long trajectory in all cases. □

Remarks As mentioned in the Introduction, Kojima, Megiddo, and Noma [33] consider a continuous $\mathbf{P}_0$-function and show, in the presence of strict feasibility and properness conditions, the existence of a trajectory

$$\{z(t) = (x(t), y(t)) : t > 0,\ y(t) = f(x(t)) + tb,\ x_i(t)y_i(t) = ta_i \text{ for all } i\}$$

where b and $a > 0$ are fixed vectors in R^n. Their analysis is carried out by considering the function

$$H(x, y) := (Xy, y - f(x))$$

on $R^n_{++} \times R^n_{++}$ where $X = diag\, x$ denotes the diagonal matrix with entries coming from x and R^n_{++} denotes the interior of R^n. We make two observations. Firstly, Moré [40] has proved that the function

$$G(x, y) := (Xy, f(x) - y)$$

is a $\mathbf{P}$-function when restricted to $R^n_{++} \times R^n_{++}$ and a $\mathbf{P}_0$-function when restricted to $R^n_+ \times R^n_+$. Secondly, the above trajectory of Kojima, Megiddo, and Noma can also be obtained, via Theorem 4, by considering the function

$$F(x, \varepsilon) := x + [f(x) + \varepsilon b] - \sqrt{x^2 + [f(x) + \varepsilon b]^2 + 2\varepsilon a}$$

where all algebraic operations are performed componentwise.

For some recent algorithmic studies of the interior point methods and their variants, see e.g., [5], [55], [56], [57].

6. Trajectories associated with the fixed point map

In this section, we describe trajectories associated with certain generalizations of the fixed point map corresponding to a box variational inequality problem. Consider a function $f : R^n \to R^n$ and a (rectangular) box K in R^n defined by

$$K = K_1 \times K_2 \times \cdots \times K_n$$

where each K_i is a closed interval in R. Recall that the box variational inequality problem BVI (f, K) is to find a vector $x^* \in K$ such that

$$\langle f(x^*), x - x^* \rangle \geq 0 \quad (\forall x \in K) \tag{22}$$

and that when $K = R^n_+$, BVI(f, K) reduces to NCP(f). It is well known that SOL(f, K), the solution set of BVI(f, K), is precisely the zero set of the fixed point map

$$x \mapsto x - \Pi_K\,(x - f(x)) \tag{23}$$

where $\Pi_K(x)$ is the (orthogonal) projection of x onto K. First, we consider some generalizations of this map. Let us say that a function $\theta : R \to R$ satisfies *Condition* $(*)$ if

$$(*) \qquad \begin{cases} (i) & \theta(t) \text{ is increasing on R and} \\ (ii) & t - \theta(t) \text{ is increasing on } R. \end{cases}$$

Consider a *diagonal function* $\Theta : R^n \to R^n$, i.e.,

$$\Theta(x) = (\theta_1(x_1), \theta_2(x_2), \ldots, \theta_n(x_n))$$

where each component function $\Theta_j(x)$ $(= \theta_j(x_j))$ is a function of x_j only. We say that Θ satisfies **Condition** $(*)$ if each component of Θ satisfies Condition $(*)$.

Example Given a closed interval L in R, it is easily seen that the mapping $t \mapsto \Pi_L(t)$ satisfies Condition $(*)$. More generally, let μ be a positive Borel measure on R such that

$$\mu(R) = 1 \quad \text{and} \quad \int_R |s| d\mu(s) < \infty. \tag{24}$$

(Since most of our integrands are continuous functions, we have taken here a Borel measure. One could consider any positive measure defined on the σ-algebra of Lebesgue measurable sets on R.) Then by simple algebraic manipulations we can show that the function

$$\theta(t) = \int_R \Pi_L(t - s)\, d\mu(s)$$

satisfies Condition $(*)$. The same goes for

$$\theta_\lambda(t) = \int_R \Pi_L(t - \lambda s)\, d\mu(s)$$

for any $\lambda \geq 0$. When $d\mu = \rho(s)ds$ where ρ is a (probability) density function on R and $\lambda > 0$, θ_λ defines a member of the Gabriel-Moré family of smooth approximations to the projection map Π_L [21]. For $L = R_+$, some well known choices of ρ are the Neural networks smooth plus function, Chen-Harker-Kanzow-Smale smooth plus function, Pinar-Zenois smooth plus function, and Zang smooth plus function [12]. Going back to a general μ, *we note that when L is either bounded above or bounded below, and μ does not vanish on any infinite interval, the function $t - \theta(t)$ is strictly increasing.* Now consider a rectangle $K = K_1 \times K_2 \times \cdots \times K_n$ and a μ satisfying (24). Then the diagonal function Θ whose jth component function is given by

$$\theta_j(t) = \int_R \Pi_{K_j}(t - s)\, d\mu(s)$$

satisfies Condition $(*)$. Note that when μ is the point mass at the origin, $\Theta(x) = \Pi_K(x)$.

The following lemma is needed to prove the $\mathbf{P}_0(\mathbf{P})$-properties of the fixed point map and its generalizations.

LEMMA 2 *Suppose that $\theta : R \to R$ is a function satisfying Condition $(*)$ above. Let α, β, γ, and δ be real numbers. Then we have the following implications:*

(i) $\alpha > \beta,\ \gamma \geq \delta \Longrightarrow \alpha - \theta(\alpha - \gamma) \geq \beta - \theta(\beta - \delta)$.

(ii) $\alpha > \beta,\ \gamma > \delta \Longrightarrow \alpha - \theta(\alpha - \gamma) > \beta - \theta(\beta - \delta)$.

(iii) $t - \theta(t)$ *is strictly increasing,* $\alpha > \beta,\ \gamma \geq \delta \Longrightarrow \alpha - \theta(\alpha - \gamma) > \beta - \theta(\beta - \delta)$.

(iv) Let $\varepsilon > 0$ *and consider the sequences* $\{\alpha^k\}$ *and* $\{\beta^k\}$ *in* R*. Suppose* $\{\alpha^k\} \to \infty$ *and* $\{\beta^k\}$ *bounded below or* $\{\alpha^k\} \to -\infty$ *and* $\{\beta^k\}$ *bounded above in* R*. Then the sequence* $\{|\zeta^k|\}$*, where* $\zeta^k := \alpha^k - \theta(\alpha^k - \beta^k - \varepsilon\alpha^k)$*, diverges to infinity along a subsequence.*

Proof. (i) Assume $\alpha > \beta,\ \gamma \geq \delta$ but $\alpha - \theta(\alpha - \gamma) < \beta - \theta(\beta - \delta)$.
From these we get

$$0 < \alpha - \beta < \theta(\alpha - \gamma) - \theta(\beta - \delta),$$

which, in view of $(*)$, gives $s := \alpha - \gamma > \beta - \delta =: t$. Rearranging the terms of the above inequality we get

$$[s - \theta(s)] - [t - \theta(t)] < \delta - \gamma.$$

By the second Condition in $(*)$ and $s > t$, we see that the left hand side of the above inequality is nonnegative while the right hand side is nonpositive. This contradiction proves (i). The proof of (ii) is similar. To see (iii), we proceed as in the proof of (i). By assuming the negation of statement in (iii) we arrive at $s := \alpha - \gamma > \beta - \delta =: t$ and

$$[s - \theta(s)] - [t - \theta(t)] \leq \delta - \gamma.$$

Using the hypothesis and the inequality $s > t$, we see that the left hand side of the above inequality is positive while the right hand side is nonpositive. This contradiction proves (iii).
(iv) Suppose that $\{\alpha^k\} \to \infty$ and $\{\beta^k\}$ is bounded below. Consider the case when $\eta^k = \alpha^k - \beta^k - \varepsilon\alpha^k$ is bounded above by, say, Δ. In this case, $\theta(\eta^k) \leq \theta(\Delta)$ for large k, since θ is increasing. It follows from $\zeta^k = \alpha^k - \theta(\alpha^k - \beta^k - \varepsilon\alpha^k) \geq \alpha^k - \theta(\Delta)$ that $\zeta^k \to \infty$.
Now consider the alternate case when $\eta^k = \alpha^k - \beta^k - \varepsilon\alpha^k \to \infty$ along a subsequence. Assuming the given sequence to be this subsequence, we have

$$\eta^k - \theta(\eta^k) \geq 0 - \theta(0)$$

for all large k, since $t - \theta(t)$ is increasing. We see that $\zeta^k = \alpha^k - \theta(\alpha^k - \beta^k - \varepsilon\alpha^k) = [\eta^k - \theta(\eta^k)] + \beta^k + \varepsilon\alpha^k \geq$ constant $+ \varepsilon\alpha^k$ and hence $\zeta^k \to \infty$. A similar proof can be given when $\{\alpha^k\} \to -\infty$ and $\{\beta^k\}$ is bounded above; we omit the details. □

THEOREM 10 *Consider* $f : R^n \to R^n$*. Let* Θ *be a continuous diagonal function on* R^n *satisfying the Condition* $(*)$*. Then the following statements hold:*

(a) If f *is a continuous* $\mathbf{P}_0$*(*$\mathbf{P}$*)-function, then so is*

$$F(x) := x - \Theta(x - f(x)). \tag{25}$$

Moreover, if $t - \theta_i(t)$ *is strictly increasing for each* i *and* f *is a* $\mathbf{P}_0$*-function, then* F *is a* $\mathbf{P}$*-function.*

(b) If f is a continuous $\mathbf{P}_0$-function and $\varepsilon > 0$, then

$$F(x, \varepsilon) := x - \Theta(x - f(x) - \varepsilon x) \tag{26}$$

is a continuous coercive $\mathbf{P}$-function.

Proof. (a) Let f be a a continuous $\mathbf{P}_0$-function, and take $x \neq y$. Then there exists an index i such that $x_i \neq y_i$ and $(x_i - y_i)[f_i(x) - f_i(y)] \geq 0$. Without loss of generality, we assume that $x_i > y_i$ and $f_i(x) \geq f_i(y)$. Writing $\alpha = x_i$, $\beta = y_i$, $\gamma = f_i(x)$, $\delta = f_i(y)$ and $\theta = \theta_i$ we apply Lemma $2(i)$ to get $F_i(x) \geq F_i(y)$. It follows that $\max_{\{j: x_j \neq y_j\}}(x_j - y_j)[F_j(x) - F_j(y)] \geq 0$. This proves that F is a $\mathbf{P}_0$-function. A similar argument (using Lemma $2(ii)$) proves the $\mathbf{P}$-property of F when f is a $\mathbf{P}$-function. Using Lemma $2(iii)$, we get the additional statement made.
(b) Let $\varepsilon > 0$ and f be a continuous $\mathbf{P}_0$-function. Since $f(x) + \varepsilon x$ is a $\mathbf{P}$-function, it follows from part (a) that $F(x, \varepsilon)$ is a continuous $\mathbf{P}$-function. We now prove the coercivity of $F(x, \varepsilon)$ in X. Let $\{x^k\}$ be a sequence in R^n such that $||x^k|| \to \infty$. By combining Lemma 1 and Lemma $2(iv)$, we see that $||F(x^k, \varepsilon)|| \to \infty$ along a subsequence. This proves the coercivity of $F(x, \varepsilon)$. □

Remark Consider a continuous $\mathbf{P}_0$-function f and define, for $z = (x, y)$,

$$H(z) = \begin{pmatrix} f(x) - y \\ x - \Theta(x - y) \end{pmatrix} \tag{27}$$

and

$$H(z, \varepsilon) := \begin{pmatrix} f(x) - y \\ x - \Theta(x - y - \varepsilon x) \end{pmatrix} \tag{28}$$

where Θ satisfies the Condition $(*)$ and $\varepsilon > 0$. We claim that $H(z)$ is a continuous $\mathbf{P}_0$-function. To see this, let $z_1 := (x_1, y_1) \neq (x_2, y_2) := z_2$.
Case 1: $x_1 \neq x_2$. In this case, we find an index i such that $(x_1 - x_2)_i[f_i(x_1) - f_i(x_2)] \geq 0$. Without loss of generality, let $(x_1)_i > (x_2)_i$ and $f_i(x_1) \geq f_i(x_2)$.
Sub-case (i): Suppose $(y_1)_i \leq (y_2)_i$. In this case,

$$(x_1 - x_2)_i[f_i(x_1) - (y_1)_i - f_i(x_2) + (y_2)_i] \geq 0$$

and hence

$$\max_{\{j: (z_1)_j \neq (z_2)_j\}} (z_1 - z_2)_j[H_j(z_1) - H_j(z_2)] \geq 0. \tag{29}$$

Sub-case (ii): Suppose $(y_1)_i > (y_2)_i$. Since $(x_1)_i > (x_2)_i$, from Lemma $2(ii)$ we see that $[x_1 - \Theta(x_1 - y_1)]_i > [x_2 - \Theta(x_2 - y_2)]_i$ and hence we have (29).
Case 2: $x_1 = x_2$. In this case, $y_1 \neq y_2$. We take any i such that $(y_1)_i \neq (y_2)_i$. Since $-\theta_i(-t)$ is increasing and $x_1 = x_2$, we have

$$(y_1 - y_2)_i\{[x_1 - \Theta(x_1 - y_1)] - [x_2 - \Theta(x_2 - y_2)]\}_i \geq 0$$

from which we get (29). Hence we have shown that $H(z)$ is a $\mathbf{P}_0$-function.
We note that $H(z)$ need not be a $\mathbf{P}$-function when f is a $\mathbf{P}$-function. A simple example is $f(x) = x$ and $\Theta(x) \equiv 0$. For $(0, e) \neq (0, 0)$ in $R^n \times R^n$, the $\mathbf{P}$-property fails to hold. However, it can be shown that when each component function of the diagonal function Θ is strictly increasing, $H(z)$ is $\mathbf{P}$ when f is $\mathbf{P}$.

We now specialize the above results to some familiar perturbations of the fixed point map (23). Let μ be a measure as defined in (24). Define for $\lambda \geq 0$, $\varepsilon \geq 0$,

$$p_\lambda(u) := \int_R \Pi_K(u - \lambda\, s\, e)d\mu(s), \tag{30}$$

$$F_\lambda(x) := x - p_\lambda(x - f(x)), \tag{31}$$

$$F_\lambda(x, \varepsilon) := x - p_\lambda(x - f(x) - \varepsilon x), \tag{32}$$

and for $z = (x, y)$,

$$H_\lambda(z) := \begin{pmatrix} f(x) - y \\ x - p_\lambda(x - y) \end{pmatrix} \tag{33}$$

where all algebraic operations and integration are performed componentwise. We note that when μ is the point mass at the origin, and/or $\lambda = 0$, the above functions reduce to Π_K,

$$F(x) := x - \Pi_K(x - f(x)), \tag{34}$$

$$F(x, \varepsilon) := x - \Pi_K(x - f(x) - \varepsilon x), \tag{35}$$

and

$$H(z) = \begin{pmatrix} f(x) - y \\ x - \Pi_K(x - y) \end{pmatrix}. \tag{36}$$

Note that when $d\mu = \rho(s)ds$ where ρ is a density function, p_λ is a smooth approximation of Π_K and moreover when f is continuously differentiable, and $\lambda > 0$, $F_\lambda(x)$ is a smooth approximation of $F(x)$ and $F_\lambda(x, \varepsilon)$ is a smooth approximation of $F(x, \varepsilon)$ [12], [21], [22].

By specializing the previous result to $\Theta(u) = p_\lambda(u)$, we get the following.

THEOREM 11 *Let $\lambda \geq 0$, $\varepsilon \geq 0$, and μ be as above. Suppose f is a continuous $\mathbf{P}_0$-function and K be a rectangular box. Then the following statements hold:*

(a) $F_\lambda(x)$ is a $\mathbf{P}_0$-function. In particular, the fixed point map (34) is a $\mathbf{P}_0$-function.

(b) If $\lambda > 0$, each closed interval K_i is either bounded above or below, and μ does not vanish on any infinite interval of R, then $F_\lambda(x)$ is a $\mathbf{P}$-function.

(c) For each $\varepsilon > 0$ and $\lambda \geq 0$, $F_\lambda(x, \varepsilon)$ is a coercive $\mathbf{P}$-function in x.

We can now combine theorems 4 and 11 to get any number of long/short trajectories for the equation $x - \Pi_K(x - f(x)) = 0$.

For some recent studies based on the fixed point map and on the smoothing and perturbation techniques for the NCP and BVI, see e.g., [6], [10], [12], [13], [14], [21], [22], [28], [29], [47], [51], [54].

7. Trajectories associated with the normal map

In this section, we describe trajectories associated with certain generalizations of the normal map corresponding to a box variational inequality problem. So consider BVI(f, K) where f is continuous on R^n and K is a rectangular box in R^n. It is well known that BVI(f, K) can be solved via the equation

$$F(x) := f(\Pi_K(x)) + x - \Pi_K(x) = 0 \tag{37}$$

where F is called the normal map. Denoting the solution set of BVI(f, K) by SOL(f, K), we note that if x^* is a solution of $F(x) = 0$, then $\Pi_K(x^*) \in$ SOL(f, K), and conversely, if $u^* \in$ SOL(f, K), then $u^* - f(u^*)$ solves $F(x) = 0$.

THEOREM 12 *Consider $f : R^n \to R^n$ and let Θ be a continuous diagonal function on R^n satisfying Condition $(*)$. Then the following statements hold:*

(a) If f is a continuous $\mathbf{P}_0(\mathbf{P})$-function, then so is

$$F(x) := f(\Theta(x)) + x - \Theta(x). \tag{38}$$

Moreover, F is a continuous $\mathbf{P}$-function when $t - \theta_i(t)$ is strictly increasing for each i and f is a continuous $\mathbf{P}_0$-function.

(b) If f is a continuous $\mathbf{P}_0$-function, then

$$F(x, \varepsilon) := F(x) + \varepsilon\Theta(x) \tag{39}$$

is a continuous coercive $\mathbf{P}$-function.

Proof. (a) Assume that f is a $\mathbf{P}_0$-function. Let $x \neq y$ in R^n. We have to show that for some index i,

$$(x_i - y_i)[F(x) - F(y)]_i \geq 0. \tag{40}$$

We consider two cases.

Case (i): $\Theta(x) \neq \Theta(y)$. Since f is a $\mathbf{P}_0$-function, there is an index i such that

$$\Theta_i(x) \neq \Theta_i(y) \quad \text{and} \quad [\Theta(x) - \Theta(y)]_i[f(\Theta(x)) - f(\Theta(y))]_i \geq 0. \tag{41}$$

Without loss of generality, we may assume that

$$\Theta_i(x) > \Theta_i(y) \quad \text{and} \quad f_i(\Theta(x)) \geq f_i(\Theta(y)). \tag{42}$$

Since θ_i is increasing, we see from $\theta_i(x_i) = \Theta_i(x) > \Theta_i(y) = \theta_i(y_i)$ that $x_i > y_i$. Since $t - \theta_i(t)$ is increasing in t, we have $x_i - \theta_i(x_i) \geq y_i - \theta_i(y_i)$. It follows that

$$F_i(x) = f_i(\Theta(x)) + x_i - \theta_i(x_i) \geq f_i(\Theta(y)) + y_i - \theta_i(y_i) = F_i(y).$$

From this inequality and $x_i > y_i$, we get

$$(x_i - y_i)[F(x) - F(y)]_i \geq 0$$

proving (40).

Case (ii): $\Theta(x) = \Theta(y)$. In this case, we take any index i such that $x_i \neq y_i$ and assume without loss of generality that $x_i > y_i$. Then

$$F_i(x) = f_i(\Theta(x)) + x_i - \Theta_i(x) = f_i(\Theta(y)) + x_i - \Theta_i(y) > f_i(\Theta(y)) + y_i - \Theta_i(y) = F_i(y).$$

It follows that

$$(x_i - y_i)[F(x) - F(y)]_i > 0$$

proving (40). Thus we see that F is a continuous $\mathbf{P}_0$-function. Careful inspection of the above proof shows that F is a $\mathbf{P}$-function if either f is a $\mathbf{P}$-function, or f is a $\mathbf{P}_0$-function and $t - \theta_i(t)$ is strictly increasing for each i.

(b) Suppose that f is $\mathbf{P}_0$. Then $f(x) + \varepsilon x$ is a $\mathbf{P}$-function. Putting this $\mathbf{P}$-function in place of f in (a) gives us the $\mathbf{P}$-property of $F(x, \varepsilon)$. Now for the coercive part. Let us fix $\varepsilon > 0$, and write

$$g(x) := F(x, \varepsilon).$$

To show that g is coercive, it is enough to show that for each sequence $\{x^k\}$ with $||x^k|| \to \infty$ there is a subsequence of $\{g(x^k)\}$ which goes to infinity in the norm. To this end, fix a sequence $\{x^k\}$ with $||x^k|| \to \infty$. Let $u^k := \Theta(x^k)$. If $\{u^k\}$ is bounded, then because of the continuity of f, we see that

$$g(x^k) = f(u^k) + \varepsilon u^k + x^k - u^k$$

is unbounded so some subsequence of $\{g(x^k)\}$ goes to infinity in the norm. Hence we may assume that $\{u^k\}$ is unbounded and without loss of generality, $||u^k|| \to \infty$. Because of Lemma 1, we may assume that either $u_1^k \to \infty$ and $\{f_1(u^k)\}$ is bounded below or $u_1^k \to -\infty$ and $\{f_1(u^k)\}$ is bounded above. Suppose the first alternative holds. Then $\theta_1(x_1^k) = [\Theta(x^k)]_1 = u_1^k \to \infty$ implies (by Condition $(*)$) that $x_1^k \to \infty$ and $x_1^k - \theta_1(x_1^k)$ is bounded below. So in the right hand side of the sum

$$g_1(x^k) = f_1(u^k) + \varepsilon u_1^k + (x_1^k - \theta_1(x_1^k)) \tag{43}$$

the first and the last terms are bounded below while the middle term is going to infinity. We see $|g_1(x^k)| \to \infty$. So in this case, we have produced a subsequence of $\{g(x^k)\}$ which goes to infinity.

When u_1^k goes to $-\infty$ we use a similar argument. (In the right hand side of (43) we will have two terms bounded above and one term going to negative infinity.) Thus we see that g is coercive. □

Remark By specializing Θ, combining Theorems 4 and 12, we can generate trajectories for the normal map corresponding to a BVI. As an illustration, consider $K = R^n_+$, $d\mu = \rho(s)ds$ where ρ is a density function which is positive everywhere. Then for each component function θ in p_λ, $t - \theta(t)$ is strictly increasing. By Item (a) in the above theorem, the function

$$F_\lambda(x) = f(p_\lambda(x)) + x - p_\lambda(x)$$

is a $\mathbf{P}$-function. Now suppose that NCP(f) has nonempty and bounded solution set. Then the normal map $x \mapsto f(x^+) + x - x^+$ will have a nonempty and bounded solution set. By Theorem 4, we have a short trajectory

$$\{x(\lambda) : 0 < \lambda < \bar{\lambda}\}$$

for the normal map. This trajectory satisfies property $2(b)$ of Theorem 4. For an algorithmic study of a variant of F_λ considered above, see [48], [45].

8. Trajectories for the vertical complementarity problem

In this section, we describe trajectories for the vertical complementarity problem. Corresponding to functions $f^1, f^2, \ldots, f^N$ from R^n into itself, the vertical nonlinear complementarity problem VNCP($f^1, f^2, \ldots, f^N$) is to find an $x \in R^n$ such that

$$F(x) := \min\{x, f^1(x), f^2(x), \ldots, f^N(x)\} = 0. \tag{44}$$

(In the literature on VNCPs, the problem is usually defined without the identity function in the above equation. However, the presence of the identity function is quite important in our study here.)

Note that when $N = 1$, the problem reduces to the usual nonlinear complementarity problem. When the functions f^i are affine, we get a vertical linear complementarity problem, see [52]. In a recent study, Peng and Lin [43], and H-D. Qi and L-Z. Liao [46] consider smoothing techniques for solving vertical complementarity problems. Their approach is based on the so-called *aggregate function*

$$g(x,t) := t\ln(\sum_0^N e^{\frac{g^i(x)}{t}}) \tag{45}$$

of a finite collection $\{g^0, g^1, \ldots, g^N\}$ of real-valued functions. This $g(x,t)$ is a smooth perturbation of $\max\{g^0, g^1, \ldots, g^N\}$ when the individual functions involved are smooth. For the VNCP, the above authors consider the perturbation

$$F(x,\varepsilon) := -\varepsilon\ln(\sum_0^N e^{-\frac{f^i(x)}{\varepsilon}}) \tag{46}$$

where $f^0(x) = x$, $\varepsilon > 0$ and all operations are performed componentwise. (We note that for $N = 1$, this perturbation is already covered in (31) by taking $K = R^n_+$ and $d\mu = \frac{e^{-s}}{(1+e^{-s})^2}\, ds$.) In this section, we show that under appropriate assumptions, $F(x)$ and $F(x, \varepsilon)$ are $\mathbf{P}_0$ ($\mathbf{P}$) thereby demonstrating the applicability of Theorem 4 in this setting.

First a definition. Given the finite set $\{f^1, f^2, \ldots, f^N\}$ of functions from R^n into itself, we say that a function f is a *(row) representative* if each component function f_j is one of the functions in the set $\{(f^1)_j, (f^2)_j, \ldots, (f^N)_j\}$. We say that the set $\{f^1, f^2, \ldots, f^N\}$ is a $\mathbf{P}_0$ ($\mathbf{P}$)-family if each representative is a $\mathbf{P}_0$($\mathbf{P}$)-function.

THEOREM 13 *Suppose that $\{f^1, f^2, \ldots, f^N\}$ is a set of continuous functions from R^n into itself. Define $F(x)$ by (44) and $F(x, \varepsilon)$ by (46). Then the following statements hold.*

(a) If $\{f^1, f^2, \ldots, f^N\}$ is a $\mathbf{P}_0$($\mathbf{P}$)-family, then $F(x)$ is a continuous $\mathbf{P}_0$ ($\mathbf{P}$)-function.

(b) If $\{f^1, f^2, \ldots, f^N\}$ is a $\mathbf{P}_0$-family and $\varepsilon > 0$, then the function $F(x, \varepsilon)$ is a continuous $\mathbf{P}$-function.

Proof. (a) Assume that $\{f^1, f^2, \ldots, f^N\}$ is a $\mathbf{P}_0$-family and $F(x)$ is not a $\mathbf{P}_0$ ($\mathbf{P}$)-function. Then there exist x and y such that $x \neq y$ and

$$\max_{\{i: x_i \neq y_i\}} (x_i - y_i)[F(x) - F(y)]_i < 0 \quad (\leq 0).$$

We construct a representative f as follows. Consider any index i. If $x_i = y_i$, then we pick any index i^* and consider f^{i^*}. When $x_i < y_i$, we have $[F(x) - F(y)]_i > 0 \quad (\geq 0)$; in this case, we choose an index i^* from $\{1, 2, \ldots, N\}$ such that $[f^{i^*}(x) - f^{i^*}(y)]_i > 0 \quad (\geq 0)$. Finally, when $x_i > y_i$, we choose an index i^* such that $[f^{i^*}(x) - f^{i^*}(y)]_i < 0 \quad (\leq 0)$. Let f the row representative of $\{f^1, f^2, \ldots, f^N\}$ whose ith component function is $(f^{i^*})_i$. We see that for any $x_i \neq y_i$, $(x_i - y_i)[f(x) - f(y)]_i < 0 \quad (\leq 0)$ contradicting the assumption. Hence we have the $\mathbf{P}_0$ ($\mathbf{P}$)-property of F.

(b) Suppose that $\{f^1, f^2, \ldots, f^N\}$ is a $\mathbf{P}_0$-family. For a fixed $\varepsilon > 0$, suppose that $F(x, \varepsilon)$ is not a $\mathbf{P}$-function. Then there exist x and y such that $x \neq y$ and for any index i,

$$(x_i - y_i)[F(x, \varepsilon) - F(y, \varepsilon)]_i \leq 0.$$

Now define index sets $I = \{i : x_i = y_i\}$ and $J = \{j : x_j \neq y_j\}$. Consider an index $i \in J$ and assume, without loss of generality, that $x_i > y_i$. Then $F_i(x, \varepsilon) \leq F_i(y, \varepsilon)$. For simplicity, let $\varepsilon a_k = (f^k(x))_i$, $\varepsilon b_k = (f^k(y))_i$ $\forall k = 1, 2, \ldots, N$. Also, let $\varepsilon a_0 = x_i$ and $\varepsilon b_0 = y_i$. We have $a_0 > b_0$ and $e^{-a_0} + e^{-a_1} + e^{-a_2} + \cdots + e^{-a_N} \geq e^{-b_0} + e^{-b_1} + e^{-b_2} + \cdots + e^{-b_N}$; hence there exists an index $i^* \neq 0$ such that $a_{i^*} < b_{i^*}$. The corresponding function f^{i^*} has the property $(x_i - y_i)[f^{i^*}(x) - f^{i^*}(y)]_i < 0$. For $i \in I$, we pick any f^{i^*}. We now construct a representative f of $\{f^1, f^2, \ldots, f^N\}$ by defining $f_i := (f^{i^*})_i$ for all i. For this f,

$$\max_{\{i: x_i \neq y_i\}} (x_i - y_i)[f(x) - f(y)]_i < 0$$

contradicting the assumption. Hence $F(x, \varepsilon)$ is a $\mathbf{P}$-function. □

Remarks The function $F(x,\varepsilon)$ defined in the above theorem need not be coercive. This can be seen by taking $N = 1$, $f^1 = 0$.
It is possible to construct a different perturbation of F that is both **P** and coercive: For $\varepsilon > 0$, we put

$$G(x,\varepsilon) := \min\{x, f^1(x) + \varepsilon x, f^2(x) + \varepsilon x, \ldots, f^N(x) + \varepsilon x\}.$$

Suppose that $\{f^1, f^2, \ldots, f^N\}$ is a $\mathbf{P}_0$-family. Then by item (a) of the above theorem, $G(x,\varepsilon)$ is a **P**-function. That $G(x,\varepsilon)$ is coercive in x for each $\varepsilon > 0$, is easily seen by applying Lemma 1 to the $\mathbf{P}_0$-function $F(x)$. Based on these perturbations and theorem 4, one could generate trajectories for the function $F(x)$. For the long trajectory generated by $G(x,\varepsilon)$, one could apply Theorem 5 when each f^j is a polynomial function on R^n.

Acknowledgments

The work of the first author is partially supported by the National Science Foundation Grant CCR-9307685.

References

1. S. Banach and S. Mazur, "Uber Mehrdeutige Stetige Abildungen," Studia Math, vol. 5, pp. 174-178, 1934.
2. M.S. Berger, Nonlinearity and Functional Analysis, Academic Press, New York, 1977.
3. R. Benedetti and J.J. Risler, Real Algebraic and Semi-Algebraic Sets, Hermann, Éditeurs des Sciences et des Arts, Paris, 1990.
4. S.C. Billups and M.C. Ferris, "QPCOMP: A Quadratic Programming Based Solver for Mixed Complementarity Problems," Mathematical Programming, vol. 76, pp. 533-562, 1997.
5. J. Burke and S. Xu, "The Global Linear Convergence of a Non-Interior Path-Following Algorithm for Linear Complementarity Problems," Technical Report, Department of Mathematics, University of Washington, Seattle, WA, 1996 (revised July 1997).
6. B. Chen and X. Chen, "A Global and Local Superlinear Continuation-Smoothing Method for $P_0 + R_0$ and Monotone NCP," SIAM J. Optimization, to appear.
7. B. Chen and X. Chen, "A Global Linear and Local Quadratic Continuation Smoothing Method for Variational Inequalities with Box Constraints," Technical Report, Department of Management and Systems, Washington State University, Pullman, WA, 1997.
8. B. Chen, X. Chen and C. Kanzow, "A Penalized Fischer-Burmeister NCP-Function: Theoretical Investigation and Numerical Results," Preprint 126, Institute of Applied Mathematics, University of Hamburg, Hamburg, Germany, 1997.
9. B. Chen and P.T. Harker, "A Non-Interior-Point Continuation Method for Linear Complementarity Problems," SIAM Journal on Matrix Analysis and Applications, vol. 14, pp. 1168-1190, 1993.
10. B. Chen and P. T. Harker, "Smooth Approximations to Nonlinear Complementarity Problems," SIAM Journal on Optimization, vol. 7, pp.403-420, 1997.
11. B. Chen and N. Xiu, "A Global Linear and Local Quadratic Non-Interior Continuation Method for Nonlinear Complementarity Problems Based on Chen-Mangasarian Smoothing Function," Technical Report, Department of Management and Systems, Washington State University, Pullman, WA, 1997.
12. C. Chen and O.L. Mangasarian, "A Class of Smoothing Functions for Nonlinear and Mixed Complementarity Problems," Computational Optimization and Applications, vol. 5, pp. 97-138, 1996.
13. X. Chen, L. Qi and D. Sun, "Global and Superlinear Convergence of the Smoothing Newton Method and its Application to General Box Constrained Variational Inequalities," Mathematics of Computation, vol. 67, pp. 519-540, 1998.
14. X. Chen and Y. Ye, "On Homotopy-Smoothing Methods for Variational Inequalities," SIAM J. Control and Optimization, to appear.

15. R.W. Cottle, J.-S. Pang and R.E. Stone, The Linear Complementarity Problem, Academic Press, Boston, 1992.
16. R.W. Cottle, F. Giannessi and J.-L. Lions, editors, Variational Inequalities and Complementarity Problems: Theory and Applications, J. Wiley: New York, 1980.
17. F. Facchinei and C. Kanzow, "Beyond Monotonicity in Regularization Methods for Nonlinear Complementarity Problems," SIAM J. on Control and Optimization, to appear.
18. M.C. Ferris and J.-S. Pang, editors, "Complementarity and Variational problems: State of the Art, SIAM, Philadelphia, 1997.
19. M.C. Ferris and J.-S. Pang, " Engineering and Economic Applications of Complementarity Problems," SIAM Review, vol. 39, pp. 669-713, 1997.
20. A. Fischer, "A Special Newton-Type Optimization Method," Optimization, vol. 24, pp. 269-284, 1992.
21. S.A. Gabriel and J.J. Moré, "Smoothing of Mixed Complementarity Problems and Variational Problems," Complementarity and Variational Problems: State of the Art, SIAM, Philadelphia, PA, pp. 105-116, 1997.
22. S.A. Gabriel, "A Hybrid Smoothing Method for Mixed Nonlinear Complementarity Problems," Computational Optimization and Applications, to appear.
23. O. Güler, "Existence of Interior Points and Interior Paths in Nonlinear Monotone Complementarity Problems," Mathematics of Operations Research, vol. 18, pp. 128-147, 1993.
24. O. Güler, "Limiting Behavior of the Weighted Central Paths in Linear Programming," Mathematical Programming, vol. 65, pp. 347-363, 1994.
25. M.S. Gowda and R. Sznajder, "The Generalized Order Complementarity Problem," SIAM Journal on Matrix Analysis and Applications, vol. 15, pp. 779-795, 1994.
26. M.S. Gowda and R. Sznajder, "Weak Univalence and the Connectedness of Inverse Images of Continuous Functions," Research Report 97-02, Department of Mathematics and Statistics, University of Maryland Baltimore County, Baltimore, Maryland, 1997.
27. P.T. Harker and J.-S. Pang, "Finite Dimension Variational Inequality and Nonlinear Complementarity Problems: A Survey of Theory, Algorithms and Applications," Mathematical Programming, vol. 48, pp. 161-220, 1990.
28. K. Hotta and A. Yoshise, "Global Convergence of a Class of Non-Interior-Point Algorithms Using Chen-Harker-Kanzow Function for Nonlinear Complementarity Problems," Technical Report 708, Institute of Policy and Planning Sciences, University of Tsukuba, Tsukuba, Ibaraki, Japan, 1996.
29. H. Jiang, "Smoothed Fischer-Burmeister Equation Methods for the Complementarity Problem," Technical Report, Department of Mathematics, the University of Melbourne, Parkville, Victoria, Australia, 1997.
30. H. Jiang and L. Qi, "A New Nonsmooth Equations Approach to Nonlinear Complementarity Problems," SIAM Journal on Control and Optimization, vol. 35, pp. 178-193, 1997.
31. C. Kanzow, "Some Noninterior Continuation Methods for Linear Complementarity Problems," Journal on Matrix Analysis and Applications, vol. 17, pp. 851-868, 1996.
32. C. Kanzow, "A New Approach to Continuation Methods for Complementarity Problems with Uniform P-Functions," Operations Research Letters, vol. 20, pp. 85-92, 1997.
33. M. Kojima, N. Megiddo and T. Noma, "Homotopy Continuation Methods for Nonlinear Complementarity Problems," Mathematics of Operations Research, vol. 16, pp. 754-774, 1991.
34. M. Kojima, N. Megiddo, T. Noma and A. Yoshise, A Unified Approach to Interior Point Algorithms for Linear Complementarity Problems, volume 538 of Lecture Notes in Computer Science, Springer-Verlag: New York, 1991.
35. M. Kojima, M. Mizuno and T. Noma, "A New Continuation Method for Complementarity Problems with Uniform P-Functions," Mathematical Programming, vol. 43, pp. 107-113, 1989.
36. M. Kojima, M. Mizuno and T. Noma, "Limiting Behavior of Trajectories Generated by a Continuation Method for Monotone Complementarity problems," Mathematics of Operations Research, vol. 43, pp. 662-675, 1990.
37. L. McLinden, "The Complementarity Problem for Maximal Monotone Multifunction," in Variational Inequalities and Complementarity Problems, John Wiley and Sons: New York, pp. 251-270, 1980.
38. N. Megiddo, "Pathways to the Optimal Set in Linear Programming," in Progress in Mathematical Programming: Interior-Point and Related Methods, Springer-Verlag: New York, pp. 131-158, 1989.
39. R.D.C. Monteiro and J.-S. Pang, " Properties of an Interior-Point Mapping for Mixed Complementarity Problems," Mathematics of Operations Research, vol. 21, pp. 629-654, 1996.
40. J.J. Moré, "Global Methods for Nonlinear Complementarity Problems," Mathematics of Operations Research, vol. 21, pp. 589-614, 1996.

41. Y. Nesterov and A. Nemirovskii, Interior-Point Polynomial Algorithms in Convex Programming, SIAM Publications, 1994.
42. J.-S. Pang, "Complementarity Problems," in Handbook on Global Optimization, Kluwer Academic publishers: Boston, 1994.
43. J.-M. Peng and Z. Lin, "A Non-interior Continuation Method for Generalized Linear Complementarity Problems," Research Report, Academica Sinica, Beijing, China, 1997.
44. G. Di Pillo and F. Giannessi, editors, Nonlinear Optimization and Applications, Plenum Press: New York, 1996.
45. H.-D. Qi, "A Regularized Smoothing Newton Method for Box Constrained Variational Inequality Problems with $\mathbf{P}_0$-Functions," Research Report, Institute of Computational Mathematics and Scientific/Engineering Computing, Chinese Academy of Sciences, Beijing, China, Revised 1997.
46. H.-D. Qi and L.-Z. Liao, "A Smoothing Newton Method for Extended Vertical Linear Complementarity Problems," Research Report, Institute of Computational Mathematics and Scientific Computing, Chinese Academy of Sciences, Beijing, China, 1997.
47. L. Qi and D. Sun, "Globally Linearly, and Globally and Locally Superlinearly Convergent Versions of the Hotta-Yoshise Non-Interior Point Algorithm for Nonlinear Complementarity Problems," Technical Report, School of Mathematics, The University of New South Wales, Sydney, Australia, 1997.
48. L. Qi, D. Sun and G. Zhou, "A New Look at Smoothing Newton Methods for Nonlinear Complementarity Problems and Box Constrained Variational Inequalities," Technical Report, School of Mathematics, The University of New South Wales, Sydney, Australia, 1997.
49. G. Ravindran and M.S. Gowda, "Regularization of $\mathbf{P}_0$-Functions in Box Variational Inequality Problems," Research Report, Department of Mathematics and Statistics, University of Maryland Baltimore County, Baltimore, MD, 1997.
50. P.K. Subramanian,"A Note on Least Two Norm Solutions of Monotone Complementarity Problems," Appl. Math. Letters, vol. 1, pp. 395-397 ,1988.
51. D. Sun, "A Regularized Newton Method for Solving Nonlinear Complementarity Problems," Technical Report, School of Mathematics, The University of New South Wales, Sydney, Australia, 1997.
52. R. Sznajder and M.S. Gowda, "On the Limiting Behavior of the Trajectory of Regularized Solutions of a $\mathbf{P}_0$-Complementarity Problem," in Reformulation-Nonsmooth, Piecewise Smooth, Semismooth and Smoothing Methods, Kluwer Academic Publishers: Boston, to appear.
53. P. Tseng, "Growth Behavior of a Class of Merit Functions for the Nonlinear Complementarity Problem," Journal of Optimization Theory and Applications, vol. 89, pp. 17-37, 1996.
54. P. Tseng, "Analysis of a Non-Interior Continuation Method Based on Chen-Mangasarian Smoothing Functions for Complementarity Problems," Technical Report, Department of Mathematics, University of Washington, Seattle, WA, 1997.
55. S. Xu, "The Global Linear Convergence of an Infeasible Non-Interior Path-Following Algorithm for Complementarity Problems with Uniform P-Functions," Technical Report, University of Washington, Seattle, WA, 1996.
56. S. Xu, "The Global Linear Convergence and Complexity of a Non-Interior Path-Following Algorithm for Monotone LCP Based on Chen-Harker-Kanzow-Smale Smooth Functions," Technical Report, University of Washington, Seattle, WA, 1997.
57. S. Xu and J.V. Burke, "A Polynomial Time Interior-Point Path-Following Algorithm for LCP Based on Chen-Harker-Kanzow Smoothing Techniques," Technical Report, University of Washington, Seattle, WA, 1997.

Computational Optimization and Applications 12, 253–273 (1999)

Stabilized Sequential Quadratic Programming*

WILLIAM W. HAGER hager@math.ufl.edu
Department of Mathematics, University of Florida, Gainesville, FL 32611

Received January, 1997; Revised January 19, 1998

Abstract. Recently, Wright proposed a stabilized sequential quadratic programming algorithm for inequality constrained optimization. Assuming the Mangasarian-Fromovitz constraint qualification and the existence of a strictly positive multiplier (but possibly dependent constraint gradients), he proved a local quadratic convergence result. In this paper, we establish quadratic convergence in cases where both strict complementarity and the Mangasarian-Fromovitz constraint qualification do not hold. The constraints on the stabilization parameter are relaxed, and linear convergence is demonstrated when the parameter is kept fixed. We show that the analysis of this method can be carried out using recent results for the stability of variational problems.

Keywords: sequential quadratic programming, quadratic convergence, superlinear convergence, degenerate optimization, stabilized SQP, error estimation

Dedication: This paper is dedicated to Olvi L. Mangasarian on the occasion of his 65th birthday.

1. Introduction

Let us consider the following inequality constrained optimization problem:

$$\text{minimize} \quad f(z) \qquad \text{subject to} \quad c(z) \le 0, \quad z \in \mathbf{R}^n, \tag{1}$$

where f is real-valued and $c : \mathbf{R}^n \to \mathbf{R}^m$. Given $\lambda \in \mathbf{R}^m$, the Lagrangian $\mathcal{L}$ is defined by

$$\mathcal{L}(z, \lambda) = f(z) + \lambda^{\mathsf{T}} c(z).$$

Let (z_k, λ_k) denote the current approximation to a local minimizer z_* and an associated multiplier λ_* for (1). In the sequential quadratic programming (SQP) algorithm, the new approximation z_{k+1} to z_* is given by $z_{k+1} = z_k + \Delta z$ where Δz is a local minimizer of the following quadratic problem:

$$\begin{aligned} &\text{minimize} \quad \nabla f(z_k)\Delta z + \frac{1}{2}\Delta z^{\mathsf{T}} \nabla_z^2 \mathcal{L}(z_k, \lambda_k)\Delta z \\ &\text{subject to} \quad c(z_k) + \nabla c(z_k)\Delta z \le 0 \end{aligned} \tag{2}$$

*This work was supported by the National Science Foundation.

There are various ways to specify the new multiplier. Often λ_{k+1} is a multiplier associated with the constraint in the quadratic problem (2).

The typical convergence theorem for (2) (for example, see Robinson's paper [16]) states that for (z_0, λ_0) in a neighborhood of a solution/multiplier pair (z_*, λ_*) associated with (1), the iteration is quadratically convergent when the following conditions hold:

(R1) The gradients of the active constraints are linearly independent.
(R2) The multipliers associated with the active constraints are strictly positive.
(R3) There exists a scalar $\alpha > 0$ such that

$$w^{\mathsf{T}}\nabla_z^2\mathcal{L}(z_*, \lambda_*)w \geq \alpha\|w\|^2 \tag{3}$$

for each w satisfying $\nabla c_i(z_*)w = 0$ for every i such that $c_i(z_*) = 0$.

When the constraint gradients are linearly dependent, quadratic convergence in the SQP algorithm is lost in even the simplest cases. For example, consider the problem

$$\text{minimize} \quad z^2 \qquad \text{subject to} \quad z^2 \leq 0. \tag{4}$$

The unique solution is $z_* = 0$ while λ_* can be any nonnegative number. If the multiplier approximation λ_k is held fixed at $\underline{\lambda} \geq 0$ and if $z_0 > 0$, then for $0 \leq \underline{\lambda} \leq 1$, the iteration reduces to $z_{k+1} = \underline{\lambda} z_k/(1 + \underline{\lambda})$, while for $\underline{\lambda} > 1$, the iteration reduces to $z_{k+1} = z_k/2$. In either case, the convergence is linear.

Wright's stabilized sequential quadratic programming algorithm [19] is obtained by applying Rockafellar's augmented Lagrangian [18] to the quadratic program (2). If $\rho_k > 0$ is the penalty parameter at iteration k, then (z_{k+1}, λ_{k+1}) is a local minimax for the problem

$$\begin{aligned}\min_z \max_{\lambda \geq 0} \; & (z - z_k)^{\mathsf{T}}\nabla f(z_k) + \frac{1}{2}(z - z_k)^{\mathsf{T}}\nabla_z^2\mathcal{L}(z_k, \lambda_k)(z - z_k) \\ & + \lambda^{\mathsf{T}}(c(z_k) + \nabla c(z_k)(z - z_k)) - \frac{1}{2}\rho_k\|\lambda - \lambda_k\|^2. \end{aligned} \tag{5}$$

Wright shows that this method is locally quadratically convergent if the following conditions hold:

(W1) The Mangasarian-Fromovitz [15] constraint qualification (MFCQ) holds. In the context of the inequality constrained problem (1), this means that there exists y such that $c(z_*) + \nabla c(z_*)y < 0$.
(W2) There exists a multiplier vector whose components associated with the active constraints are strictly positive.
(W3) For some fixed $\alpha > 0$, the coercivity condition (3) holds for all choices of λ_* satisfying the following first-order conditions:

$$\nabla_z\mathcal{L}(z_*, \lambda_*) = 0, \quad \lambda_* \geq 0, \quad c(z_*)^{\mathsf{T}}\lambda_* = 0. \tag{6}$$

(W4) The parameter ρ_k tends to zero proportional to the error in (z_k, λ_k).

Notice that (W1) is weaker than (R1) since there may exist y such that

$$c(z_*) + \nabla c(z_*)y < 0$$

even when the constraint gradients are linearly dependent. On the other hand, the MFCQ does not hold for the example (4), or in cases where an equality constraint is written as a pair of inequalities.

Let us consider the stabilized iteration (5) for the example (4) with z_0 near $z_* = 0$. There are two cases to consider, depending on the choice of $\lambda_k = \underline{\lambda}$. If $\underline{\lambda}$ is sufficiently large (for example, $\underline{\lambda} \geq z_k^2/(\rho_k + z_k^2)$), then at the solution of (5), the maximizing λ is positive and the successive iterates are given by

$$z_{k+1} = \frac{z_k^3}{(1+\underline{\lambda})\rho_k + 2z_k^2}. \tag{7}$$

Hence, if $\rho_k = z_k - z_* = z_k$ (the error at step k), then we have

$$z_{k+1} = \frac{z_k^2}{1+\underline{\lambda} + 2z_k},$$

which implies local quadratic convergence to the solution $z_* = 0$.

The second case corresponds to the situation where the maximizing λ in (5) vanishes. For this to happen, we must have $\underline{\lambda} \leq z_k^2/(\rho_k + z_k^2)$, and the new iterate is expressed:

$$z_{k+1} = \frac{z_k\underline{\lambda}}{1+\underline{\lambda}}.$$

Again, if $\rho_k = z_k - z_* = z_k$, then

$$\underline{\lambda} \leq z_k^2/(\rho_k + z_k^2) = z_k/(1+z_k). \tag{8}$$

and we have

$$z_{k+1} = \frac{z_k\underline{\lambda}}{1+\underline{\lambda}} \leq \frac{z_k^2}{(1+\underline{\lambda})(1+z_k)}. \tag{9}$$

In each of these cases, the convergence to the solution $z_* = 0$ is locally quadratic.

Also notice in this example that choosing ρ_k much smaller than the error at step k can slow the convergence. In particular, if $\rho_k = 0$ and the max is changed to sup in (5), then the scheme (5) reduces to the usual SQP iteration (2) for which the convergence in the example (4) is linear. On the other hand, we still obtain fast convergence even when ρ_k is much larger than the error at step k. For example, if $\rho_k > 0$ is fixed, then (7) gives cubic convergence. Likewise, the initial inequality in (8) implies that $\underline{\lambda} \leq z_k^2/\rho_k$, which combines with (9) to give $z_{k+1} \leq z_k^3/\rho_k$. In either case, when $\rho_k > 0$ is fixed, we obtain cubic convergence near the solution $z_* = 0$. Hence, from an implementational viewpoint, a large ρ_k is safer than a small one.

In this example, quadratic convergence is preserved with the stabilized SQP scheme even though strict complementarity and the MFCQ are violated. In fact, when $\underline{\lambda} = 0$ and strict complementarity is violated, we have convergence in one step. In this paper, we show in general that Wright's stabilized scheme is locally, quadratically convergent even though both the MFCQ and strict complementarity are violated. In contrast to Wright's assumption (W3) that the second-order condition holds for all multipliers, we give in this paper a local analysis where a second-order condition is required to hold only at a given solution/multiplier pair (z_*, λ_*). When strict complementarity is violated, our second-order condition is slightly stronger than the usual second-order condition in that we assume

$$w^{\mathsf{T}} \nabla_z^2 \mathcal{L}(z_*, \lambda_*) w \geq \alpha \|w\|^2 \tag{10}$$

for all w satisfying $\nabla c_i(z_*)w = 0$ for every i such that $c_i(z_*) = 0$ and $(\lambda_*)_i > 0$. This strengthened form of the second-order sufficient condition first appears in Robinson's study [17] of Lipschitz stability of optimization problems. Dontchev and Rockafellar [7] show that this condition along with linear independence of the active constraint gradients are necessary and sufficient for Lipschitz stability of the solution and multipliers under canonical perturbations of the problem constraints and cost function.

The strong second-order sufficient condition is stable in the sense that it holds when $\nabla_z^2 \mathcal{L}(z_*, \lambda_*)$ and $\nabla c_i(z_*)$ are replaced by nearby matrices, while the usual second-order condition is unstable under problem perturbations. The usual second-order sufficient condition imposes on w in (10) the additional constraint $\nabla c_i(z_*)w \leq 0$ for every i such that $c_i(z_*) = 0 = (\lambda_*)_i$. That is, (10) must hold for all w in the set

$$\{w \in \mathbf{R}^n : \nabla c_i(z_*)w = 0 \ \forall i \in \mathcal{A}_+, \quad \nabla c_i(z_*)w \leq 0 \ \forall i \in \mathcal{A}_0\},$$

where

$$\mathcal{A}_+ = \{i : (\lambda_*)_i > 0\} \quad \text{and} \quad \mathcal{A}_0 = \{i : c_i(z_*) = 0 = (\lambda_*)_i\}.$$

If the usual second-order condition holds for some pair (z_*, λ_*), then we can perturb the constraint $c(z) \leq 0$ to $c(z) + \xi \leq 0$ where $\xi_i < 0$ if $(\lambda_*)_i = 0 = c_i(z_*)$, and $\xi_i = 0$ otherwise. For this perturbed problem, (z_*, λ_*) again satisfies the first-order conditions, however, the active constraints for the perturbed problem are precisely the constraints in the unperturbed problem with positive multipliers. Therefore, even though the usual second-order sufficient condition holds at (z_*, λ_*), small perturbations in the constraints can yield a problem whose stationary point does not satisfy this condition.

Our analysis of (5) is based on the application of tools from stability analysis. That is, we introduce parameters in the iteration map and we study how the map depends on the parameters using a stability result established in [6, Lemma 2.1]. Once we understand how the iteration map depends on the parameters, we can write down a convergence theorem. Other applications of stability theory to the convergence of algorithms and to the analysis of discretizations appear in [3–6], and [11]. Our analysis of (5) also leads to a new expression for the error in each iterate. In particular, we show that linear convergence is achieved when ρ_k is fixed, but small. This paper is a revised version of the report [12].

Another approach for dealing with degeneracy in nonlinear programming is developed by Fischer in [9]. In his approach, the original quadratic program (2) is retained, however, the multiplier estimate is gotten by solving a separate quadratic program. Fischer obtains quadratic convergence assuming the MFCQ, the second-order sufficient optimality condition, a constant rank condition for the active constraint gradients in a neighborhood of z_*, and a condition concerning the representation of the cost function gradient in terms of the constraint gradients. Although these assumptions seem more stringent than those used in our analysis of Wright's method, there are no parameters like ρ_k in Fischer's method that must be specified in each iteration.

2. Convergence theory

Let z_* denote a local minimizer for (1) and let λ_* be an associated multiplier satisfying the first-order conditions (6). To state our assumptions, we partition c and λ into (g, h) and (μ, π) where the components of h correspond to components of c associated with strictly positive components π_* of λ_*, while the components of g are the remaining components of c for which the associated components μ_* of λ_* could be zero. Let $\mathcal{M}$ denote the set of all multipliers associated with a local minimizer z_* for (1):

$$\lambda \in \mathcal{M} \quad \text{if and only if} \quad \nabla_z \mathcal{L}(z_*, \lambda) = 0, \quad \lambda \geq 0, \quad \text{and} \quad \lambda^{\mathsf{T}} c(z_*) = 0.$$

Letting $\mathcal{B}_\delta(z)$ denote the ball with center z and radius δ, our main result is the following:

Theorem 1. *Suppose that f and c are twice Lipschitz continuously differentiable in a neighborhood of a local minimizer z_* of (1), that $\lambda_* = (\mu_*, \pi_*)$ is an associated multiplier in $\mathcal{M}$ with $\pi_* > 0$, and that*

$$w^{\mathsf{T}} \nabla_z^2 \mathcal{L}(z_*, \lambda_*) w \geq \alpha \|w\|^2 \tag{11}$$

for each w such that $\nabla h(z_*) w = 0$. Then for any choice of the constant σ_0 sufficiently large, there exist constants σ_1, δ, and $\bar{\beta}$ with the property that $\sigma_0 \delta \leq \sigma_1$ and for each starting guess $(z_0, \lambda_0) \in \mathcal{B}_\delta(z_*, \lambda_*)$, there are iterates (z_k, λ_k) contained in $\mathcal{B}_\delta(z_*, \lambda_*)$, where each z_{k+1} is a strict local minimizer in the stabilized problem (5), λ_{k+1} is the unique maximizer in (5) associated with $z = z_{k+1}$, and ρ_k is any scalar that satisfies the condition

$$\sigma_0 \|z_k - z_*\| \leq \rho_k \leq \sigma_1. \tag{12}$$

Moreover, the following estimate holds:

$$\|z_{k+1} - z_*\| + \|\lambda_{k+1} - \hat{\lambda}_{k+1}\| \leq \bar{\beta}(\|z_k - z_*\|^2 + \|\lambda_k - \hat{\lambda}_k\|^2 + \rho_k \|\lambda_k - \hat{\lambda}_k\|), \tag{13}$$

where $\hat{\lambda}_k$ and $\hat{\lambda}_{k+1}$ are the closest elements of $\mathcal{M}$ to λ_k and λ_{k+1} respectively.

By Theorem 1, letting ρ_k go to zero proportional to the total error

$$\|z_k - z_*\| + \|\lambda_k - \hat{\lambda}_k\|$$

leads to local quadratic convergence. Techniques for estimating the error in the current iterate can be found in [13, 19]. Since Theorem 1 is a local convergence result, we assume (without loss of generality), that $c(z_*) = 0$. That is, if some constraint is inactive at z_*, we simply discard this constraint and apply Theorem 1 to the reduced problem, obtaining a neighborhood where the iterations converge and (13) holds. When this constraint is included in c, it can be shown that for (z_k, λ_k) near (z_*, λ_*), the associated component of the maximizing multiplier in (5) vanishes. Hence, the iterates obtained either with or without this inactive constraint included in c are identical.

Although an equality constraint does not appear explicitly in (1), we can include the equality constraint $e(z) = 0$ by writing it as a pair of inequalities: $e(z) \le 0$ and $-e(z) \le 0$. One of these constraint functions should be included in g and the other in h. There are an infinite number of multipliers associated with this pair of constraint functions with linearly dependent gradients, and it can always be arranged so that the associated component in π_* is strictly positive.

Throughout this paper, $\|\cdot\|$ denotes the Euclidean norm and β denotes a generic positive constant that has different values in different equations, and which can be bounded in terms of the derivatives through second order of f and c in a neighborhood of (z_*, λ_*) and in terms of fixed constants like α in (11). In order to prove Theorem 1, we recast (5) in the form of a perturbed variational inequality. Let T be the function defined by

$$T(z, \lambda, \underline{z}, \underline{\lambda}_1, \underline{\lambda}_2) = \begin{pmatrix} \nabla_z \mathcal{L}(\underline{z}, \lambda) + \nabla_z^2 \mathcal{L}(\underline{z}, \underline{\lambda}_1)(z - \underline{z}) \\ c(\underline{z}) + \nabla c(\underline{z})(z - \underline{z}) - \rho(\lambda - \underline{\lambda}_2) \end{pmatrix}, \tag{14}$$

where ρ and $p = (\underline{z}, \underline{\lambda}_1, \underline{\lambda}_2)$ are regarded parameters. Since we later impose a constraint on ρ in terms of p, as in (12), we do not make ρ an explicit argument of T. We study properties of solutions to the following inclusion relative to the parameters: Find (z, λ) such that

$$T(z, \lambda, p) \in \begin{pmatrix} 0 \\ N(\lambda) \end{pmatrix}, \quad \lambda \ge 0, \tag{15}$$

where N is the usual normal cone: If $\lambda \ge 0$, then $y \in N(\lambda)$ if and only if $y \le 0$ and $y^{\mathsf T}\lambda = 0$. By analyzing how the solutions to (15) depend on p, we will establish Theorem 1.

If (z_{k+1}, λ_{k+1}) is a local solution to (5), then for $p = p_k = (z_k, \lambda_k, \lambda_k)$, $(z, \lambda) = (z_{k+1}, \lambda_{k+1})$ is a solution to (15), and in this case, (15) represents the first-order optimality conditions associated with (5). More explicitly, (15) implies that

$$\nabla_z \mathcal{L}(z_k, \lambda_{k+1}) + \nabla_z^2 \mathcal{L}(z_k, \lambda_k)(z_{k+1} - z_k) = 0, \tag{16}$$

$$c(z_k) + \nabla c(z_k)(z_{k+1} - z_k) - \rho(\lambda_{k+1} - \lambda_k) \le 0, \quad \lambda_{k+1} \ge 0, \tag{17}$$

$$\lambda_{k+1}^{\mathsf T}(c(z_k) + \nabla c(z_k)(z_{k+1} - z_k) - \rho(\lambda_{k+1} - \lambda_k)) = 0. \tag{18}$$

Conditions (17) and (18) are equivalent to saying that λ_{k+1} achieves the maximum in (5) corresponding to $z = z_{k+1}$. By the standard rules for differentiating under a maximization (see [2]), the derivative of the extremand in (5) with respect to z is obtained by computing the partial derivative with respect to z and evaluating the resulting expression at that λ

where the extremand is maximized. Hence, (16) is equivalent to saying the derivative of the extremand with respect to z vanishes at $z = z_{k+1}$.

Observe that when $p = (z_*, \lambda_*, \tilde{\lambda})$, where $\tilde{\lambda}$ is an arbitrary element of $\mathcal{M}$, then $(z, \lambda) = (z_*, \tilde{\lambda})$ is a solution to (15). In this section, we apply the following stability result, describing how the solution to (15) changes as p changes, to obtain Theorem 1. The proof of this stability result is given in the next section.

Lemma 1. *Under the hypotheses of Theorem 1, for any choice of the constant σ_0 sufficiently large and for any $\sigma_1 > 0$, there exist constants β and δ such that $\sigma_0\delta \le \sigma_1$ and for each $p = (\underline{z}, \underline{\lambda}_1, \underline{\lambda}_2) \in \mathcal{B}_\delta(p_*)$ and for each ρ satisfying*

$$\Lambda(p) \le \rho \le \sigma_1 \quad \textit{where } \Lambda(p) = \sigma_0\|\underline{z} - z_*\|, \tag{19}$$

(15) has a unique solution $(z, \lambda) = (z(p), \lambda(p)) \in \mathcal{N}(\rho)$ where

$$\mathcal{N}(\rho) = \{(z, \lambda) : \|z - z_*\| + \rho\|\lambda - \lambda_*\| \le \rho\}.$$

Moreover, for every p_1 and $p_2 \in \mathcal{B}_\delta(p_*)$, and ρ satisfying (19) for $p = p_1$ and $p = p_2$, if (z_1, λ_1) and (z_2, λ_2) are the associated solutions to (15), then we have

$$\|z_1 - z_2\| + \rho\|\lambda_1 - \lambda_2\| \le \beta\|T(z_1, \lambda_1, p_1) - T(z_1, \lambda_1, p_2)\|. \tag{20}$$

There are three parts to the proof of Theorem 1. Initially, we show that the estimate (13) holds for each (z_k, λ_k) near (z_*, λ_*), where (z_{k+1}, λ_{k+1}) is a solution to (15) associated with $p = p_k = (z_k, \lambda_k, \lambda_k)$. Next, we show that for (z_0, λ_0) sufficiently close to (z_*, λ_*), we can construct a sequence (z_1, λ_1), (z_2, λ_2), ..., contained in a fixed ball centered at (z_*, λ_*), where (z_{k+1}, λ_{k+1}) is the unique solution in $\mathcal{N}(\rho_k)$ to (15) for $p = (z_k, \lambda_k, \lambda_k)$. Finally, we show that for this unique solution (z_{k+1}, λ_{k+1}) to (15), z_{k+1} is a local minimizer of (5).

Part 1 (Error estimate). Let $\sigma_1 > 0$ be any fixed scalar (independent of k) and let σ_0 and δ be chosen in accordance with Lemma 1. By Lemma 1, there exists a ball $\mathcal{B}_\delta(p_*)$ with the property that for each $p_k = (z_k, \lambda_k, \lambda_k) \in \mathcal{B}_\delta(p_*)$, (15) has a unique solution $(z, \lambda) = (z_{k+1}, \lambda_{k+1})$ in $\mathcal{N}(\rho)$ where ρ is any scalar that satisfies the condition

$$\sigma_0\|z_k - z_*\| \le \rho \le \sigma_1. \tag{21}$$

We apply Lemma 1 taking

$$p_1 = (z_*, \lambda_*, \hat{\lambda}_k), \quad (z_1, \lambda_1) = (z_*, \hat{\lambda}_k),$$
$$p_2 = (z_k, \lambda_k, \lambda_k), \quad (z_2, \lambda_2) = (z_{k+1}, \lambda_{k+1}).$$

If λ_k is near λ_*, then $\hat{\lambda}_k$ is near λ_* since $\|\lambda_k - \hat{\lambda}_k\| \le \|\lambda_k - \lambda_*\|$ and

$$\|\hat{\lambda}_k - \lambda_*\| \le \|\hat{\lambda}_k - \lambda_k\| + \|\lambda_k - \lambda_*\| \le 2\|\lambda_k - \lambda_*\|.$$

Suppose that $p_2 = p_k \in \mathcal{B}_\delta(p_*)$ is close enough to p_* that $p_1 = (z_*, \lambda_*, \hat{\lambda}_k) \in \mathcal{B}_\delta(p_*)$ and $(z_1, \lambda_1) = (z_*, \hat{\lambda}_k) \in \mathcal{N}(\rho)$. Note that (19) holds for $p = p_1$ since $\Lambda(p_1) = 0$. Assuming that ρ is chosen so that (19) holds for $p = p_k = (z_k, \lambda_k, \lambda_k)$, it follows from (20) that

$$\|z_{k+1} - z_*\| + \rho\|\lambda_{k+1} - \hat{\lambda}_k\| \leq \beta E_k, \tag{22}$$

where

$$E_k = \left\| \begin{pmatrix} \nabla_z \mathcal{L}(z_k, \hat{\lambda}_k) + \nabla_z^2 \mathcal{L}(z_k, \lambda_k)(z_* - z_k) \\ c(z_k) + \nabla c(z_k)(z_* - z_k) - \rho(\hat{\lambda}_k - \lambda_k) \end{pmatrix} \right\|. \tag{23}$$

Expanding E_k in a Taylor series around z_* gives

$$E_k \leq \bar{E}_k := \beta(\|z_k - z_*\|^2 + \|\lambda_k - \hat{\lambda}_k\|\|z_k - z_*\| + \rho\|\lambda_k - \hat{\lambda}_k\|) \tag{24}$$

$$\leq \beta(\|z_k - z_*\|^2 + \|\lambda_k - \hat{\lambda}_k\|^2 + \rho\|\lambda_k - \hat{\lambda}_k\|), \tag{25}$$

where β is a generic positive constant. The second inequality (25) is obtained using the relation $ab \leq (a^2 + b^2)/2$. Combining (22), (23), and (25) establishes the estimate for z_{k+1} in Theorem 1.

Dividing (24) by ρ gives

$$\bar{E}_k/\rho \leq \beta((\|z_k - z_*\| + \|\lambda_k - \hat{\lambda}_k\|)\|z_k - z_*\|/\rho + \|\lambda_k - \hat{\lambda}_k\|).$$

Utilizing the lower bound $\rho \geq \sigma_0\|z_k - z_*\|$, it follows that

$$\bar{E}_k/\rho \leq \beta(\|z_k - z_*\| + \|\lambda_k - \hat{\lambda}_k\|). \tag{26}$$

Hence, dividing (22) by ρ and referring to (26), we deduce that

$$\|\lambda_{k+1} - \hat{\lambda}_k\| \leq \beta(\|z_k - z_*\| + \|\lambda_k - \hat{\lambda}_k\|). \tag{27}$$

By the triangle inequality, we have

$$\|\lambda_{k+1} - \lambda_*\| \leq \|\lambda_{k+1} - \hat{\lambda}_k\| + \|\hat{\lambda}_k - \lambda_k\| + \|\lambda_k - \lambda_*\|,$$

and combining this with (27) gives

$$\|\lambda_{k+1} - \lambda_*\| \leq \|\lambda_k - \lambda_*\| + \beta(\|z_k - z_*\| + \|\lambda_k - \hat{\lambda}_k\|)$$

$$\leq \|\lambda_k - \lambda_*\| + \beta(\|z_k - z_*\| + \|\lambda_k - \lambda_*\|). \tag{28}$$

This shows that λ_{k+1} is near λ_* when (z_k, λ_k) is near (z_*, λ_*).

We now show that

$$\|\nabla_z \mathcal{L}(z_*, \lambda_{k+1})\| \leq \beta\bar{E}_k. \tag{29}$$

In order to establish this, we exploit the Lipschitz continuity of $\nabla_z \mathcal{L}$, the bound (22), and our observation that λ_{k+1} is near λ_* to obtain

$$\|\nabla_z \mathcal{L}(z_*, \lambda_{k+1}) - \nabla_z \mathcal{L}(z_{k+1}, \lambda_{k+1})\| \leq \beta \|z_{k+1} - z_*\| \leq \beta E_k. \tag{30}$$

Expanding $\nabla_z \mathcal{L}(z_{k+1}, \lambda_{k+1})$ in a Taylor series around z_k and substituting from (16) gives

$$\begin{aligned}
&\|\nabla_z \mathcal{L}(z_{k+1}, \lambda_{k+1})\| \\
&\quad \leq \left\|\nabla_z \mathcal{L}(z_k, \lambda_{k+1}) + \nabla_z^2 \mathcal{L}(z_k, \lambda_{k+1})(z_{k+1} - z_k)\right\| + \beta \|z_{k+1} - z_k\|^2 \\
&\quad = \left\|\left(\nabla_z^2 \mathcal{L}(z_k, \lambda_{k+1}) - \nabla_z^2 \mathcal{L}(z_k, \lambda_k)\right)(z_{k+1} - z_k)\right\| + \beta \|z_{k+1} - z_k\|^2 \\
&\quad \leq \beta(\|z_{k+1} - z_k\| + \|\lambda_{k+1} - \lambda_k\|)\|z_{k+1} - z_k\| \\
&\quad \leq \beta(\|z_{k+1} - z_k\|^2 + \|\lambda_{k+1} - \lambda_k\|^2).
\end{aligned} \tag{31}$$

By the triangle inequality, we have

$$\|z_{k+1} - z_k\| \leq \|z_{k+1} - z_*\| + \|z_k - z_*\| \leq \beta E_k + \|z_k - z_*\|.$$

Squaring this gives

$$\|z_{k+1} - z_k\|^2 \leq \beta E_k^2 + 2\|z_k - z_*\|^2 \leq \beta \bar{E}_k. \tag{32}$$

If it can be shown that

$$\|\lambda_{k+1} - \lambda_k\| \leq \beta(\|z_k - z_*\| + \|\lambda_k - \hat{\lambda}_k\|), \tag{33}$$

then by squaring, we have

$$\|\lambda_{k+1} - \lambda_k\|^2 \leq \beta(\|z_k - z_*\|^2 + \|\lambda_k - \hat{\lambda}_k\|^2) \leq \beta \bar{E}_k. \tag{34}$$

Combining (31) with (32) and (34) gives

$$\|\nabla_z \mathcal{L}(z_{k+1}, \lambda_{k+1})\| \leq \beta \bar{E}_k,$$

and combining this with (30) yields

$$\begin{aligned}
\|\nabla_z \mathcal{L}(z_*, \lambda_{k+1})\| &\leq \|\nabla_z \mathcal{L}(z_{k+1}, \lambda_{k+1})\| \\
&\quad + \|\nabla_z \mathcal{L}(z_*, \lambda_{k+1}) - \nabla_z \mathcal{L}(z_{k+1}, \lambda_{k+1})\| \leq \beta \bar{E}_k,
\end{aligned}$$

which completes the proof of (29).

To prove (33), we focus on the individual components of $\lambda_{k+1} - \lambda_k$ and establish the relation

$$|(\lambda_{k+1} - \lambda_k)_i| \leq \beta(\|z_k - z_*\| + \|\lambda_k - \hat{\lambda}_k\|) \tag{35}$$

for each i. There are three cases to consider:

(C1) $(\lambda_{k+1})_i = 0 = (\lambda_k)_i$. For these components, (35) is a triviality.

(C2) $(\lambda_{k+1})_i > 0$. By complementary slackness (18), we have

$$(c(z_k) + \nabla c(z_k)(z_{k+1} - z_k))_i = \rho(\lambda_{k+1} - \lambda_k)_i. \tag{36}$$

Expanding $c(z_k)$ in a Taylor expansion around z_{k+1}, utilizing (32), and taking absolute values yields

$$\begin{aligned} |(c(z_k) + \nabla c(z_k)(z_{k+1} - z_k))_i| &\leq |c_i(z_{k+1})| + \beta\|z_{k+1} - z_k\|^2 \\ = |c_i(z_{k+1}) - c_i(z_*)| + \beta\|z_{k+1} - z_k\|^2 &\leq \beta\|z_{k+1} - z_*\| + \beta\bar{E}_k \leq \beta\bar{E}_k. \end{aligned} \tag{37}$$

Dividing (36) by ρ and utilizing (37) and (26) gives (35).

(C3) $(\lambda_{k+1})_i = 0$ and $(\lambda_k)_i > 0$. By (17), we have

$$0 < \rho(\lambda_k)_i = -\rho(\lambda_{k+1} - \lambda_k)_i \leq |(c(z_k) + \nabla c(z_k)(z_{k+1} - z_k))_i|.$$

Dividing this by ρ and again utilizing (37) and (26) gives (35).

This completes the proof of both (33) and (29).

Consider the following system of linear equations and inequalities in λ:

$$\nabla_z \mathcal{L}(z_*, \lambda) = 0, \quad \lambda \geq 0. \tag{38}$$

This system is feasible since any $\lambda \in \mathcal{M}$ is a solution. By (29) and a result of Hoffman [14], the closest solution $\hat{\lambda}_{k+1}$ of (38) to λ_{k+1} satisfies

$$\|\lambda_{k+1} - \hat{\lambda}_{k+1}\| \leq \beta\bar{E}_k. \tag{39}$$

That is, Hoffman's result states that if a linear system of inequalities is feasible, then the distance from any given point to the set of feasible points is bounded by a constant times the norm of the constraint violation at the given point. By (29), the norm of the constraint violation is at most $\beta\bar{E}_k$ at λ_{k+1}, from which it follows that the distance from λ_{k+1} to the closest solution of (38) is bounded by a constant times $\bar{E}_k$. Since $c(z_*) = 0$, this solution of (38) is contained in $\mathcal{M}$ and it is the closest element of $\mathcal{M}$ to λ_{k+1}. Relations (25) and (39) combine to complete the proof of (13).

Part 2 (Containment). Collecting results, we have shown that if

$$p = p_k = (z_k, \lambda_k, \lambda_k)$$

is sufficiently close to $p_* = (z_*, \lambda_*, \lambda_*)$, then (15) has a unique solution $(z_{k+1}, \lambda_{k+1}) \in \mathcal{N}(\rho)$ where ρ is any scalar satisfying (21), where z_{k+1} and λ_{k+1} satisfy (13), and where λ_{k+1} also satisfies (28). As σ_1 or δ in Lemma 1 decreases, the constant β in (20) can be kept fixed since the set of ρ and p that satisfies the constraints of the lemma becomes smaller. That is, if (20) holds for one set of ρ and p values, then it holds for all subsets. Let $\bar{\beta}$ be the constant appearing in (13) that we estimated in Part 1 using Lemma 1. Given any positive $\epsilon < 1$, let us choose σ_1 and δ of Lemma 1 small enough that

$$\bar{\beta}(\|z_k - z_*\|^2 + \|\lambda_k - \hat{\lambda}_k\|^2 + \rho_k\|\lambda_k - \hat{\lambda}_k\|) \leq \epsilon(\|z_k - z_*\| + \|\lambda_k - \hat{\lambda}_k\|)$$

for all $p_k \in \mathcal{B}_\delta(p_*)$ and $\rho_k \leq \sigma_1$. From the analysis of Part 1, both (13) and (28), there exists, for all $p_k \in \mathcal{B}_\delta(p_*)$ and ρ_k satisfying

$$\sigma_0\|z_k - z_*\| \leq \rho_k \leq \sigma_1, \tag{40}$$

a unique solution $(z_{k+1}, \lambda_{k+1}) \in \mathcal{N}(\rho_k)$ to (15), and we have

$$\|z_{k+1} - z_*\| + \|\lambda_{k+1} - \hat{\lambda}_{k+1}\| \leq \epsilon(\|z_k - z_*\| + \|\lambda_k - \hat{\lambda}_k\|), \tag{41}$$

and

$$\|\lambda_{k+1} - \lambda_*\| \leq \|\lambda_k - \lambda_*\| + \beta_0(\|z_k - z_*\| + \|\lambda_k - \hat{\lambda}_k\|), \tag{42}$$

where β_0 denotes the specific constant β appearing in (28).

We now show in an inductive fashion that for (z_0, λ_0) sufficiently close to (z_*, λ_*), there exists a sequence (z_k, λ_k), $k = 0, 1, \ldots$, where (z_{k+1}, λ_{k+1}) is the unique solution to (15) in $\mathcal{N}(\rho_k)$ corresponding to $p = (z_k, \lambda_k, \lambda_k)$, and to ρ_k satisfying (40). In particular, let r_0 be chosen small enough that

$$r_1 := 2r_0\left(1 + \frac{\beta_0}{1-\epsilon}\right) \leq \delta/2.$$

Starting from any $(z_0, \lambda_0) \in \mathcal{B}_{r_0}(z_*, \lambda_*)$, we proceed by induction and suppose that (z_0, λ_0), $(z_1, \lambda_1), \ldots, (z_j, \lambda_j)$ are all contained in $\mathcal{B}_{r_1}(z_*, \lambda_*)$. Since $r_1 \leq \delta/2$, there exists a unique solution $(z_{j+1}, \lambda_{j+1}) \in \mathcal{N}(\rho_j)$ to (15) for $p = (z_j, \lambda_j, \lambda_j)$. By (41), it follows that for $0 \leq k \leq j+1$,

$$\begin{aligned}\|z_k - z_*\| + \|\lambda_k - \hat{\lambda}_k\| &\leq \epsilon^k(\|z_0 - z_*\| + \|\lambda_0 - \hat{\lambda}_0\|) \\ &\leq \epsilon^k(\|z_0 - z_*\| + \|\lambda_0 - \lambda_*\|) \leq r_0 \leq r_1/2.\end{aligned} \tag{43}$$

By (42) and (43), we have

$$\begin{aligned}\|\lambda_{j+1} - \lambda_*\| &\leq \|\lambda_j - \lambda_*\| + \beta_0\epsilon^j(\|z_0 - z_*\| + \|\lambda_0 - \hat{\lambda}_0\|) \\ &\leq \|\lambda_0 - \lambda_*\| + \beta_0(\|z_0 - z_*\| + \|\lambda_0 - \hat{\lambda}_0\|)\sum_{k=0}^{j}\epsilon^k \\ &\leq \|\lambda_0 - \lambda_*\| + \frac{\beta_0}{1-\epsilon}(\|z_0 - z_*\| + \|\lambda_0 - \hat{\lambda}_0\|) \\ &\leq \|\lambda_0 - \lambda_*\| + \frac{\beta_0}{1-\epsilon}(\|z_0 - z_*\| + \|\lambda_0 - \lambda_*\|) \\ &\leq r_0 + \frac{\beta_0 r_0}{1-\epsilon} \leq r_1/2.\end{aligned} \tag{44}$$

Combining (43) and (44) yields

$$\|z_{j+1} - z_*\| + \|\lambda_{j+1} - \lambda_*\| \le r_1/2 + r_1/2 = r_1.$$

Hence, $(z_{j+1}, \lambda_{j+1}) \in \mathcal{B}_{r_1}(z_*, \lambda_*)$ and the induction is complete.

Part 3 (Local minimizer). Finally, we show that z_{k+1} is a local minimizer for (5). Since $\lambda_* = (\mu_*, \pi_*)$ with $\pi_* > 0$, it follows that by taking r_0 sufficiently small, $\pi_{k+1} > 0$ for all k. By complementary slackness (18), we have

$$\pi_{k+1} = \pi_k + (h(z_k) + \nabla h(z_k)(z_{k+1} - z_k))/\rho.$$

As noted after (18), if $(z, \lambda) = (z_{k+1}, \lambda_{k+1})$ is a solution of (15), then $\lambda_{k+1} = (\mu_{k+1}, \pi_{k+1})$ achieves the maximum in (5) for $z = z_{k+1}$. Since the maximizing λ in (5) is a continuous function of z (see [3, Lemma 4]), we conclude that for z near z_{k+1}, the maximizing $\lambda = (\mu, \pi)$ has $\pi > 0$; hence, by complementary slackness and for z near z_{k+1}, the maximizing π is given by

$$\pi = \pi_k + (h(z_k) + \nabla h(z_k)(z - z_k))/\rho.$$

After making this substitution in (5), the cost function of the minimax problem can be decomposed into the sum of a convex function of z:

$$\max_{\mu \ge 0} \ \mu^{\mathsf{T}}(g(z_k) + \nabla g(z_k)(z - z_k)) - \frac{1}{2}\rho_k \|\mu - \mu_k\|^2,$$

and a strongly convex part

$$\begin{aligned}&(z - z_k)^{\mathsf{T}}\nabla f(z_k) + \frac{1}{2}(z - z_k)^{\mathsf{T}}\nabla_z^2 \mathcal{L}(z_k, \lambda_k)(z - z_k)\\ &\quad + (h(z_k) + \nabla h(z_k)(z - z_k))^{\mathsf{T}}\left(\pi_k + \frac{1}{2\rho_k}(h(z_k) + \nabla h(z_k)(z - z_k))\right).\end{aligned}$$

The first part is convex since the extremand is a linear function of z and the max of a sum is less than or equal to the sum of the maxs. The second part is strongly convex since the Hessian matrix

$$\nabla_z^2 \mathcal{L}(z_k, \lambda_k) + \frac{1}{\rho_k}\nabla h(z_k)^{\mathsf{T}}\nabla h(z_k)$$

is positive definite for ρ_k and r_0 sufficiently small by Lemma 3 in the Appendix. Hence, the cost function of (5) is a strongly convex function of z in a neighborhood of z_{k+1}, and since the derivative vanishes at z_{k+1} by (16), z_{k+1} is a local minimum. This completes the proof of Theorem 1. □

3. Stability for the linearized system

The proof of Lemma 1 is based on the following result, which is a variation of Lemma 2.1 in [6].

Lemma 2. *Let X be a subset of $\mathbf{R}^n$ and let $\|\cdot\|_\rho$ denote the norm on X. Given $w_* \in X$ and $\tau > 0$, define*

$$W = \{x \in \operatorname{cl} X : \|x - w_*\|_\rho \le \tau\}.$$

In other words, W is the intersection of the closure of X and the ball with center w_* and radius τ. Suppose that F maps W to the subsets of $\mathbf{R}^m$, and $T : W \times P \to \mathbf{R}^m$, where P is a set. Let $p_* \in P$ with $T(w_*, p_*) \in F(w_*)$, let L be an $m \times n$ matrix, and let τ, η, ϵ, and γ denote any positive numbers for which $\epsilon\gamma < 1$, $\tau \ge \eta\gamma/(1 - \epsilon\gamma)$, and the following properties hold:

(P1) $\|T(w_*, p_*) - T(w_*, p)\| \le \eta$ *for all* $p \in P$.

(P2) $\|T(w_2, p) - T(w_1, p) - L(w_2 - w_1)\| \le \epsilon\|w_2 - w_1\|_\rho$ *for all* $w_1, w_2 \in W$ *and* $p \in P$.

(P3) *For some set* $\mathcal{N} \supset \{T(w, p) - Lw : w \in W,\ p \in P\}$, *the following problem has a unique solution for each* $\psi \in \mathcal{N}$:

$$\text{Find } x \in X \text{ such that } Lx + \psi \in F(x) \tag{45}$$

and if $x(\psi)$ denotes the solution corresponding to ψ, we have

$$\|x(\psi_2) - x(\psi_1)\|_\rho \le \gamma\|\psi_2 - \psi_1\| \tag{46}$$

for each $\psi_1, \psi_2 \in \mathcal{N}$.

Then for each $p \in P$, there exists a unique $w \in W$ such that $T(w, p) \in F(w)$. Moreover, for every $p_i \in P$, $i = 1, 2$, if w_i denotes the w associated with p_i, then we have

$$\|w_2 - w_1\|_\rho \le \frac{\gamma}{1 - \gamma\epsilon}\|T(w_1, p_2) - T(w_1, p_1)\|. \tag{47}$$

Proof: Fix $p \in P$ and for $w \in W$, let $\Phi(w)$ denote the solution to (45) corresponding to $\psi = T(w, p) - Lw$. That is, $\Phi(w) = x[T(w, p) - Lw]$. For $w_i \in W$, $i = 1, 2$, define $\psi_i = T(w_i, p) - Lw_i$. Observe that

$$\begin{aligned}\|\Phi(w_1) - \Phi(w_2)\|_\rho &= \|x(\psi_1) - x(\psi_2)\|_\rho \le \gamma\|\psi_1 - \psi_2\| \\ &= \gamma\|T(w_1, p) - T(w_2, p) - L(w_1 - w_2)\| \\ &\le \gamma\epsilon\|w_1 - w_2\|_\rho\end{aligned}$$

for each $w_1, w_2 \in W$. Since $\gamma\epsilon < 1$, Φ is a contraction on W with contraction constant $\gamma\epsilon$. From the assumption $T(w_*, p_*) \in F(w_*)$, it follows that

$$w_* = x[T(w_*, p_*) - Lw_*].$$

Given $w \in W$, we have

$$\begin{aligned}\|\Phi(w) - w_*\|_\rho &= \|x[T(w, p) - Lw] - x[T(w_*, p_*) - Lw_*]\|_\rho \\ &\le \gamma\|T(w, p) - T(w_*, p_*) - L(w - w_*)\| \\ &\le \gamma(\|T(w, p) - T(w_*, p) - L(w - w_*)\| \\ &\quad + \|T(w_*, p) - T(w_*, p_*)\|) \\ &\le \gamma(\epsilon\|w - w_*\|_\rho + \eta) \le \gamma(\epsilon\tau + \eta) \le \tau\end{aligned}$$

since $\|w - w_*\|_\rho \le \tau$ for all $w \in W$ and $\tau \ge \gamma\eta/(1 - \gamma\epsilon)$. Thus Φ maps W into itself. By the Banach contraction mapping principle, there exists a unique $w \in W$ such that $w = \Phi(w)$. Since $w = \Phi(w)$ is equivalent to $T(w, p) \in F(w)$ for $w \in W$ and $p \in P$, we conclude that for each $p \in P$, there is a unique $w \in W$ such that $T(w, p) \in F(w)$. For $p = p_i \in P$, $i = 1, 2$, let $w = w_i$ denote the associated solutions to $T(w, p) \in F(w)$. We have

$$\begin{aligned}\|w_1 - w_2\|_\rho &= \|x[T(w_1, p_1) - Lw_1] - x[T(w_2, p_2) - Lw_2]\|_\rho \\ &\le \gamma\|T(w_1, p_1) - T(w_2, p_2) - L(w_1 - w_2)\| \\ &\le \gamma\|T(w_1, p_1) - T(w_1, p_2)\| \\ &\quad + \gamma\|T(w_1, p_2) - T(w_2, p_2) - L(w_1 - w_2)\| \\ &\le \gamma\|T(w_1, p_1) - T(w_1, p_2)\| + \gamma\epsilon\|w_1 - w_2\|_\rho.\end{aligned}$$

Rearranging this inequality, the proof is complete. □

Proof of Lemma 1: In order to apply Lemma 2 to T defined in (14), we identify w or x with the pair (z, λ), we identify p with the triple $(\underline{z}, \underline{\lambda}_1, \underline{\lambda}_2)$, and we choose

$$F(w) = F(z, \lambda) = \begin{pmatrix} 0 \\ N(\lambda) \end{pmatrix},$$

and

$$X = \{(z, \lambda) \in \mathbf{R}^n \times \mathbf{R}^m : \lambda = (\mu, \pi),\ \mu \ge 0,\ \pi > 0\}.$$

The set P, chosen later, is a neighborhood of $(z_*, \lambda_*, \lambda_*)$. In presenting the linearization L of Lemma 2, we partition both the constraint function c and the multiplier λ into their components (g, h) and (μ, π) respectively. The linearization L of $T(\cdot, p_*)$ around w_* is given by

$$L\begin{pmatrix} z \\ \mu \\ \pi \end{pmatrix} = \begin{pmatrix} Qz + A^{\mathsf T}\mu + B^{\mathsf T}\pi \\ Az - \rho\mu \\ Bz - \rho\pi \end{pmatrix},$$

where

$$Q = \nabla_z^2\mathcal{L}(z_*, \mu_*, \pi_*), \quad A = \nabla g(z_*), \quad B = \nabla h(z_*).$$

In order to apply Lemma 2 to the function T in (14), we need to establish the Lipschitz property (46). This leads us to consider the problem: Find $x \in X$ such that $L(x) + \psi \in F(x)$. Since L has three components, we partition $\psi = (\varphi, r, s)$, and the linearized problem takes the form: Find $(z, \mu, \pi) \in X$ such that

$$Qz + A^{\mathsf{T}}\mu + B^{\mathsf{T}}\pi + \varphi = 0, \tag{48}$$

$$\mu \geq 0, \ Az - \rho\mu + r \in N(\mu), \tag{49}$$

$$Bz - \rho\pi + s = 0, \tag{50}$$

where in the last equation (50), we exploit the fact that $\pi > 0$ for all $(z, \mu, \pi) \in X$.

In order to analyze the linearization (48)–(50), we introduce the following auxiliary problem:

$$\min_{z} \max_{\mu \geq 0} \ z^{\mathsf{T}}\varphi + \frac{1}{2}z^{\mathsf{T}}Qz + \frac{1}{2\rho}\|Bz + s\|^2 + \mu^{\mathsf{T}}(Az + r) - \frac{\rho}{2}\|\mu\|^2. \tag{51}$$

By (11) and Lemma 3 in the Appendix, the matrix $Q + B^{\mathsf{T}}B/\rho$ is positive definite with smallest eigenvalue at least $\alpha/2$ for ρ sufficiently small, where α appears in (11). Hence, the extremand in (51) is strongly convex in z and strongly concave in μ. By [8, Proposition 2.2, p. 173], the max and the min can be interchanged. For fixed μ, the min in (51) is attained by the solution z of the following linear equation:

$$\left(Q + \frac{1}{\rho}B^{\mathsf{T}}B\right)z + \varphi + A^{\mathsf{T}}\mu + B^{\mathsf{T}}s/\rho = 0. \tag{52}$$

After substituting this z in (51), we obtain an *equivalent* strongly concave maximization problem in the variable μ and the parameters φ, r, and s appear linearly in the cost function. Since strongly concave maximization problems are Lipschitz continuous functions of linear parameters in the cost (for example, see [3, Lemma 4]), the maximizing μ is a Lipschitz continuous function of the parameter ψ, and by (52), the minimizing z is also a Lipschitz continuous function of ψ.

Since (48)–(50) are the first-order conditions for a solution of (51), and since the first-order conditions are necessary and sufficient for optimality in this convex/concave setting, we conclude that (48)–(50) have a unique solution $(z(\psi), \lambda(\psi))$ depending Lipschitz continuously on the parameters φ, r, and s. We now apply [10, Theorem 2.1] in order to determine more precisely how the Lipschitz constant of $(z(\psi), \lambda(\psi))$ depends on ρ. Defining the set

$$c(\psi) = \{i : (Az(\psi) - \rho\mu(\psi) + r)_i = 0\}, \tag{53}$$

where $\lambda(\psi) = (\mu(\psi), \pi(\psi))$, it follows from [10, Theorem 2.1], that if γ_1 and γ_2 satisfy

$$\|z(\psi_2) - z(\psi_1)\| \leq \gamma_1\|\psi_1 - \psi_2\|, \quad \|\lambda(\psi_2) - \lambda(\psi_1)\| \leq \gamma_2\|\psi_2 - \psi_1\|,$$

whenever $c(\psi_1) = c(\psi_2)$, then these same Lipschitz constants work for all ψ_1 and ψ_2.

After substituting for μ in (52), using the relation $(Az(\psi) - \rho\mu(\psi) + r)_i = 0$ for $i \in c(\psi)$, we see that $z = z(\psi)$ satisfies

$$\left(Q + \frac{1}{\rho}C^{\mathsf{T}}C\right)z + \varphi + C^{\mathsf{T}}t/\rho = 0, \tag{54}$$

where C and t are gotten by augmenting B and s by the rows of A and the components of r associated with $i \in c(\psi)$. Let UR denote an orthogonal decomposition of C where R is right triangular (that is, $R_{ij} = 0$ if $i > j$) with linearly independent rows and U has orthonormal columns. After substituting $C = UR$ in (54), we obtain the equivalent system

$$\begin{pmatrix} Q & R^{\mathsf{T}} \\ R & -\rho I \end{pmatrix}\begin{pmatrix} z \\ \chi \end{pmatrix} = \begin{pmatrix} -\varphi \\ -U^{\mathsf{T}}t \end{pmatrix}.$$

The second equation $\chi = (Rz + U^{\mathsf{T}}t)/\rho$ in this system is the definition of χ and the first equation in this system is (54). Since the coefficient matrix is nonsingular for ρ sufficiently small (see [1, Lemma 1.27]), both $z(\psi)$ and $\chi(\psi)$ are Lipschitz continuous functions of ψ, where the Lipschitz constant is independent of ρ for ρ sufficiently small:

$$\|z(\psi_2) - z(\psi_1)\| + \|\chi(\psi_2) - \chi(\psi_1)\| \le \beta\|\psi_1 - \psi_2\| \tag{55}$$

Let V have orthonormal columns chosen so that the matrix $(U \mid V)$ is orthogonal. The vector $\pi(\psi)$ satisfies (50) and the components $\mu_0(\psi)$ of $\mu(\psi)$ associated with $i \in c(\psi)$ satisfy an analogous relation in (53). Hence, we have

$$\begin{pmatrix} \mu_0(\psi) \\ \pi(\psi) \end{pmatrix} = (Cz(\psi) + t)/\rho.$$

Multiplying by $(U \mid V)^{\mathsf{T}}$ yields:

$$\begin{aligned} (U \mid V)^{\mathsf{T}}\begin{pmatrix} \mu_0(\psi) \\ \pi(\psi) \end{pmatrix} &= (U \mid V)^{\mathsf{T}}(Cz(\psi) + t)/\rho \\ &= \frac{1}{\rho}\begin{pmatrix} U^{\mathsf{T}}(Cz(\psi) + t) \\ V^{\mathsf{T}}t \end{pmatrix} = \begin{pmatrix} \chi(\psi) \\ V^{\mathsf{T}}t/\rho \end{pmatrix} \end{aligned}$$

Multiplying again by $(U \mid V)$ gives

$$\begin{pmatrix} \mu_0(\psi) \\ \pi(\psi) \end{pmatrix} = U\chi(\psi) + VV^{\mathsf{T}}t/\rho. \tag{56}$$

Since $\chi(\psi)$ is a Lipschitz continuous function of ψ, it follows from (56) that $\mu_0(\psi)$ and $\pi(\psi)$ are Lipschitz continuous functions of ψ, while the remaining components of $\mu(\psi)$

vanish. Therefore, when $c(\psi_1) = c(\psi_2)$, (55) and (56) give us the estimates

$$\| z(\psi_2) - z(\psi_1)\| \le \beta \|\psi_2 - \psi_1\|, \tag{57}$$

$$\|\lambda(\psi_2) - \lambda(\psi_1)\| \le \beta \|\psi_2 - \psi_1\| + \|q_2 - q_1\|/\rho, \tag{58}$$

where $q = (r, s)$ and β is independent of ρ for ρ sufficiently small. By [10, Theorem 2.1], this estimate is valid for arbitrary choices of the parameters.

Given a fixed positive scalar σ_1, we assume that ρ is always $\le \sigma_1$. Hence, after multiplying (58) by ρ and adding to (57), we conclude that

$$\|w(\psi_1) - w(\psi_2)\|_\rho \le \gamma \|\psi_1 - \psi_2\|, \quad w(\psi) = (z(\psi), \lambda(\psi)), \tag{59}$$

for some constant γ independent of ρ, where

$$\|w\|_\rho = \|(z, \lambda)\|_\rho = \|z\| + \rho\|\lambda\|.$$

For the choice $\psi = \psi_* = (\varphi_*, r_*, s_*) = T(w_*, p_*) - L(w_*)$ where

$$\varphi_* = -Qz_* - A^\mathsf{T}\mu_* - B^\mathsf{T}\pi_*, \quad r_* = -Az_* + \rho\mu_*, \quad s_* = -Bz_* + \rho\pi_*,$$

(48)–(50) have the solution $z = z_*$ and $\lambda = \lambda_* = (\mu_*, \pi_*)$. Defining the parameter

$$\Delta = \frac{1}{2\gamma} \min_i \, (\pi_*)_i, \tag{60}$$

it follows from (59) that for all $\psi \in \mathcal{B}_{\rho\Delta}(\psi_*)$ and for all j,

$$\begin{aligned} |(\pi(\psi) - \pi_*)_j| &\le \|\pi(\psi) - \pi_*\| \\ &= \|\pi(\psi) - \pi(\psi_*)\| \le (\gamma/\rho)\|\psi - \psi_*\| \le \min_i \, (\pi_*)_i/2. \end{aligned}$$

Hence, $\pi(\psi) > 0$ for all $\psi \in \mathcal{B}_{\rho\Delta}(\psi_*)$, from which it follows that $(z(\psi), \lambda(\psi)) \in X$ for all $\psi \in \mathcal{B}_{\rho\Delta}(\psi_*)$. Combining this with (59), we conclude that (45) has a unique solution and (46) holds for all $\psi \in \mathcal{B}_{\rho\Delta}(\psi_*)$.

Given an arbitrary scalar $\sigma_1 > 0$, and positive scalars σ_0 and δ, chosen shortly, we define

$$P = \{(\underline{z}, \underline{\lambda}_1, \underline{\lambda}_2) \in \mathcal{B}_\delta(p_*) : \sigma_0\|\underline{z} - z_*\| \le \rho\}, \tag{61}$$

where $p_* = (z_*, \lambda_*, \lambda_*)$. By choosing σ_0 sufficiently large and δ sufficiently small, we will satisfy the condition $\epsilon\gamma < 1$ of Lemma 2, and by choosing δ smaller if necessary, the remaining conditions of Lemma 2 will be satisfied.

(P1) Observe that $T(w_*, p_*) = 0$ where $w_* = (z_*, \lambda_*)$. Defining $p = (\underline{z}, \underline{\lambda}_1, \underline{\lambda}_2)$, we have

$$T(w_*, p_*) - T(w_*, p) = \begin{pmatrix} \nabla_z \mathcal{L}(\underline{z}, \lambda_*) + \nabla_z^2 \mathcal{L}(\underline{z}, \underline{\lambda}_1)(z_* - \underline{z}) \\ c(\underline{z}) + \nabla c(\underline{z})(z_* - \underline{z}) - \rho(\lambda_* - \underline{\lambda}_2) \end{pmatrix}.$$

Expanding in a Taylor series around p_* gives

$$\|T(w_*, p_*) - T(w_*, p)\| \leq \beta(\|\underline{z} - z_*\|^2 + \|\underline{\lambda}_1 - \lambda_*\| \|\underline{z} - z_*\| + \rho\|\underline{\lambda}_2 - \lambda_*\|) \tag{62}$$

for all $p \in P$. Since the right side of (62) is bounded by $\beta\delta$, the constant η in (P1) can be made arbitrarily small by taking δ small.

(P2) Let ϵ be any positive number small enough that $\epsilon\gamma < 1$ where γ appears in (59). Observe that

$$T(w_1, p) - T(w_2, p) - L(w_1 - w_2) = \begin{pmatrix} \left(\nabla_z^2 \mathcal{L}(\underline{z}, \underline{\lambda}_1) - \nabla_z^2 \mathcal{L}(z_*, \lambda_*)\right)(z_2 - z_1) + (\nabla c(\underline{z}) - \nabla c(z_*))^{\mathsf{T}}(\lambda_2 - \lambda_1) \\ (\nabla c(\underline{z}) - \nabla c(z_*))(z_2 - z_1) \end{pmatrix},$$

where $w_1 = (z_1, \lambda_1)$ and $w_2 = (z_2, \lambda_2)$. By the assumed Lipschitz continuity of the derivatives, and by (61), we have, for all $p \in P$ and for any choice of w_1 and w_2,

$$\begin{aligned} &\|T(w_1, p) - T(w_2, p) - L(w_1 - w_2)\| \\ &\quad \leq \beta\|\underline{z} - z_*\| \|w_1 - w_2\| + \beta\|\underline{\lambda}_1 - \lambda_*\| \|z_1 - z_2\| \\ &\quad \leq \frac{\beta\rho}{\sigma_0}\|w_1 - w_2\| + \beta\delta\|z_1 - z_2\| \\ &\quad \leq \frac{\beta}{\sigma_0}(\rho\|z_1 - z_2\| + \rho\|\lambda_1 - \lambda_2\|) + \beta\delta\|z_1 - z_2\| \\ &\quad \leq \frac{\beta}{\sigma_0}(\sigma_1\|w_1 - w_2\|_\rho + \|w_1 - w_2\|_\rho) + \beta\delta\|w_1 - w_2\|_\rho \\ &\quad = \beta\left(\frac{(1 + \sigma_1)}{\sigma_0} + \delta\right)\|w_1 - w_2\|_\rho. \end{aligned} \tag{63}$$

Choose σ_0 large enough and δ small enough that the factor multiplying $\|w_1 - w_2\|_\rho$ in (63) is $\leq \epsilon$. This establishes (P2) and $\epsilon\gamma < 1$.

(P3) Choosing $\tau = \rho$, the set W of Lemma 2 is

$$W = \{w = (z, \lambda) \in \mathbf{R}^n \times \mathbf{R}^m : \lambda = (\mu, \pi),\ \mu \geq 0,\ \pi > 0,\ \|w - w_*\|_\rho \leq \rho\}.$$

By (62) and (63), we have for all $w \in W$ and $p \in P$,

$$\begin{aligned} &\|T(w, p) - L(w) - (T(w_*, p_*) - L(w_*))\| \\ &\quad \leq \|(T(w, p) - T(w_*, p)) - L(w - w_*)\| + \|T(w_*, p) - T(w_*, p_*)\| \\ &\quad \leq \epsilon\|w - w_*\|_\rho + \beta(\|\underline{z} - z_*\|^2 + \|\underline{\lambda}_1 - \lambda_*\| \|\underline{z} - z_*\| + \rho\|\underline{\lambda}_2 - \lambda_*\|) \\ &\quad \leq \rho(\epsilon + \beta\delta) \end{aligned} \tag{64}$$

since $\|w - w_*\|_\rho \le \tau = \rho$, $\|\underline{z} - z_*\| \le \rho/\sigma_0$, and $(\underline{z}, \underline{\lambda}_1, \underline{\lambda}_2) \in \mathcal{B}_\delta(p_*)$. Choose ϵ and δ smaller if necessary so that

$$\epsilon + \beta\delta \le \Delta,$$

where Δ is defined in (60). Hence, by (64), we have

$$\|T(w, p) - L(w) - (T(w_*, p_*) - L(w_*))\| \le \Delta\rho$$

for all $w \in W$ and $p \in P$. Since $\psi_* = T(w_*, p_*) - L(w_*)$, it follows that

$$T(w, p) - L(w) \in \mathcal{B}_{\rho\Delta}(\psi_*)$$

for all $w \in W$ and $p \in P$. This completes the proof of (P3) since we already showed that (45) has a unique solution satisfying (46) for all $\psi \in \mathcal{B}_{\rho\Delta}(\psi_*)$.

Finally, let us consider the condition

$$\tau \ge \gamma\eta/(1 - \epsilon\gamma) \tag{65}$$

of Lemma 2, where $\eta = \sup\{\|T(w_*, p_*) - T(w_*, p)\| : p \in P\}$. Recalling that $\tau = \rho$, and utilizing (62), we see that (65) is satisfied if

$$\rho \ge \beta(\|\underline{z} - z_*\|^2 + \|\underline{\lambda}_1 - \lambda_*\|\|\underline{z} - z_*\| + \rho\|\underline{\lambda}_2 - \lambda_*\|) \tag{66}$$

for each $(\underline{z}, \underline{\lambda}_1, \underline{\lambda}_2) \in P$; here the factor $\gamma/(1 - \epsilon\gamma)$ of (65) is absorbed into β. Assuming δ is small enough that $\beta\|\underline{\lambda}_2 - \lambda_*\| < 1$, we rearrange (66) to obtain the equivalent relation

$$\rho \ge \frac{\beta(\|\underline{z} - z_*\| + \|\underline{\lambda}_1 - \lambda_*\|)\|\underline{z} - z_*\|}{1 - \beta\|\underline{\lambda}_2 - \lambda_*\|}. \tag{67}$$

By the definition of P, $\rho \ge \sigma_0\|\underline{z} - z_*\|$ for all $p = (\underline{z}, \underline{\lambda}_1, \underline{\lambda}_2) \in P$. Hence, if

$$\sigma_0 \ge \frac{\beta(\|\underline{z} - z_*\| + \|\underline{\lambda}_1 - \lambda_*\|)}{1 - \beta\|\underline{\lambda}_2 - \lambda_*\|}. \tag{68}$$

(67) will be satisfied. Choosing δ small enough that (68) is satisfied, it follows that (67) holds, which implies in turn (65). Since all the assumptions of Lemma 2 are satisfied, Lemma 1 follows almost directly. The neighborhood $\mathcal{N}(\rho)$ of Lemma 1 coincides with W of Lemma 2, while the ball $\mathcal{B}_\delta$ of Lemma 1 is the same ball appearing in the definition of P in (61). The constant β of Lemma 1 is the expression $\gamma/(1 - \gamma\epsilon)$ of (47). □

Appendix: A matrix bound

Lemma 3. *Given matrices Q_* and B_* where Q_* is symmetric, suppose that*

$$w^{\mathsf{T}} Q_* w \ge \alpha\|w\|^2 \quad \textit{whenever } B_* w = 0, \quad w \in \mathbf{R}^n. \tag{69}$$

Then given any $\delta > 0$, there exists $\sigma > 0$ and neighborhoods $\mathcal{B}$ of B_* and $\mathcal{Q}$ of Q_* such that

$$v^{\mathsf{T}}\left(Q + \frac{1}{\rho}B^{\mathsf{T}}B\right)v \geq (\alpha - \delta)\|v\|^2$$

for all $v \in \mathbf{R}^n$, $0 < \rho \leq \sigma$, $B \in \mathcal{B}$, and $Q \in \mathcal{Q}$.

Proof: If w lies in the null space of B_*, then

$$w^{\mathsf{T}}(Q_* + B_*^{\mathsf{T}}B_*/\rho)w \geq \alpha\|w\|^2$$

by (69). There exists a scalar $\tau > 0$ such that $\|B_*u\| \geq \tau\|u\|$ for all u in the row space of B_*. Hence, for u in the row space of B_*, we have

$$u^{\mathsf{T}}(Q_* + B_*^{\mathsf{T}}B_*/\rho)u = u^{\mathsf{T}}Q_*u + \|B_*u\|^2/\rho \geq (\tau^2/\rho - \|Q_*\|)\|u\|^2.$$

An arbitrary vector in $v \in \mathbf{R}^n$ has the orthogonal decomposition $v = u + w$ where u is in the row space of B_* and w is in the null space of B_*. Since $B_*w = 0$, it follows that

$$\begin{aligned} v^{\mathsf{T}}(Q_* + B_*^{\mathsf{T}}B_*/\rho)v &= (u+w)^{\mathsf{T}}(Q_* + B_*^{\mathsf{T}}B_*/\rho)(u+w) \\ &= w^{\mathsf{T}}Q_*w + u^{\mathsf{T}}(Q_* + B_*^{\mathsf{T}}B_*/\rho)u + 2u^{\mathsf{T}}Q_*w \\ &\geq \alpha\|w\|^2 + \left(\frac{\tau^2}{\rho} - \|Q_*\|\right)\|u\|^2 - 2\|u\|\|w\|\|Q_*\|. \end{aligned} \tag{70}$$

Utilizing the inequality

$$ab \leq \epsilon a^2 + b^2/4\epsilon,$$

with $a = \|w\|$ and $b = 2\|u\|\|Q_*\|$ gives

$$2\|u\|\|w\|\|Q_*\| \leq \epsilon\|w\|^2 + \|Q_*\|^2\|u\|^2/\epsilon.$$

Inserting this in (70), we have

$$v^{\mathsf{T}}(Q_* + B_*^{\mathsf{T}}B_*/\rho)v \geq (\alpha - \epsilon)\|w\|^2 + \left(\frac{\tau^2}{\rho} - \|Q_*\| - \frac{\|Q_*\|^2}{\epsilon}\right)\|u\|^2.$$

Let us choose σ small enough that

$$\frac{\tau^2}{\sigma} - \|Q_*\| - \frac{\|Q_*\|^2}{\epsilon} \geq \alpha - \epsilon.$$

Since $\|w\|^2 = \|u\|^2 + \|v\|^2$, it follows that

$$v^{\mathsf{T}}(Q_* + B_*^{\mathsf{T}}B_*/\rho)v \geq (\alpha - \epsilon)\|v\|^2$$

for all v and for all $0 < \rho \leq \sigma$. Since the expression $Q + B^{\mathsf{T}}B/\sigma$ is a continuous function of B and Q, there exists neighborhoods $\mathcal{Q}$ of Q_* and $\mathcal{B}$ of B_* such that

$$v^{\mathsf{T}}(Q + B^{\mathsf{T}}B/\sigma)v \geq (\alpha - 2\epsilon)\|v\|^2$$

for all v and for all $Q \in \mathcal{Q}$ and $B \in \mathcal{B}$. When $0 < \rho \le \sigma$, we have

$$v^{\mathsf{T}}(Q + B^{\mathsf{T}}B/\rho)v \ge v^{\mathsf{T}}(Q + B^{\mathsf{T}}B/\sigma)v \ge (\alpha - 2\epsilon)\|v\|^2.$$

Taking $\delta = 2\epsilon$, the proof is complete. □

References

1. D.P. Bertsekas, Constrained Optimization and Lagrange Multiplier Methods, Academic Press: New York, 1982.
2. F.H. Clarke, "Generalized gradients and applications," Trans. of the Amer. Math. Soc., vol. 205, pp. 247–262, 1975.
3. A.L. Dontchev and W.W. Hager, "Lipschitzian stability in nonlinear control and optimization," SIAM J. Control Optim., vol. 31, pp. 569–603, 1993.
4. A.L. Dontchev, W.W. Hager, A.B. Poore, and B. Yang, "Optimality, stability and convergence in nonlinear control," Appl. Math. Optim., vol. 31, pp. 297–326, 1995.
5. A.L. Dontchev and W.W. Hager, "The Euler approximation in state constrained optimal control," Department of Mathematics, University of Florida, Gainesville, FL 32611, November, 1997.
6. A.L. Dontchev and W.W. Hager, "Lipschitzian stability for state constrained nonlinear optimal control," SIAM J. Control Optim., vol. 35, pp. 698–718, 1998.
7. A.L. Dontchev and R.T. Rockafellar, "Characterizations of strong regularity for variational inequalities over polyhedral convex sets," SIAM J. Optim., vol. 6, pp. 1087–1105, 1996.
8. I. Ekeland and R. Temam, Convex Analysis and Variational Problems, North-Holland: Amsterdam, 1976.
9. A. Fischer, "Modified Wilson method for nonlinear programs with nonunique multipliers," Technische Universität Dresden, Germany, February, 1997.
10. W.W. Hager, "Lipschitz continuity for constrained processes," SIAM J. Control Optim., vol. 17, pp. 321–338, 1979.
11. W.W. Hager, "Approximations to the multiplier method," SIAM J. Numer. Anal., vol. 22, pp. 16–46, 1985.
12. W.W. Hager, "Convergence of Wright's stabilized SQP algorithm," Mathematics Department, University of Florida, Gainesville, FL 32611, January, 1997.
13. W.W. Hager and M.S. Gowda, "Stability in the presence of degeneracy and error estimation," Mathematics Department, University of Florida, Gainesville, FL 32611, November 22, 1997 (to appear in Math. Programming).
14. A.J. Hoffman, "On approximate solutions of systems of linear inequalities," J. Res. Nat. Bur. Standards, vol. 49, pp. 263–265, 1952.
15. O.L. Mangasarian and S. Fromovitz, "The Fritz-John necessary optimality conditions in the presence of equality and inequality constraints," J. of Math. Anal. and Appl., vol. 17, pp. 37–47, 1967.
16. S.M. Robinson, "Perturbed Kuhn-Tucker points and rates of convergence for a class of nonlinear-programming algorithms," Math. Programming, vol. 7, pp. 1–16, 1974.
17. S.M. Robinson, "Strongly regular generalized equations," Math. Oper. Res., vol. 5, pp. 43–62, 1980.
18. R.T. Rockafellar, "The multiplier method of Hestenes and Powell applied to convex programming," J. Optim. Theory Appl., vol. 12, pp. 555–562, 1973.
19. S.J. Wright, "Superlinear convergence of a stabilized SQP method to a degenerate solution," Comput. Optim. Appl., vol. 11, pp. 253–275, 1998.

MIX
Papier aus verantwortungsvollen Quellen
Paper from responsible sources
FSC® C105338

If you have any concerns about our products,
you can contact us on
ProductSafety@springernature.com

In case Publisher is established outside the EU,
the EU authorized representative is:
Springer Nature Customer Service Center GmbH
Europaplatz 3, 69115 Heidelberg, Germany

Printed by Libri Plureos GmbH
in Hamburg, Germany